SOCIETY FOR THE STUDY OF HUMAN BIOLOGY SYMPOSIUM SERIES: 30

Diet and disease

PUBLISHED SYMPOSIA OF THE SOCIETY FOR THE STUDY OF HUMAN BIOLOGY

1 The Scope of Physical Anthropology and its Place in Academic Studies ED. D.F. ROBERTS AND J.S. WEINER
2 Natural Selection in Human Populations ED. D.F. ROBERTS AND G.A. HARRISON
3 Human Growth ED. J.M. TANNER
4 Genetical Variation in Human Populations ED. G.A. HARRISON
5 Dental Anthropology ED. D.R. BROTHWELL
6 Teaching and Research in Human Biology ED. G.A. HARRISON
7 Human Body Composition, Approaches and Applications ED. D.R. BROTHWELL
8 Skeletal Biology of Earlier Human Populations ED. J. BROZEK
9 Human Ecology in the Tropics ED. J.P. GARLICK AND R.W.J. KEAY
10 Biological Aspects of Demography ED. W. BRASS
11 Human Evolution ED. M.H. DAY
12 Genetic Variation in Britain ED. D.F. ROBERTS AND E. SUNDERLAND
13 Human Variation and Natural Selection ED. E.F. ROBERTS (Penrose Memorial Volume reprint)
14 Chromosome Variation in Human Evolution ED. A.J. BOYCE
15 Biology of Human Foetal Growth ED. D.F. ROBERTS
16 Human Ecology in the Tropics ED. J.P. GARLICK AND R.W.J. DAY
17 Physiological Variation and its Genetic Base ED. J.S. WEINER
18 Human Behaviour and Adaptation ED. N.J. BLURTON JONES AND V. REYNOLDS
19 Demographic Patterns in Developed Societies ED. R.W. HIORNS
20 Disease and Urbanisation ED. E.J. CLEGG AND J.P. GARLICK
21 Aspects of Human Evolution ED. C.B. STRINGER
22 Energy and Effort ED. G.A. HARRISON
23 Migration and Mobility ED. A.J. BOYCE
24 Sexual Dimorphism ED. F. NEWCOMBE *et al.*
25 The Biology of Human Ageing ED. A.H. BITTLES AND K.J. COLLINS
26 Capacity for Work in the Tropics ED. K.J. COLLINS AND D.F. ROBERTS
27 Genetic Variation and its Maintenance ED. D.F. ROBERTS AND G.F. DE STEFANO
28 Human Mating Patterns ED. C.G.N. MASCIE-TAYLOR AND A.J. BOYCE
29 The Physiology of Human Growth ED. J.M. TANNER AND M.A. PREECE
30 Diet and Disease ED. G.A. HARRISON AND J.C. WATERLOW

Numbers 1–9 were published by Pergamon Press, Headington Hill Hall, Headington, Oxford OX3 0BY. Numbers 10–24 were published by Taylor & Francis Ltd, 10–14 Macklin Street, London WC2B 5NF. Further details and prices of back-list numbers are available from the Secretary of the Society for the Study of Human Biology.

Diet and disease

In traditional and developing societies

30th Symposium Volume of the Society for the Study of Human Biology

EDITED BY

G.A. HARRISON
Department of Biological Anthropology
University of Oxford

AND

J.C. WATERLOW
Department of Community Health
London School of Hygiene and Tropical Medicine

Organised in collaboration with
The International Commission on the Anthropology of Food
(I.U.A.E.S.)

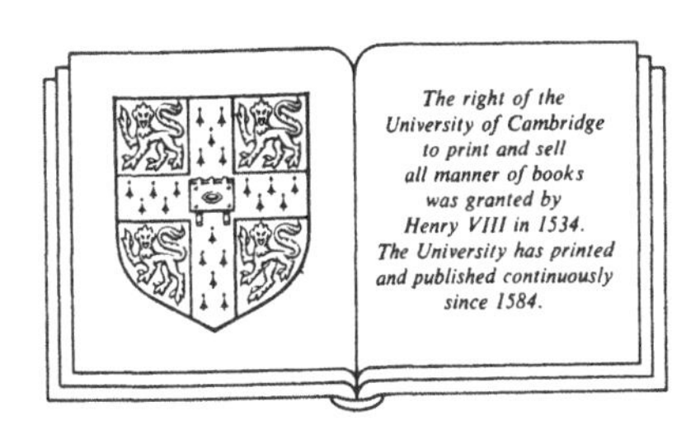

CAMBRIDGE UNIVERSITY PRESS
Cambridge New York Port Chester
Melbourne Sydney

CAMBRIDGE UNIVERSITY PRESS
Cambridge, New York, Melbourne, Madrid, Cape Town, Singapore, São Paulo, Delhi

Cambridge University Press
The Edinburgh Building, Cambridge CB2 8RU, UK

Published in the United States of America by Cambridge University Press, New York

www.cambridge.org
Information on this title: www.cambridge.org/9780521104722

First published 1990
This digitally printed version 2009

A catalogue record for this publication is available from the British Library

ISBN 978-0-521-38454-4 hardback
ISBN 978-0-521-10472-2 paperback

Contents

List of contributors

Professor G.A. Harrison
Department of Biological Anthropology, 58 Banbury Road, Oxford OX2 6QS.

Professor J.C. Waterlow, FRS
Department of Community Health, London School of Hygiene and Tropical Medicine, Keppel Street, London WC1E 7HT.
Correspondence to 15 Hillgate Street, London W8 7SP.

Dr G.B. Spurr
Department of Physiology, Medical College of Wisconsin, 8701 Watertown Plank Road, Milwaukee, Wis. 5326, USA.

Dr S. Grantham-McGregor
Tropical Metabolism Research Unit, University of the West Indies, Kingston, Jamaica.

Professor A.A. Jackson
Department of Human Nutrition, University of Southampton, Bassett Crescent East, SO9 3TU.

Dr B.S. Hetzel
CSIRO Division of Human Nutrition, Kintore Avenue, Adelaide, S.A. 5000, Australia.

Dr S.J. Ulijaszek
Department of Biological Anthropology, Downing Street, Cambridge CB2 3DZ.

Professor M. Hodkinson
Department of Geriatric Medicine, University College and Middlesex School of Medicine, St. Pancras Way, London NW1 0PE.

Dr J.D. Haas
Division of Nutritional Sciences, Cornell University, Ithaca, New York 14853, USA.

Dr P.S. Nestel
Harvard Institute for International Development, Cambridge, Mass. 02138, USA

Dr S.S. Strickland
London School of Hygiene and Tropical Medicine, Keppel Street, London WC1E 7HT.

Dr I. de Garine
C.R.N.S. Pargade, Lasseube, 64290 Gan, France.

Dr B. Harriss
University of Oxford, Queen Elizabeth House, St. Giles, Oxford OX1 3LA.

Dr R. Burghart
Department of Anthropology, School of Oriental and African Studies, Malet Street, London WC1. Currently at Südasien Institut, Universität Heidelberg, In Neuenheimer Feld 330, Heidelberg, W. Germany.

Acknowledgements

We wish to thank Professor D.F. Roberts and Professor J.V.G.A. Durnin, who shared with the Editors the pleasure of chairing the Symposium sessions.

Grateful acknowledgment is also made to the following organisations for grants towards the costs of the meeting: The Royal Society; Cadbury Schweppes Charitable Trust; Express Foods Group; J. Lyons & Co. Ltd; Milupa; RHM; Rowntree Mackintosh; Unilever p.l.c.

1 *Introduction*

G. AINSWORTH HARRISON

Much has been written about the relationship between the quantity and quality of the food available to people and their health and disease susceptibility. Even for traditional and developing societies, which are the concern of this book and where nutritional problems are essentially ones of shortage, an enormous amount of work has been undertaken looking at the morbid effects of these shortages. What then has this book to offer that is new?

In the main, nutritionists have tackled the problems of malnutrition in physiological and medical terms, concentrating on the mechanisms by which nutritional diseases develop and the requirements to prevent or remedy them. This has rightly been a top academic as well as practical priority, but it deals only with the proximal situation of human nutritional requirements for physiological homeostasis. Were human beings domestic animals such a focus might be sufficient in itself, but because people live in variable and changing environments, and because they are called upon to participate in varying activities, skills and behaviours, and because, most importantly of all, they are parts of complex social structures with varying social demands and privileges, the full understanding of human morbidity from nutritional causes needs a wider approach than the exclusively physiological one. Food is as much a social resource as a collection of nutrients, and it can dominate social organisation as much as biological need. The way it is produced, and distributed, stored and preserved, prepared and cooked can as profoundly affect the nutritional health of a community as its total quantity. To understand, therefore, many features of human nutritional disease, requires an anthropological as well as a physiological insight. The aim of this book, and the symposium upon which it is based, is to encourage such a dual approach.

In the symposium (July 18th & 19th, 1988), a group of nutritionists with wide experience of nutritional problems in the Third World, and with a sympathy for holistic approaches, were brought together with some biological and social anthropologists who have an interest in nutrition from an evolutionary, ecological and social point of view, in the hope that a fruitful dialogue would develop. The book, we believe, demonstrates that this indeed happened, but also indicates that there remain useful opportunities from continuing it.

The first contribution, by Professor J.C. Waterlow, deals with the very important issue of human nutritional needs, and particularly with the question of possible variability in these needs. Surprisingly, nutritionists have only quite recently been concerned with whether or not there is nutritional adaptability, i.e. whether nutritional needs respond to nutritional environments in ways that safeguard health. Perhaps this is partly because they became isolated from evolutionary thinking, but it probably also reflects a deep assumption that nutrition was too fundamental to permit variation. It is gratifying that this assumption is being questioned but, as Waterlow's paper makes clear, resolving the issue is empirically very difficult.

The second and third chapters deal respectively with the effects of malnutrition on work capacity and mental development. Dr G.B. Spurr reviews his own extensive researches, mainly in Columbia, on the relationship between nutritional status, physical development and an individual's ability to perform physical work, and Dr S. Grantham-McGregor considers the ways in which poor nutrition can affect the mental activities of West Indian children, mainly by restricting their ability to play. Although neither of these contributions deal explicitly with the morbid state, they have profound relevance to it. In human beings, capacities to perform physical and mental work have ramifying consequences on economic and social status, and through this to nutritional health. The financially poor form the nutritionally and environmentally deprived; and ill-health can lead to a vicious spiral of increasing deprivation, through its reduction in ability to work and thus greater ill-health.

The two contributions by Professor A.A. Jackson and Dr B.S. Hetzel deal explicitly with important diseases of nutritional origin. Kwashiorkor, so common among children in many developing countries, has been known now for a number of years. Its aetiology, however, is still far from fully understood, as Professor Jackson clearly shows, and it is a much more complex entity than the simple protein deficiency disease it was once supposed to be. The situation with goitre and cretinism is much clearer, and treatment with iodine is one of the great successes in nutritional epidemiology. Such success, however, depends not only on having full scientific knowledge but also the logistic means, such as those provided by Dr Hetzel's International Council for Control of Iodine Deficiency Disorders, to ensure that treatment actually reaches those in need of it.

It is almost axiomatic that malnutrition predisposes to infectious disease; if not to infection itself, certainly to severity. The proposition has, however, been little researched empirically, and there are at least some cases, e.g. malaria, where the well-nourished may suffer more than the malnourished. What is known is reviewed very systematically by Dr S.J. Ulijaszek.

It may seem odd to find in a book concerned with Third World countries a paper on the elderly in Britain. However, as Professor Hodkinson shows, there are a number of interesting parallels between the nutritional status in the two situations. The causes for inadequate food intake are, of course, not the same, but both have important financial and social elements. It is also salutary to see the problems of one's own elderly within the broader framework of world malnutrition.

On the world scale, however, it is the child who is most vulnerable to food shortage. This is for biological and social reasons, and both are no doubt involved in contributing to the enormous variability in child malnutrition within a single population and, indeed, within a single family. The situation is deserving of much more attention, particularly in post-weaning children who have been largely ignored. Here two contributions are devoted specifically to children (as well as the analysis of mental development by Grantham-McGregor). Drs J.D. Haas and J-P. Habicht deal with the difficult problem of assessing levels of malnutrition in children; a particularly pressing concern in intervention programmes with limited resources, and Dr P.S. Nestel provides a case study of child health and growth in the Sudan.

It is easy to forget, particularly in modern industrial societies, that food is not only a starting point for human biological needs, but also an end point of human economic activities. These activities are themselves determined by innumerable factors of environment, technological development and social custom, and are endlessly variable. We may therefore expect that nutritional disease will be closely related to economy, but this is a matter little considered by human biologists. The valuable potential of this relationship is clearly demonstrated in the review of Dr S.S. Strickland.

It has been argued that widespread malnutrition is rare in fully traditional societies; that such societies have acquired systems that regulate population size to the resources and technology available. Whether this be so or not many such societies experience severe seasonal hunger, and food shortage can then be so great as to dominate social processes as well as biological ones. Because of their regularity seasonal shortages can, however, be prepared for, and the way this is done is not only intrinsically interesting but may well be of wider practical value. The topic is discussed here by Drs I. de Garine and S. Koppert, mainly from their extensive experience of the 'hungry season' among West African cultivators.

De Garine and Koppert bring us firmly into the social domain, and the two final chapters are concerned exclusively with this. Dr Barbara Harriss considers the extremely important issue of food distribution within the household and its relation to disease in South Asia. She amply demonstrates the marked sex bias there against women in the ways food is usually shared, and she traces the consequences of this in varying gender morbidity and mortality. She also, however, sees other important cultural factors contributing to this variation.

Dr R. Burghart pursues such cultural factors even further; specifically in relation to varying conceptions that different people have of their body images and self-hood. He stresses how important these are to a person's sense of well-being. He is also concerned with the way the 'social body' is identified and conceptualised in different societies, and the consequences of this in food distribution priorities. Much of what he writes is of vital relevance for intervention programmes, since he well demonstrates that 'the nutritional knowledge of the health educator does not replace the ignorance of local people'.

And so from fundamentals of physiology to fundamentals of social anthropology; all brought together most meaningfully in a proper discussion of diet and disease. It surely demonstrates that no field of human biology demands a more interdisciplinary approach than that of human nutrition.

2 *Mechanisms of adaptation to low energy intakes*

J.C. WATERLOW

Summary

No attempt is made in this paper to define adaptation in any precise way. The aim is simply to describe actual and possible responses of man to life-long intakes of energy that by Western standards often seem inadequate. The responses are discussed under three headings:

(i) Maintenance of a low body weight, which can be achieved by one or both of two ways: to be small in height as a consequence of retarded growth in childhood, and to have a low body weight in relation to height. An attempt is made to define the acceptable limits of low body weight.

(ii) Economy in the cost of physical activity. It is probable that there are ergonomic adaptations acquired by tradition and practice, of which perhaps the most important is that given tasks should be performed slowly rather than fast. This is an area where more research is clearly needed.

(iii) Biochemical and metabolic changes that promote economy. Here we can do little more than speculate. Possible mechanisms are reductions in the rates of energy-dissipating reactions such as protein turnover and ion-pumping. Since slow muscle fibres are more economical of energy than fast fibres, a preponderance of slow fibres would in theory confer an advantage, provided that, as indicated in (ii) above, a pattern of moving slowly is acceptable. Many of the biochemical mechanisms are controlled, or at least influenced, by the activity of the thyroid gland.

At all levels it is necessary to consider costs or trade-offs that balance possible advantages.

Introduction

I shall not try to define adaptation in any precise way or to distinguish it from concepts such as accommodation, homeostasis, regulation, etc. (for discussions see Waterlow, 1985a, b, 1986). Many years ago some of us wasted a great deal of time in sterile arguments about the exact distinction between the clinical forms of malnutrition known as kwashiorkor and marasmus, and I do not want to fall into that trap again. However, there is one aspect of adaptation that does need to be stressed in the present context: the fact that the word as ordinarily used carries a value judgement. In any given environment, such as a hot climate or a high altitude, it is by definition 'better' to be adapted than not adapted. But what if two people, A and N, are living in different environments? How can we compare them? N is tempted to say 'I am normal and you are adapted'. This tendency is shown by the way in which we apply universally our Western standards of growth, body size, basal metabolic rate (BMR), etc.

It is probably a reasonable generalization to suppose that every adapted state carries both gains and losses, and it is difficult, if not impossible, to avoid subjective value-judgements in comparing the end-results. For this reason, after 25 years of thinking and writing about adaptation, I am almost inclined to give up using the word and to confine myself to objective questions of the form: 'how does this particular function respond to this particular stress or change?'.

There is one other point worth making at this stage. Physiological characteristics seem to be of three general kinds: first, those that are rather closely fixed, as defined by Claude Bernard's concept of the constancy of the internal environment. Secondly, there are characteristics that seem to have an upper limit, from which they can only vary downwards. In this category come, for example, the plasma albumin concentration or the amount of protein per unit DNA in any particular type of cell. However much protein one eats, it does not seem to be possible to pass these upper limits, and the argument then is whether they should be regarded as 'preferred' points (see Payne, 1987). Thirdly, there are characteristics which in any population vary about a mean, so that a statistical definition can be made of a 'normal' range. For many of the characteristics in which we are interested the coefficient of variation seems to be of the order of 10–15 per cent, so that the range may cover a span of almost two-fold, from 70 to 130 per cent of the mean.* The BMR would fall into this category. I suggest that without physiological variability of this kind there would be little or no capacity for adaptation. Obviously, a characteristic that is fixed cannot adapt, though other things may adapt to keep it fixed.

* Since this paper was written Dr L. Garby of the Institute of Physiology, University of Odense, informs me that when conditions are rigidly standardized and results related to body composition, the between subject variation in energy expenditure may be as low as five per cent.

Table 2.1. *Effect of body weight on energy expenditure*

	Subject	
	A	B
Weight, kg	50	65
Height, m	1.6	1.6
Body mass index, wt/ht^2	19.5	25.4
Basal metabolic rate kcal/d (MJ/d)	1445 (6.04)	1670 (6.98)
Maintenance expenditure (1.4 × BMR)	2020 (8.44)	2340 (9.78)
Expenditure of a moderately active man (1.8 × BMR)	2600 (10.87)	3010 (12.28)

Basal metabolic rates calculated from Schofield *et al.* (1985) for male subjects aged 18–30. A's expenditure is 14 per cent less than that of B.

There are three general strategies for economizing energy expenditure:

(i) a low body weight;
(ii) reduction in the amount or cost of physical activity;
(iii) increased biochemical efficiency, i.e. 'metabolic' adaptation.

The word efficiency is used here as a general term; it is essential to define it precisely in each particular context. Much of what I have to say about metabolic adaptation is little more than speculation – a discussion of possibilities. In most cases it is likely that only small savings can be made, but together they could add up to a significant economy. As Barcroft (1934) has said: 'Every adaptation is an integration'.

The relationship of body size to energy expenditure

This subject has been well discussed by Ferro-Luzzi (1988) in a previous symposium of this series. It is obvious that a smaller person will have a lower energy expenditure (EE), but there are some interesting implications. BMR, which accounts for more than 50 per cent of an average person's total EE, is related to body weight, and so also is the cost of physical activities that involve moving the body. Table 2.1 shows estimates of EE based on the prediction equations for BMR in relation to body weight produced by Schofield, Schofield & James (1985) and used in the most recent UN report on energy requirements (FAO/WHO/UNU, 1985). These equations were derived mainly from measurements in Western countries and cover a rather narrow range of body weights. At lower weights, as found in Third World countries, they may overestimate the BMR (McNeill *et al.*, 1987).

In Table 2.1 total daily EE is expressed as a multiple of BMR. The expenditure of a sedentary person with very little physical activity would be 1.4 BMR; that of a moderately active man, such as a peasant farmer in the Third World, would be 1.8 BMR. Subject A saves 14 per cent of B's expenditure, and if, as mentioned above, A's BMR is overestimated, the saving will be even greater. Quantitatively this is probably the largest single contribution to the adaptation that we are considering.

Suppose that B's intake becomes the same as A's, so that he is short of 300–400 kcal per day: if the energy equivalent of body tissue is taken as 6000 kcal per kg, in about nine months, B will have come down to the same weight as A and will again be in energy balance. If, as is conventional, the acceptable limits of body mass index (BMI) are taken as 20–25, B, who was a little too fat has now become a little too thin (Table 2.1). FAO (1987) calls this process of adapting to a low energy intake by losing body weight a *costless biological adaptation*. Is it really costless? Although the hibernating bear can apparently lose some 20 kg of weight without loss of lean tissue (Nelson *et al.*, 1975), this is not so for man. From the data collected by Forbes (1985), a weight loss of 15 kg, which is admittedly rather extreme, would involve a loss of some 3 kg of lean tissue, of which about half or more would be muscle. This is not a very great loss – about six per cent of the initial muscle mass; whether it should be regarded as a negligible cost will depend on the extent to which muscular work is important for the individual's way of life.

One may ask: what is the limit of this process of adaptation? Much is known about the upper limit of 'acceptable' BMI, above which the risks attached to being overweight become significant. The upper limit is usually set at 25. At the present time the average BMI of young adults in the UK is about 24 (Knight, 1984) so that a high proportion are overweight. By contrast, in Third World countries the average BMI of apparently healthy groups is usually in the range 21–19 (Eveleth & Tanner, 1976). Not very much is known about the lower limits of BMI compatible with health and normal function. A recent review suggests that people with BMI between 18.5 and 17 may be regarded as borderline; between 17 and 16 as probably having a deficient energy intake; and below 16 as frankly malnourished (James, Ferro-Luzzi & Waterlow, 1988).

Another way of achieving a low body weight is to be short rather than tall, but with the same body proportions and the same BMI. If C and D both have a BMI of 22, C, at 1.72 m tall will weigh 65 kg and D, at 1.5 m, will weigh 50 kg. D has many advantages in being small: he or she not only needs less food but also needs less material for clothing and less living space. Therefore it might be argued that it is smart to be small. However, I would prefer to apply this argument only to adults. The suppression of linear growth in children (stunting) by nutritional and environmental handicaps is another matter altogether (see Waterlow, 1988). Here I want to consider only the effects of differences in height in adults, regardless of the way in which those differences may result from the interplay of genetic and environmental factors during growth.

There are disadvantages in being small, to set against the benefits. To the extent that people have to do external work, such as carrying loads or cutting sugar cane, it will be a disadvantage to have a smaller muscle mass. A small person has a smaller absolute working capacity ($\dot{V}_{O_2}$ max); at a given load our subject D, weighing 50 kg, has to operate at a higher proportion of his $\dot{V}_{O_2}$ max than C, weighing 65 kg, and therefore needs to have a higher level of cardio-respiratory fitness. This disadvantage is less at low work loads and disappears when people are mainly sedentary.

Stature also influences the energetics of movement. Leg length is related to stature, and in walking on the level in a free and natural way, common observation shows that a taller person takes longer strides. Alexander (1984) has introduced the concept of isodynamic patterns of movement. Walkers of different sizes will walk in an isodynamic fashion if their relative velocities are proportional to the square roots of their heights. If the body weights are the same, the activities will not only be isodynamic but also isoenergetic. It follows that at equal body weights a tall person will walk faster than a short one for the same energy expenditure, and will cover a fixed distance at a lower total cost. A man 1.7 m tall would be expected to save about seven per cent of the energy expended by a man of 1.5 m. In reality tall people tend to be heavier than short people, so the greater weight cancels out the advantage of greater height. The advantage will only be significant for people who are tall but very thin, such as the Dinka of the Sudan, whose average height is 1.82 m and weight 58 kg (Eveleth & Tanner, 1976). These people are nomads who presumably have to walk great distances, and one might speculate that their great height represents a biological/genetic adaptation.

I do not know how these relationships with stature would apply under other conditions, such as walking with loads or on a gradient. Mountain people are generally small, and it would be interesting to know whether gradient reverses the advantage of tall stature for walking on the level. It is likely that body weight is the dominant consideration, since the effect of weight on absolute energy expenditure increases with walking up hill.

Reduction in the amount or cost of physical activity

The level of activity

The most obvious way of economizing energy is to reduce one's level of physical activity. In the last UN report (FAO/WHO/UNU, 1985) two categories of voluntary physical activity were defined, economic and discretionary. Economic activity is that on which a person's livelihood depends, and for many of us in the Western world is virtually nil. Discretionary activities are those which are not immediately essential, but which improve the quality of life or perhaps help to prolong it. In our society we could put sports, gardening, jogging, etc. in this category. In the Third World an example often cited is the energy cost for a mother to walk several miles to take her child to a health centre.

We know little about the extent to which people actually do reduce their activity in response to a shortage of energy supply. Such an adaptation would normally be undesirable if valuable activities were curtailed, but in extreme cases it may be necessary to preserve life. Levi (1987) has given examples from his experience in a concentration camp of extraordinary ingenuity in cutting down work without appearing to do so. In the wartime experiment on volunteers on a low energy intake, by 24 weeks the subjects had become lethargic, with a greatly reduced level of spontaneous physical activity (Keys *et al.*, 1950). Under more ordinary conditions Nature has done the experiment by subjecting people to a 'hungry season', when food from the previous harvest is exhausted and at the same time a great deal of work is needed for cultivating the new crop. For example, in the hungry season farmers in Upper Volta increased their energy expenditure at the expense of a loss of 3–4 kg in body weight. This seems to be a fairly general pattern in Africa (Ferro-Luzzi, Pastore & Sette, 1988). In the light of the previous discussion such a loss can probably be regarded as functionally costless.

Children seem to respond in a different way to a low energy intake, maintaining body weight and even growth at the expense of a reduction in physical activity (Rutishauser & Whitehead, 1972; Torun & Viteri, 1981). We do not know very much about this, because of the extreme difficulty of measuring free-living activity in children. A reduction in physical and exploratory activity in early life may be expected to impair psychological and social development and is therefore not an acceptable adaptation.

Economy of energy expenditure in work

Many physiologists have set out to measure the mechanical efficiency of muscular work and it is tempting to suppose that an increase in this efficiency might be an element in adaptation. I shall return to this subject later. Measurements of mechanical efficiency *in vivo* are full of pitfalls (see e.g. Whipp & Wasserman, 1969; Stainsby *et al.*, 1980) and I shall not consider them here. Of greater interest to the nutritionist, as opposed to the physiologist, is the energy cost of carrying out a given task, of the sort that is performed in real life. I suggest that, as a first step, we need to give more attention to the ergonomic aspects of work and its cost. The appropriate measure of efficiency here is the gross energy cost.

As everyone knows, there are efficient and inefficient ways of doing any piece of physical work. It is reasonable to suppose that tradition and experience have enabled people living on marginal food intakes to find the most economical methods of doing the jobs that they have to do. A great deal of work has been done on the energy cost of different ways of carrying a load (e.g. Legg & Mahanty, 1985). On first principles, for greatest efficiency the load should be carried as close as possible to the centre of gravity of the body (Parkes, 1869). Much interest attaches to the carriage of loads on the head or supported from the head by a frontal yoke. This is the method used by porters in countries such as India and Nepal, who are able to carry very large loads, up to twice their body weight, up steep hills and over long distance. Maloiy *et al.* (1986) found that women in Kenya were able to carry loads up to 20 per cent of their body weight on their heads or with a frontal yoke without any extra energy cost above that of walking unloaded. Jones *et al.* (1987) extended these studies to take account of the fact that body fat represents a 'load'. They found that in Gambian women with varying amounts of body fat there was no increase in the energy expenditure of walking so long as the total load (external load + body fat) did not exceed 40 per cent of the lean body mass. Thus a woman weighing 44 kg with 16 per cent fat (7 kg) and a lean body mass of 37 kg, would have a 'free' load of $(0.4 \times 37) - 7 = 8$ kg, whereas a woman weighing 64 kg, with 29 per cent fat, would have virtually no free load at all. Maloiy *et al.* (1986) suggested that the ability of Kenyan women to carry loads at no cost may result from an anatomical modification of the spinal column, whereas Jones *et al.* (1987) argued in effect that they are able to do this because, compared with people from Western countries, they are carrying such a small load of body fat.

Reanalysis of the Gambian data, kindly supplied by Dr Jones, suggests to me that this is not a sufficient explanation. Table 2.2 shows the

Table 2.2. *Incremental energy cost of load carrying by Gambian women[a] compared with expected cost[b], calculated for US men expressed as increase in rate of energy expenditure above that at zero load (W/kg load)*

Load, per cent of body wt Load, kg	15 7–9.5	35 17–22	45 20–29
Gambian women ($n=8$)	0.81	1.43	1.81
US men	1.61	2.20	2.55
significance of difference	$0.005>p>0.01$	$p<0.01$	$p<0.01$

[a]Data on energy expenditure while walking at 3.2 kg/h kindly supplied by Dr Colette Jones.
[b]From the prediction equation of Pimental & Pandolf (1979), assuming that gradient = 0, terrain factor = 1, velocity = 3.2 km/h.

incremental costs of carrying increasing loads, compared with the expected incremental costs in North American subjects, calculated from the equation of Pimental & Pandolf (1979). This method of expression to a large extent eliminates the effects of differences in body fat. At all three levels of load the incremental cost in the Gambian women was significantly lower than expected, although the advantage decreased with increasing load.

In the US studies the loads were carried on a standard back-pack. The difference is too great to be attributed simply to greater efficiency of load-carrying on the head.* Another factor that might account for the difference is the gait. Many people have commented on the 'gliding' motion of women carrying loads on the head (e.g. Lawrence & Whitehead, 1988). According to R.McN. Alexander (personal communication), from a strictly kinematic point of view up and down movements during walking should not affect the energy cost, since they will simply represent interconversions of kinetic and potential energy. However, it remains to be seen whether this expectation is fulfilled in practice.

The third possible explanation of the findings in Table 2.2 is increased metabolic efficiency of physical work, but this hypothesis must be regarded with great reserve, since up to the present such studies as have been made in Third World subjects have not provided any unequivocal evidence of increased mechanical efficiency (summarized by Waterlow, 1986).

* I have not found a direct comparison between the energy costs of carrying a load on the head or the back, but an indirect comparison can be made between the results of Balogun *et al.* (1986) (load on head) and those of Legg & Mahanty (1985) (load on back). Data are available at similar speeds (4.4 and 4.5 km/h and similar total weights of body+load (94 and 96 kg). In terms of $\dot{V}_{O_2}$ the results are: load on head, 1.21 l/min; load on back, 1.17 l/min.

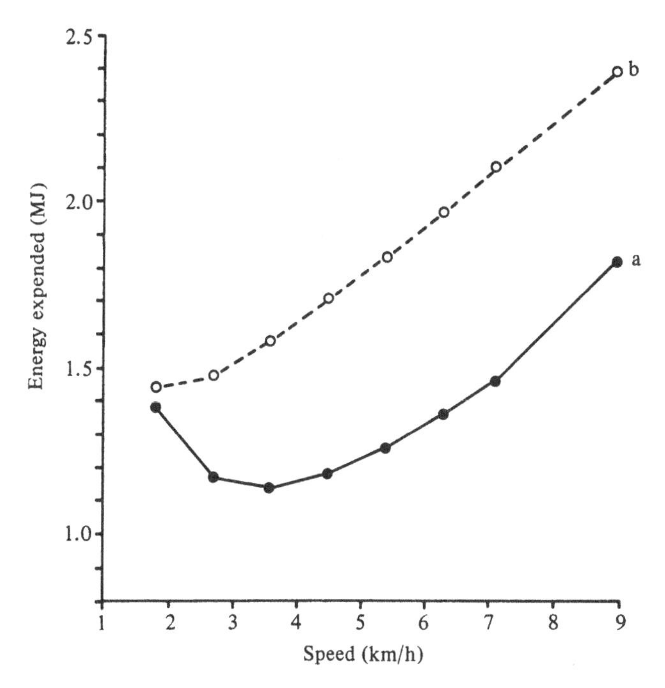

Fig. 2.1 Gross energy expenditure of a 60 kg man carrying a 20 kg load at zero gradient, calculated from Pimental & Pandolf (1979): (a) cost of covering 5 km at different speeds; (b) cost over 3 h, made up as cost of 5 km walk at different speeds plus cost of resting for the remaining time at 280 kJ/h.

Perhaps the most important factor of all in determining the energy cost of walking is the speed. The cost per minute increases as the square of the speed (van der Walt & Wyndham, 1973; Pimental & Pandolf, 1979). It follows, as is very well known, that the cost of walking a given distance, like the cost per step (Workman & Armstrong, 1963), has a U-shaped relation to speed, with a minimum or optimum at 3–3.5 km/h (Fig. 2.1). Suppose one now calculates the total energy expenditure over 3 h, the time being partly spent in walking 5 km and partly resting at the basal rate; a rather different picture now emerges. The curve is no longer U-shaped, and the slower the speed the less the total expenditure. This situation is a good example of the pros and cons of an adaptive process. The fast walker is left with time to spare in which he can do something else, *if* he has the energy to do it. The slow walker has lost this opportunity; on Payne's concept, that the most successful adaptation is one that allows the greatest range of opportunities (Payne, 1985), the slow walker has lost out. On the other hand, from the more limited point of view of economy of energy, which may determine the capacity to survive, it is evidently smart not only to be small but also to be slow.

I have discussed the energy cost of walking in some detail because, at least in theory, it provides opportunities for economy which have not so far received much attention from nutritionists. With the exception of a body of work in India (see Legg & Mahanty, 1985; Balogun *et al.*, 1986), most of these ergonomic studies have been done on Western subjects, very often soldiers. I think it would be very rewarding to extend such studies to people in the Third World who are subsisting on marginal energy intakes.

Possibilities for metabolic or biochemical adaptation

If ATP is the unit of energy transduction from food to work, the problem resolves into how to economize ATP. By analogy with income and expenditure, if one is forced to economize by a reduction in income, there are three possible strategies:

(1) to spend less money (ATP) overall, which means, among other things, avoiding waste;
(2) somehow to earn more money for the same amount of effort;
(3) to find ways of buying more for the same amount of money.

Reduction in the amount of ATP used

In metabolic terms this means reduction in fluxes that utilize ATP. The term 'flux', as used here, means the rate of flow of material through a chemical pathway, or the rate of transfer from one site to another, with units amount/time. Since ATP is used up in muscular contraction, a reduction in muscular work, as discussed above, would be one way of reducing total ATP flux. We must now consider the very large component of energy expenditure that is not involved in physical work, namely the basal or resting metabolic rate. The BMR does, of course, include the cost of some muscular work, such as that needed for respiration and circulation of the blood.

Values for the BMR collected from all over the world show that in tropical countries the rate per kilogram body weight tends to be some ten per cent lower than in Europe or North America (Schofield *et al.*, 1985; Henry & Rees, 1988). It is not absolutely clear how far, if at all, this is a climatic effect. Mason, an American missionary, found that her BMR fell every time she moved from Boston to India (Mason & Jacob, 1972), and Sir Charles Martin, who recorded his BMR every day during a 35 day voyage to Australia, found a distinct fall when his ship entered the Indian Ocean and a rise to the original level when she reached the south-easterly

Trades (Martin, 1930). On the other hand, Eijkman (1924) stated that his BMR was the same in Batavia (Dutch East Indies) as in Holland.

Shetty (1984), working in Bangalore, India, found that labourers with a very low BMI (average 17.6) and a low energy intake (average 1600 kcal per day), had BMRs per kg that were about 10 per cent lower than those of well-nourished controls in the same environment and 17 per cent lower than the rates expected from the prediction equations of Schofield *et al.* (1985). McNeill *et al.* (1987), in studies on peasant farmers in Tamil Nadu (South India), with an average BMI of 19, also found BMRs lower than the predicted rates. They attributed this to differences in body weight and body composition, rather than to a true reduction in the metabolic rate of the tissues. The difficulty with this explanation is that people with a low BMI would have less body fat; moreover, judging by the experimental evidence, any reduction of lean body mass is likely to be mainly at the expense of slowly metabolizing tissues, particularly muscle. One would therefore expect small lean people to have higher BMRs per kilogram than their heavier and fatter Western counterparts.

These studies were on people who had probably been exposed to low energy intakes throughout their lives. The classical experiment of Keys and coworkers (1950) on semistarvation in North American volunteers is, however, relevant because it was continued for six months. James & Shetty (1982), reanalysing the results of that experiment, concluded that 'the small (*sic*) fall in BMR per unit active tissue mass (by 16 per cent) in the first 2 weeks remains essentially unchanged for the subsequent 22 weeks of semistarvation'. A 16 per cent fall is, however, not so small, particularly when one takes account of the probable preferential loss of muscle. Therefore it is logical to look at two of the main components of BMR, protein turnover and ion-pumping, to see how far they might contribute to such a reduction.

Protein turnover

As I said at the beginning of this paper, where there is no variability there is no capacity for adaptation. It is therefore relevant that quite a large range of variation has been observed, perhaps two-fold, in the rates of protein turnover of different individuals (Waterlow, 1988). We do not know the advantages or handicaps of having a high or low rate of whole body protein turnover, nor do we know the effects on it of a habitually low energy intake. We are currently engaged in measuring these rates in Professor Shetty's subjects. It is generally accepted that protein turnover may account for 15–20 per cent of the BMR (Waterlow, 1986), although Jackson (1985) suggested that it may be more. A reduction of one quarter, which is well within the observed range of inter-individual variation, could thus produce a saving of five per cent of the BMR. It should be noted also that the rate of protein turnover is influenced by the activity of the thyroid gland. This has been shown for whole body turnover in obese patients treated with triiodothyronine (Nair *et al.*, 1981; Wolman *et al.*, 1985), and for muscle protein turnover in the rat (Jepson, Bates & Millward, 1988).

Ion-pumping

The concentrations of sodium, potassium and calcium inside and outside cells are very different, and it takes energy to maintain these concentration gradients. What are colloquially called 'pumps' keep sodium out and potassium in. We know the energy cost of pumping one mole, but it is unknown how much energy is used for this process per unit time in the body as a whole. Early estimates (Edelman, 1974) put the cost of Na^+–K^+ pumping at 20–45 per cent of BMR. More recent studies (Biron *et al.*, 1979) suggest a much lower figure – a maximum of 12 per cent. The concentration of free calcium inside cells is lower than outside by about two orders of magnitude, and it is of crucial importance for many vital processes that the internal concentration should be maintained at the correct level. The cost of calcium pumping may be substantial, but I know of no values for the whole body.

As in the case of protein turnover, Na^+–K^+pumping is affected by thyroid hormones. Apparently they do not alter the efficiency of the pumps, but their number (Biron *et al.*, 1979; Kjeldsen *et al.*, 1984); in other words, they produce changes in flux rate. Here again, the capacity to vary provides opportunities for adaptation.

Futile cycles

There are a number of metabolic reactions which go backwards and forwards, producing no net chemical work but only heat. It has been calculated that six of these reactions that have been fairly well characterized might normally account for 10–15 per cent of the BMR (Reeds, Fuller & Nicholson, 1985). There is evidence in at least one case of the activity being greatly reduced when an animal is starved (Challiss, Arch & Newsholme, 1985). It has been suggested that these cycles, which appear simply to waste energy, have in fact a regulatory function (Newsholme, Challiss & Crabtree, 1984). If so, reducing the waste of energy as an adaptation to starvation would be at the cost of losing some of the capacity for fine control.

Efficiency of ATP production

The amount of ATP produced per unit of food energy used or of oxygen consumed varies a little with the nature of the foodstuff oxidized and the route of oxidation. Elia & Livesey (1988) give a very comprehensive and up-to-date review. A typical Third World diet, being low in fat and protein, would be about three per cent more economical than a Western diet. The difference is not large, but every little helps. The difference will be increased if, on the Western diet, some of the carbohydrate is cycled through fat before being oxidized. This process involves an energy loss of some 20 per cent (Milligan, 1971).

These calculations are based on conventional values obtained *in vitro* for the P:O ratio (moles ATP formed per mole oxygen used). In recent years it has become apparent that this process may not always be maximally efficient and that there may be some degree of 'uncoupling' of oxygen utilization and ATP formation. The result would be a greater loss of oxidative energy as heat. The extreme example of uncoupling is shown by brown adipose tissue, which produces heat without any formation of ATP. This tissue probably does not have any important function in adult human subjects. However, the basis exists for speculating that in some people oxidation might be more tightly coupled than in others, so that food energy would be used more efficiently. Stücki (1980), in experiments *in vitro*, has shown that maximum rate and maximum efficiency are incompatible. He uses the analogy of a car: to cover a distance at the least cost of fuel you do not drive as fast as you can. We reach exactly the same conclusion as in our consideration of the energy cost of walking (Fig. 2.1).

Efficiency of ATP utilization

For the reactions considered so far the relationship between the amount of ATP used and the amount of chemical work done is probably fixed. Thus it is thought to take one mole of ATP to pump out three moles of sodium ions, two being exchanged for potassium, and four moles to bind together two moles of amino acids in the synthesis of protein. Even though we may not know the numbers exactly, they are fixed by the nature of the chemical reactions. To return to our analogy: there is probably no way of getting better value for the money spent.

However, muscular contraction provides an exception that is related to mechanical rather than chemical work. The energy that makes the components of a muscle fibre slide along each other when the muscle exerts a pull is ultimately derived from the breakdown of ATP. The relation between the force developed or the amount of mechanical work done to the amount of ATP used is called the contraction coupling efficiency. It has long been known that muscle fibres can be divided into two principle types: the slow-twitch (ST), oxidative or red fibres and the fast-twitch (FT) glycolytic or white fibres, with some intermediate types (see Saltin & Gollnick, 1983). Some muscles are functionally slow and have a predominance of ST fibres; others are fast and contain mainly FT fibres. We can call these ST and FT *muscles*. Many muscles, however, contain a mixture of both types of fibre. People differ in the pattern of fibres in their muscles. For example, the proportion of ST fibres in the *vastus lateralis* muscle, measured in biopsies, has been found to vary from 10 to over 80 per cent (Grindrod, Round & Rutherford, 1987). It is generally considered that these patterns are genetically determined.

The relevance of this to the problem of adaptation is that ST fibres have a higher contraction coupling efficiency than FT fibres. In muscles contracting isometrically, so that no external work is done, ST muscles use less ATP per unit tension developed (Gibbs & Gibson, 1972; Wendt & Gibbs, 1973; Awan & Goldspink, 1972; Goldspink, 1975; Crow & Kushmerik, 1982). In isotonic contractions, where the muscle is allowed to shorten, the efficiency depends not only on the fibre type but also on the speed of shortening. When muscles of different species were compared an inverse relationship was found between intrinsic speed of shortening and biochemical efficiency (Nwoye & Goldspink, 1981). In the intact animal, as the rate of work increases, more and more FT fibres are recruited (Armstrong & Laughlin, 1985). One would expect this to be accompanied by a decrease in biochemical efficiency.

In man a positive correlation has been found between the percentage of ST fibres in the *vastus lateralis* muscle of the thigh and the force developed in an isometric contraction (Young, 1984; Grindrod *et al.*, 1987). In what can only be described as a heroic experiment Suzuki (1979) identified two groups, each of three subjects; the ST group had 70–90 per cent of ST fibres in the *vastus lateralis*; the other group (FT) had only 15–30 per cent. Efficiency of work was measured as Δ work/Δ oxygen consumption on the bicycle ergometer. At 60 rpm there was no difference in efficiency between the two groups; at higher speeds the FT group was more efficient. It is a pity that in this experiment no tests were made at slower speeds and greater work loads.

In conditions of low energy intake and in hypothyroidism a reduction has been reported in the proportion and diameter of FT fibres (Russell *et al.*, 1984; Wiles *et al.*, 1979). In hypothyroidism this appears to be accompanied by an increase in the contraction coupling efficiency (Wiles *et al.*, 1979; Leijendekker, van Hardeveld & Kassenaar, 1983). It is of interest in this connection that experimentally thyroidectomy has been found to cause a much greater depression of protein synthesis in the FT muscle gastrocnemius than in the ST soleus (Brown & Millward, 1983). Perhaps slow muscles and fibres are selectively preserved in conditions in which the metabolic rate is reduced. This would certainly make sense from the point of view of adaptation.

I have discussed this subject at some length because as a physiologist I find it very interesting and very relevant to the question of possible mechanisms of adaptation to a shortage of energy. Goldspink (1975) has suggested that in some cases in the course of evolution efficiency may have been gained at the cost of speed. Is it possible that it is indeed smart to be slow, and that natural selection has favoured people with a particular pattern of muscle fibres? This is an area that remains to be explored by physiologists and physical anthropologists.

Conclusion

It is not difficult to think of ways in which the body may respond to a lifetime of living on what we would regard as a marginal or frankly inadequate energy intake. Apart from the obvious responses of keeping down the body weight and reducing physical activity, I have speculated about more subtle physiological and biochemical mechanisms for economizing energy. The question will inevitably arise: would it serve any useful purpose to invest in research aimed at identifying and characterizing such mechanisms? We know that people survive and work on apparently low energy intakes. What does it matter how they manage to do it? I believe that a better understanding of these mechanisms is needed to throw light on the constraints under which people operate, the so-called costs of adaptation. From the point of view of food policy it is necessary to define the range of acceptable costs.

References

Alexander, R.McN. (1984). Stride length and speed for adults, children and fossil hominids. *American Journal of Physical Anthropology*, **63**, 23–7.

Armstrong, R.B. & Laughlin, M.H. (1985). Metabolic indicators of fibre recruitment in mammalian muscles during locomotion. *Journal of Experimental Biology*, **115**, 201–13.

Awan, M.Z. & Goldspink, G. (1972). Energetics of the development and maintenance of isometric tension by mammalian fast and slow muscles. *Journal of Mechanochemical Cell Motility*, **1**, 97–108.

Balogun, J.A., Robertson, R.J., Goss, F.L., Edwards, M.A., Cox, R.C. & Metz, K.F. (1986). Metabolic and perceptual responses while carrying external loads on the head and by yoke. *Ergonomics*, **29**, 1623–35.

Barcroft, H. (1934). *Features in the Architecture of Physiological Function*. Cambridge University Press, Cambridge.

Biron, R., Burger, A., Chinet, A., Clausen, T. & Dubois-Ferriere, R. (1979). Thyroid hormones and the energetics of active sodium-potassium transport in mammalian skeletal muscles. *Journal of Physiology*, **297**, 47–60.

Brown, J.G. & Millward, D.J. (1983). Dose response of protein turnover in rat skeletal muscle to triiodothyronin treatment. *Biochemica Biophysica Acta*, **757**, 182–90.

Challiss, R.A.J., Arch, J.R.S. & Newsholme, E.A. (1985). Starvation for 24h decreases fructose 6-phosphate/fructose 1, 6-biphosphate substrate cycling in skeletal muscle. *Biochemical Society Transactions*, **13**, 269–70.

Crow, M.T. & Kushmerick, M.J. (1982). Chemical energetics of mammalian muscle. *Journal of General Physiology*, **79**, 147–66.

Edelman, I.S. (1974). Thyroid thermogenesis. *New England Journal of Medicine*, **290**, 1303–8.

Eijkman, C. (1924). Some questions concerning the influence of tropical climate on man. *Lancet*, i, 887–93.

Elia, M. & Livesey, G. (1988). Theory and validity of indirect calorimetry during net lipid synthesis. *American Journal of Clinical Nutrition*, **47**, 591–607.

Eveleth, P.B. & Tanner, J.M. (1976). *World-wide Variation in Human Growth*. Cambridge University Press, Cambridge.

FAO (1987). *The Fifth World Food Survey*. FAO, Rome.

FAO/WHO/UNU (1985). *Energy and Protein Requirements*. Report of a Joint FAO/WHO/UNU Consultation. Technical Report Series 724. WHO, Geneva.

Ferro-Luzzi, A. (1988). Marginal energy malnutrition: some speculations on primary energy-sparing mechanisms. In: *Capacity for Work in the Tropics*, ed. K.J. Collins & D.F. Roberts, pp. 141–64. Cambridge University Press, Cambridge.

Ferro-Luzzi, A., Pastore, G. & Sette, S. (1988). Seasonality in energy metabolism. In: *Chronic Energy Deficiency: Consequences and Related Issues*, ed. B. Schürch & N.S. Scrimshaw, pp. 37–58.

Forbes, G.B. (1985). Body composition as affected by physical activity and nutrition. *Federation Proceedings*, **44**, 343–7.

Gibbs, C.L. & Gibson, W.R. (1972). Energy production of rat soleus muscle. *American Journal of Physiology*, **223**, 864–71.

Goldspink, G. (1975). Biochemical energetics for fast and slow muscles. In: *Comparative Physiology – Functional Aspects of Structural Materials*, ed. L. Bolis, H.P. Maddrell & K. Schmidt-Nielsen. North-Holland, Amsterdam.

Grindrod, S., Round, J.M. & Rutherford, O. (1987). Type 2 fibre composition and force per cross-sectional area in the human quadriceps. *Journal of Physiology*, **390**, 154.

Henry, C.J.K. & Rees, D.G. (1988). A preliminary analysis of basal metabolic rate and race. In: *Comparative Nutrition*, ed. K.L. Blaxter & I.A. Macdonald, pp. 149–59. John Libbey, London.

Jackson, A.A. (1985). Nutritional adaptation in disease and recovery. In: *Nut-*

ritional Adaptation in Man, ed. K.L. Blaxter & J.C. Waterlow, pp. 111–26. John Libbey, London.
James, W.P.T., Ferro-Luzzi, A. & Waterlow, J.C. (1988). Definition of chronic energy deficiency in adults. Report of a working party of the Dietary Energy Consumption Group. *European Journal of Clinical Nutrition*, **42**, 969–82.
James, W.P.T. & Shetty, P.S. (1982). Metabolic adaptations and energy requirements in developing countries. *Human Nutrition: Clinical Nutrition*, **36C**, 331–6.
Jepson, M.M., Bates, P.C. & Millward, D.J. (1988). The role of insulin and thyroid hormones in the regulation of muscle growth and protein turnover in response to dietary protein in the rat. *British Journal of Nutrition*, **59**, 397–415.
Jones, C.D.R., Jarjon, M.S., Whitehead, R.G. & Jequier, E. (1987). Fatness and the energy cost of carrying loads in African women. *Lancet*, ii, 1331–2.
Keys, A., Brozek, J., Henschel, A., Michelson, O. & Taylor, H.L. (1950). *The Biology of Human Starvation*. University of Minnesota Press, Minneapolis.
Kjeldsen, K., Norgaard, A., Gotzsche, C.O., Thomassen, A. & Clausen, T. (1984). Effect of thyroid function on number of Na-K pumps in human skeletal muscle. *Lancet*, ii, 8–10.
Knight, I. (1984). *The Heights and Weights of Adults in Great Britain*. HMSO, London.
Lawrence, M. & Whitehead, R.G. (1988). Physical activity and total energy expenditure of child-bearing Gambian women. *European Journal of Clinical Nutrition*, **42**, 145–60.
Legg, S.J. & Mahanty, A. (1985). Comparison of five modes of carrying a load close to the trunk. *Ergonomics*, **28**, 1653–60.
Leijendekker, W.J., van Hardeveld, C. & Kassenaar, A.A.H. (1983). The influence of the thyroid state on energy turnover during tetanic stimulation in the fast-twitch (mixed type) muscle of rats. *Metabolism*, **32**, 615–27.
Levi, P. (1987). *If This is a Man*. Sphere Books, London.
Maloiy, G.M.O., Heglund, N.C., Prager, L.M., Cavagna, G.A. & Taylor, C.R. (1986). Energetic costs of carrying loads: have African women discovered an economic way? *Nature (London)*, **319**, 668–9.
Martin, C.J. (1930). Thermal adjustments of man and animals to external conditions. *Lancet*, ii, 617–21.
Mason, E.D. & Jacob, M. (1972). Variations in basal metabolic rate: responses to changes between tropical and temperate climates. *Human Biology*, **44**, 141–72.
McNeill, G., Rivers, J.P.W., Payne, P.R., de Britto, J.J. & Abel, R. (1987). Basal metabolic rate of Indian men: no evidence of metabolic adaptation to a low plane of nutrition. *Human Nutrition: Clinical Nutrition*, **41C**, 473–84.
Milligan, L.P. (1971). Energetic efficiency and metabolic transformations. *Federation Proceedings*, **30**, 1454–8.
Nair, K.S., Halliday, D., Lalloz, M. & Garrow, J.S. (1981). Rate of protein turnover in obese women on energy restriction and its relationship to extra-thyroidal T4 metabolism. *Proceedings of the Nutrition Society*, **40**, 94A.
Nelson, R.A., Jones, J.D., Wahner, H.W., McGill, D.B. & Cole, C.F. (1975). Nitrogen metabolism in bears. *Mayo Clinic Proceedings*, **50**, 141–6.
Newsholme, E.A., Challiss, R.A.J. & Crabtree, B. (1984). Substrate cycles:

their role in improving sensitivity in metabolic control. *Trends in Biochemical Science*, **9**, 277–80.

Nwoye, L.O. & Goldspink, G. (1981). Biochemical efficiency and intrinsic shortening speed in selected vertebrate fast and slow muscles. *Experientia*, **37**, 856.

Parkes, E.A. (1869). *A Manual of Practical Hygiene for Use in the Medical Service of the Army*. 3rd edition. Churchill, London.

Payne, P.R. (1985). Energy and protein requirements. In: *Agricultural Development and Nutrition*, ed. A. Pacey & P.R. Payne, pp. 51–72. Hutchinson, London.

Payne, P.R. (1987). Malnutrition and human capital: problems of theory and practice. In: *Poverty, Development and Food*, ed. E. Clay & J. Shaw, pp. 22–41. Macmillan, London.

Pimental, N.A. & Pandolf, K.B. (1979). Energy expenditure while standing or walking slowly uphill or downhill with loads. *Ergonomics*, **22**, 963–73.

Reeds, P.J., Fuller, M.F. & Nicholson, B.A. (1985). Metabolic basis of energy expenditure with particular reference to protein. In: *Substrate and Energy Metabolism in Man*, ed. J.S. Garrow & D. Halliday, pp. 46–57. John Libbey, London.

Russell, D.McR., Walker, P.M., Leiter, L.A., Sima, A.A.F., Tanner, W.K., Mickle, D.A.G., Whitwell, J., Marliss, E.B. & Jeejeebhoy, K.N. (1984). Metabolic and structural changes in skeletal muscle during hypocaloric dieting. *American Journal of Clinical Nutrition*, **39**, 503–13.

Rutishauser, I.H.E. & Whitehead, R.G. (1972). Energy intake and expenditure in 1–3 year old Ugandan children living in a rural environment. *British Journal of Nutrition*, **28**, 145–52.

Saltin, B. & Gollnick, P.D. (1983). Skeletal muscle adaptability: significance for metabolism and performance. In: *Handbook of Physiology*, ed. L.D. Peachey, R.H. Adrian & S.R. Geiger, pp. 555–631. American Physiological Society, Bethesda, Md. USA.

Schofield, W.N., Schofield, C. & James, W.P.T. (1985). Basal metabolic rate – review and prediction, together with an annotated bibliography of source material. *Human Nutrition: Clinical Nutrition*, **39C**, Suppl. 1, 5–41.

Shetty, P.S. (1984). Adaptive changes in basal metabolic rate and lean body mass in chronic undernutrition. *Human Nutrition: Clinical Nutrition*, **38C**, 443–52.

Stainsby, W.N., Gladden, B.L., Barclay, J.K. & Wilson, B.A. (1980). Exercise efficiency: validity of base-line subtractions. *Journal of Applied Physiology: Respiratory, Environmental and Exercise Physiology*, **48**, 518–22.

Stücki, J.W. (1980). The optimal efficiency and the economic degrees of coupling of oxidative phosphorylation. *European Journal of Biochemistry*, **109**, 269–83.

Suzuki, Y. (1979). Mechanical efficiency of fast- and slow-twitch muscle fibers in man during cycling. *Journal of Applied Physiology: Respiratory, Environmental and Exercise Physiology*, **47**, 263–7.

Torun, B. & Viteri, F.D. (1981). Energy requirements of pre-school children and effects of varying energy intakes on protein metabolism. In: *Protein–Energy Requirements of Developing Countries: Evaluation of New Data*, ed. B. Torun, V.R. Young & W.M. Rand, pp. 229–41. UNU Food & Nutrition Bulletin, Supplement 5. United Nations University, Tokyo.

van der Walt, W.H. & Wyndham, C.H. (1973). An equation for prediction of energy expenditure of walking and running. *Journal of Applied Physiology*, **34**, 559–63.

Waterlow, J.C. (1985a). What do we mean by adaptation? In: *Nutritional Adaptation in Man*, ed. K.L. Blaxter & J.C. Waterlow, pp. 1–12. John Libbey, London.

Waterlow, J.C. (1985b). Postscript. In: *Nutritional Adaptation in Man*, ed. K.L. Blaxter & J.C. Waterlow, pp. 233–6. John Libbey, London.

Waterlow, J.C. (1986). Metabolic adaptation to low intakes of energy and protein. *Annual Reviews of Nutrition*, **6**, 495–526.

Waterlow, J.C. (1988). Observations on the variability of man. In: *Comparative Nutrition*, ed. K.L. Blaxter & I.A. Macdonald, pp. 133–40. John Libbey, London.

Wendt, I.R. & Gibbs, C.L. (1973). Energy production of rat extensor digitorum longus muscle. *American Journal of Physiology*, **224**, 1081–6.

Whipp, B.J. & Wasserman, K. (1969). Efficiency of muscular work. *Journal of Applied Physiology*, **26**, 644–8.

Wiles, C.M., Young, A., Jones, D.A. & Edwards, R.H.T. (1979). Muscle relaxation rate, fibre-type composition and energy turnover in hyper- and hypo-thyroid patients. *Clinical Science*, **57**, 375–84.

Wolman, S.J., Sheppard, H., Fern, M. & Waterlow, J.C. (1985). The effect of tri-iodothyronine (T3) on protein turnover and metabolic rate. *International Journal of Obesity*, **9**, 459–63.

Workman, J.M. & Armstrong, B.W. (1963). Oxygen cost of treadmill walking. *Journal of Applied Physiology*, **18**, 798–803.

Young, A. (1984). The relative isometric strength of type I and type II fibers in the human quadriceps. *Clinical Physiology*, **4**, 23–32.

3 *The impact of chronic undernutrition on physical work capacity and daily energy expenditure*

G.B. SPURR

In many developing countries of the world there is still a major dependence on human physical labor because of poor mechanization (Smil, 1979). It is precisely these same countries which exhibit deficits in the adult size of their populations due to lack of dietary calories and/or other specific nutrient(s) during the period of growth. Table 3.1 shows the percentage of men and women engaged in moderate to heavy physical work as well as the prevalence of stunting in the children two years of age in several countries. It is widely agreed that the primary causes of stunting in populations are environmental (nutritional) not genetic in nature (Martorell, 1985). Furthermore, the growth of the children in a population is a good indicator of the nutritional stress on that population. A recent publication details the results of deliberations on the causes and effects of linear growth retardation in less developed countries (Waterlow, 1988).

Because of the association of heavy physical labor and poor nutrition in developing countries (Table 3.1) there has been interest in the relationship between the physical work capacity (PWC) and nutritional status and between PWC and productivity in moderate to heavy physical work (Spurr, 1983).

One response to a chronic deficiency in dietary energy availability is a reduced energy expenditure in the form of decreased physical activity. Indeed, there have been reports of diminished activity in pre-school children (Rutishauser & Whitehead, 1972; Chavez & Martinez, 1979; Viteri & Torún, 1981) and in adults (Keys *et al.*, 1950; Viteri & Torún, 1975; Gorsky & Calloway, 1983) who were malnourished or on restricted calorie intakes. There are few reports of daily energy expenditure in school-aged children who were at nutritional risk (Spurr, Reina & Barac-Nieto, 1986; Spurr & Reina, 1987). We have recently reported on the pattern of daily energy expenditure in nutritionally normal and marginally malnourished children (Spurr & Reina, 1988a) and on the effects of a dietary intervention on artificially increased activity levels in

Table 3.1. *Percentages of economically active populations engaged in moderate to heavy physical work (agriculture, hunting, fishing, forestry, mining and construction) and the prevalence of stunting at two years of age in some developing countries*

Country	Percentage engaged in moderate to heavy work[a] Male	Female	Prevalence of stunting at two yrs of age[b] (%)
Honduras	75.1	7.5	40
Guatemala	70.5	7.1	43–50
Brazil	57.9	20.7	25–30
Costa Rica	52.7	4.3	20
Sri Lanka	50.8	62.1	40
Phillipines	72.0	35.8	60
Cameroon	67.6	87.4	40

[a]UN Demographic Year Book, 1980.
[b]Keller, 1988.

the same nutritional groups of boys 10–12 years of age (Spurr & Reina, 1988b).

Because there is a relationship between daily activity in children and cognitive, social and motor development (Malina, 1984; Sameroff & McDonough, 1984), as well as an apparent influence on the growth of PWC as measured by the maximum oxygen consumption ($\dot{V}_{O_2}$ max) (Mirwald & Bailey, 1981), the present report will summarize the work on the effects of nutritional status on PWC and productivity in adult males, of nutritional status on the growth of PWC in school-aged children and of nutritional status on activity levels in children 6–16 years of age. Although many investigators have contributed importantly to the development of knowledge in this field, this discussion is not meant to be a comprehensive review since others are recently available (Spurr, 1983, 1988). Rather, a brief summary of our older work on PWC will be presented together with more recent investigations on the total daily energy expenditure and the pattern of its expenditure in school-aged children, which have originated in our laboratory in Cali, Colombia.

Colombia is a country considered to be at a middle level of development. Cali is an industrial city of 1.7 million inhabitants, the third largest in Colombia, located 3° 22′ north of the equator at an altitude of 976 m. In common with other Latin American cities, during the past 20–25 years it has undergone rapid growth due, in part, to an influx of population from rural areas. It enjoys a year-round average temperature of some 24 °C (high 29 °C, low 18 °C) which varies little throughout the year, so that wide seasonal differences in ambient temperature were not a factor in the studies to be described. There are two rainy seasons (March to June and October to December) during which average monthly rainfall may reach a maximum of 18 cm, while maximum rainfall during the dry seasons is 6–7 cm/month (CVC, 1984).

Physical work capacity

The maximum oxygen consumption ($\dot{V}_{O_2}$ max) *and applied work physiology*

The V_{O_2} max can be defined as the highest oxygen uptake an individual attains during physical work while breathing air at sea level (Åstrand & Rodahl, 1986). It is a measure of the maximum energy output by the aerobic processes of the muscles involved and the functional capacity of the circulation, since a high correlation exists between $\dot{V}_{O_2}$ max and cardiac output. As a matter of fact, one expression of $\dot{V}_{O_2}$ max is; V_{O_2} max $(l \cdot min^{-1})$ = cardiac output $(l \cdot min^{-1}) \times$ arterial–venous O_2 difference $(l\ O_2 \cdot l\ blood^{-1})$. As will be seen, the $\dot{V}_{O_2}$ max is also indirectly a measure of the endurance capacity of the individual, i.e. the ability to sustain a given work load. It is usually measured on a treadmill or bicycle ergometer with the former giving values somewhat higher than the latter (Åstrand & Rodahl, 1986). Some factors which might influence the $\dot{V}_{O_2}$ max (PWC) in normal or malnourished individuals are:

- (i) *Somatic*
  - (a) sex and age
  - (b) body size
- (ii) *Training and adaptation*
- (iii) *Psychic*
  - (a) attitude
  - (b) maturation
- (iv) *Metabolic and cardio-vascular*
  - (a) fuel
    - (1) intake
    - (2) storage
    - (3) mobilization
  - (b) oxygen uptake
    - (1) pulmonary ventilation
    - (2) cardiac output – stroke volume and heart rate
    - (3) oxygen transport by the blood–anemia
    - (4) oxygen extraction – arterial–venous O_2 difference

These are discussed in detail by Åstrand & Rodahl (1986, p. 296 and p. 488).

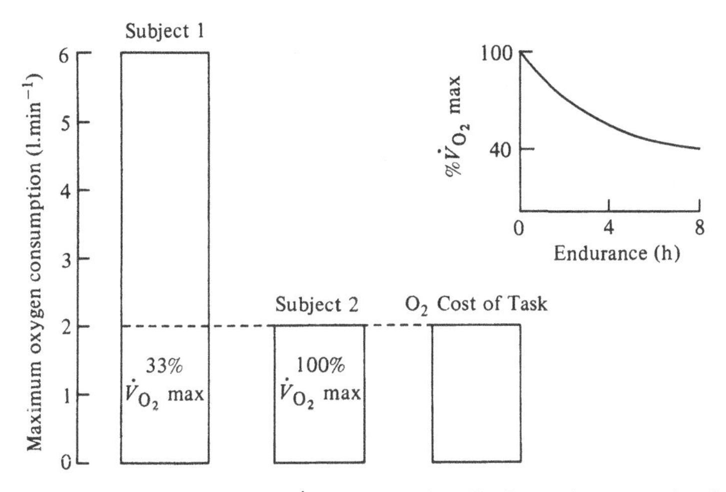

Fig. 3.1 The relationship of % $\dot{V}_{O_2}$ max (relative effort) to endurance and to the oxygen cost of a task.

The relationships among $\dot{V}_{O_2}$ max, endurance and relative effort in work, as expressed by the % $\dot{V}_{O_2}$ max utilized in a work task, are illustrated in Fig. 3.1. The graph in the upper right of Fig. 3.1 points out the negative exponential relationship which exists between relative effort (% $\dot{V}_{O_2}$ max) and endurance time (Åstrand & Rodahl, 1970; Barac-Nieto, 1987) and the fact that there is a sizeable body of information which shows that about 40% $\dot{V}_{O_2}$ max is the maximum effort that can be sustained during an eight-hour working day (Michael, Hutton & Horvath, 1961; Åstrand, 1967; Spurr, Barac-Nieto & Maksud, 1975).

Fig. 3.1 shows the $\dot{V}_{O_2}$ max ($l \cdot min^{-1}$) in two subjects in relation to the oxygen cost of an unspecified work task amounting to $2\ l \cdot min^{-1}$. One individual with a high $\dot{V}_{O_2}$ max of $6\ l \cdot min^{-1}$ would be able to accomplish the task while working at 33% of his maximum and therefore be able to easily sustain it during an eight hour working day and perhaps even longer. On the other hand, the second subject with a $\dot{V}_{O_2}$ max of only $2\ l \cdot min^{-1}$ would have to use 100% of his maximum effort and therefore would be able to sustain the task for only a few minutes. While the examples exploit two extremes of what might occur in real life, they serve to demonstrate the relationships between $\dot{V}_{O_2}$ max ($l \cdot min^{-1}$) and relative effort (% $\dot{V}_{O_2}$ max) and between the latter and endurance time (Fig. 3.1).

Growth of physical work capacity ($\dot{V}_{O_2}$ max) *in children*

Most of the studies of exercise and work capacity in malnourished

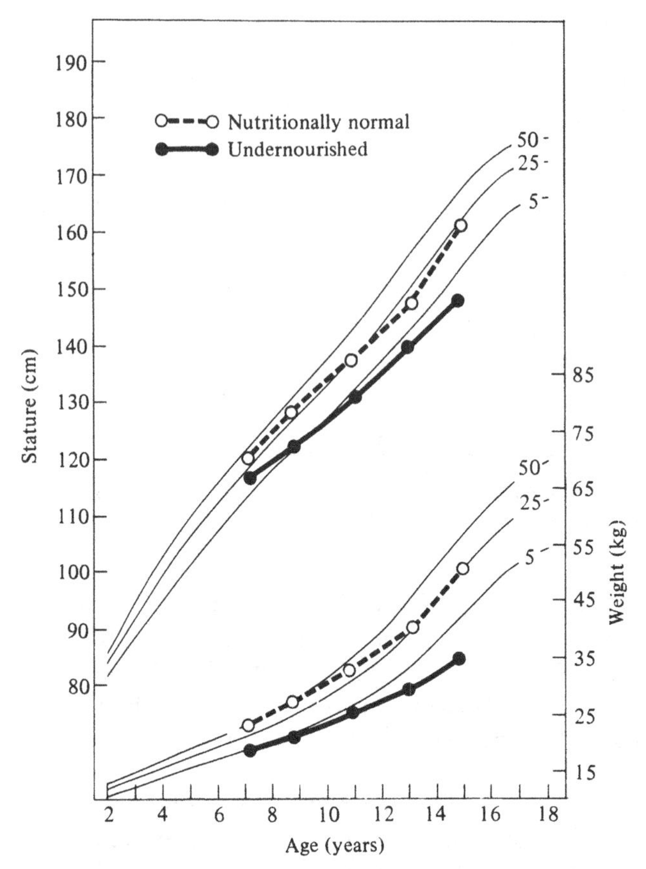

Fig. 3.2 Average values of height and weight of control and undernourished school-aged Colombian boys plotted as functions of average group ages on NCHS percentiles (from Spurr, 1988).

children have been carried out using sub-maximal exercise testing (Åreskog, Selinus & Vahlquist, 1969; Davies, 1973a; Satyanarayana, Nadamuni Naidu & Narasinga Rao, 1979). Our studies on the direct measurements of $\dot{V}_{O_2}$ max are the only ones of which we are aware in children from third world countries. In the earlier work from our laboratory, all subjects were boys who had to present their official birth certificates as a first condition for inclusion in the study. They were grouped into five age groups at two year intervals from 6 to 16 years of age. Using the Colombian standards established by Rueda-Williamson *et al.* (1969) children were selected who had weight-for-age and weight-for-height >95% (but <110%) of those predicted as being nutritionally normal and without a history of undernutrition. Those with both weight-for-age and weight-for-height <95% of the standard were considered to be undernourished at the time of the study. The reason for choosing 95% as the cut-off point was entirely arbitrary with the expectation that the group averages would be considerably below this point (Spurr *et al.*, 1983a, b). The details of the selection process and the methodology employed in the anthropometric and maturation (Spurr *et al.*, 1983a), $\dot{V}_{O_2}$ max (Spurr *et al.*, 1983b), body composition (Barac-Nieto, Spurr & Reina, 1984), and work efficiency measurements (Spurr *et al.*, 1984) have been described in the original publications.

The average heights and weights of the five age groups of nutritionally normal and undernourished boys are plotted on the US NCHS (Hamill *et al.*, 1979) percentile grids in Fig. 3.2. The number of subjects in each group varied from 24 to 60. The nutritionally normal boys followed the fiftieth percentile in the younger age groups and deviated towards the twenty-fifth percentile in the older groups (Fig. 3.2) for both height and weight. The tendency towards shorter stature in these boys is probably the result of the high percentage of mestizos (74%) who have shorter stature than other children (Zavaleta & Malina, 1980). Both height-for-age and weight-for-age of the undernourished boys were near or below the fifth percentile during this period of growth. The weight-for-height of the normal subjects fell slightly above the fiftieth percentile throughout this period, while the undernourished boys followed approximately the tenth percentile (Spurr *et al.*, 1983a). In addition to the depressed growth pattern seen in Fig. 3.2, the undernourished subjects had significantly lower values for skinfolds, a significantly delayed growth spurt and sexual maturation (Spurr *et al.*, 1983a) and, in a sub-group, increased fasting levels of circulating growth hormone (G.B. Spurr and J.C. Reina, unpublished). Consequently, the selection process resulted in the separation of undernourished boys, who were smaller and thinner than normal boys. The latter were following an essentially normal growth development when compared to either national or international (NCHS; Fig. 3.2) norms. However, the physiologic data (slowed growth velocity, delayed sexual maturation and high circulating growth hormone concentrations) make it clear that the *reason* for their smallness and thinness is that they are undergoing a process of chronic malnutrition which is no doubt 'marginal' in nature but nevertheless real. Furthermore, the fact that there is a progressive deviation from predicted values of height- and weight-for-age from younger to older boys (Fig. 3.2; Spurr *et al.*, 1983a) indicates that the process is cumulative with age.

The mean $\dot{V}_{O_2}$ max values of the boys in Fig. 3.2, together with the results of a two-way analysis of variance, are plotted in Fig. 3.3. There are statistically significant increases with age and significantly lower values in the undernourished boys than those classified as nutritionally

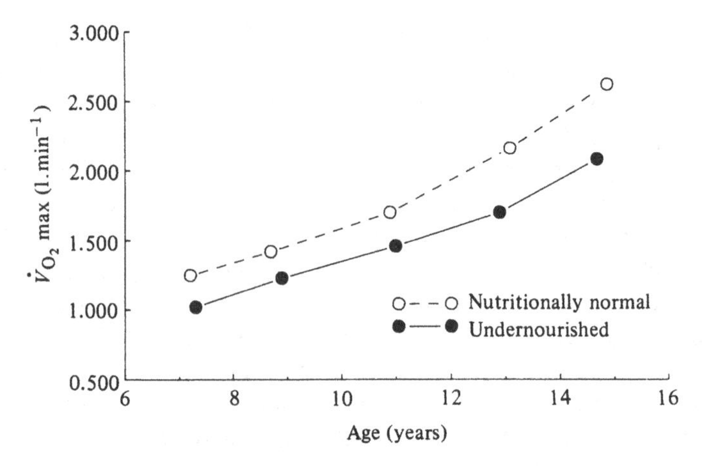

Fig. 3.3 Average maximum oxygen consumption of boys in Fig. 3.2. Results from a two-way analysis of variance showed significant ($p<0.001$) differences between age groups and between nutritional classes and a significant interaction between them (data from Spurr *et al.*, 1983b).

normal throughout the age range studied. Also, no doubt as a reflection of the progressively greater differences in body weight seen with age in the two nutritional groups (Fig. 3.2), there was also a statistically significant age–nutritional group interaction. The meaning of these data in terms of adult work capacity, and when expressed per unit of body weight and lean body mass in comparison with similar values for adult men, is discussed below.

We have recently reported (Spurr & Reina, 1989) a similar study on both boys and girls with essentially similar findings for age and nutritional group effects for both sexes. The $\dot{V}_{O_2}$ max values for girls were significantly lower than those for boys as has been reported by a number of investigators for children from more advantaged countries (Krahenbuhl, Skinner & Kohrt, 1985).

Malnutrition, physical work capacity and productivity in adult males

Since the malnourished individual is usually not working (a reason for the malnourished state), particularly in moderate or heavy work tasks, it has not been possible to relate malnourished states directly to productivity. Rather, the attempt has been made to relate both nutritional status and productivity (measured in nutritionally normal, employed subjects) to a common measurement ($\dot{V}_{O_2}$ max) and from these relationships to infer the association between nutritional status and productivity in moderate to heavy work. Most reports in the literature are the result of measurements in male subjects.

We have studied three groups of chronically malnourished adult males who were selected for their existing degree of undernutrition (Barac-Nieto *et al.*, 1978a). The most severely malnourished of these subjects were also studied during a 45 day basal period in the hospital and during 79 days of a dietary repletion regime (Barac-Nieto *et al.*, 1980). Subjects were classified into those with mild (M), intermediate (I) and severe (S) malnutrition based on their weight: height (W: H) ratios, serum albumin (AL) concentrations and daily creatinine excretions per meter of height (Cr/H) (Barac-Nieto *et al.*, 1978a). Detailed body composition and biochemical measurements of the three groups were made shortly after admission to the hospital metabolic ward (Barac-Nieto *et al.*, 1978a) and during the dietary repletion regime of Group S (Barac-Nieto *et al.*, 1979). Upon entry into the hospital, the subjects were placed on an energy intake (2240 kcal $\cdot$ day^{-1}; 9.4 MJ $\cdot$ day^{-1}) adequate for the sedentary conditions of the metabolic ward, but were maintained on the same protein intake (27 g $\cdot$ day^{-1}) they were ingesting prior to entry. Studies of work capacity and endurance in the severely malnourished men were made at the beginning and end of the 45 day basal period on the diet. The protein intake was then increased to 100 g $\cdot$ day^{-1} for the 79 day repletion regime; the increased calorie intake from proteins was balanced by reducing carbohydrate intake to maintain isoenergetic diets. Measurements of $\dot{V}_{O_2}$ max and endurance were repeated at 90 and 124 days after admission to hospital. The results for the three groups and the changes in the severely malnourished men during dietary repletion are presented in Fig. 3.4 and Fig. 3.5 and compared with data on 107 nutritionally normal control subjects who were sugar cane cutters (Spurr *et al.*, 1975), loaders (Spurr, Maksud & Barac-Nieto, 1977a), or general farm laborers (Maksud, Spurr & Barac-Nieto, 1976). There were progressive differences in body weight, W: H ratio, AL and total proteins in the control (C), M, I and S groups (Fig. 3.4). Groups C and M were not significantly different in regard to hematocrit and blood hemoglobin (Hb) but I and S were significantly and progressively depressed in these measurements. There was a slight gain in body weight of Group S during the basal period, but otherwise the variables did not change. Weight, W: H ratio and the serum proteins showed progressive improvement during the repletion regime, but the hematological values did not show improvement until the final round of measurements (Fig. 3.4).

Fig. 3.5 presents the results for maximal heart rate (f_H max), maximal aerobic power ($\dot{V}_{O_2}$ ml $\cdot$ min^{-1} $\cdot$ kg body weight^{-1}) and $\dot{V}_{O_2}$ max (l $\cdot$ min^{-1}) for the control and malnourished subjects. Average f_H max

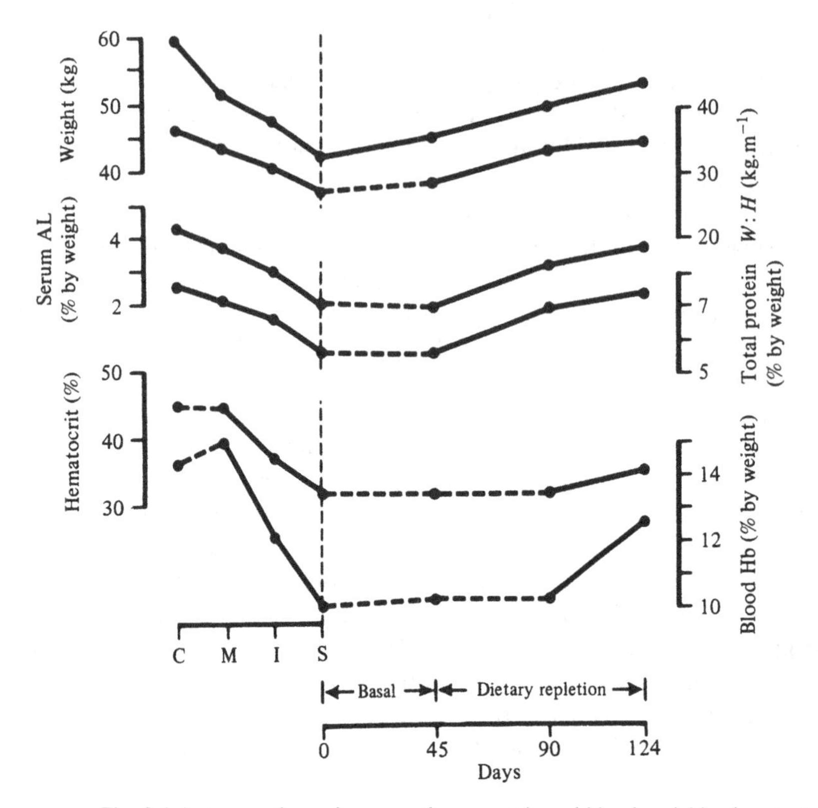

Fig. 3.4 Average values of some anthropometric and blood variables (see text for details) in nutritionally normal (C) subjects and men with mild (M), intermediate (I), and severe (S) malnutrition. The severely undernourished were studied during a basal period on adequate calories and low protein followed by a dietary repletion period on an isoenergetic but high protein diet. Solid lines connect points which are significantly different from each other (from Spurr 1983).

values were not different in the various groups nor did they change during dietary repletion. However, $\dot{V}_{O_2}$ max and maximal aerobic power were progressively less in M, I and S than in C subjects. They did not change in the S Groups during the basal period, and then progressively improved during dietary repletion, although they did not return to even the level of Group M during the period of study. Fig. 3.5 also expresses a theoretical sub-maximal work load of 0.75 $1 \cdot min^{-1}$ $\dot{V}_{O_2}$ in terms of % $\dot{V}_{O_2}$ max for each of the groups. From Fig. 3.5, it is clear that $\dot{V}_{O_2}$ max and maximal aerobic power are markedly depressed in chronic malnutrition and that the degree of reduction is related to the severity of depression in nutritional status. Using the three groups of malnourished

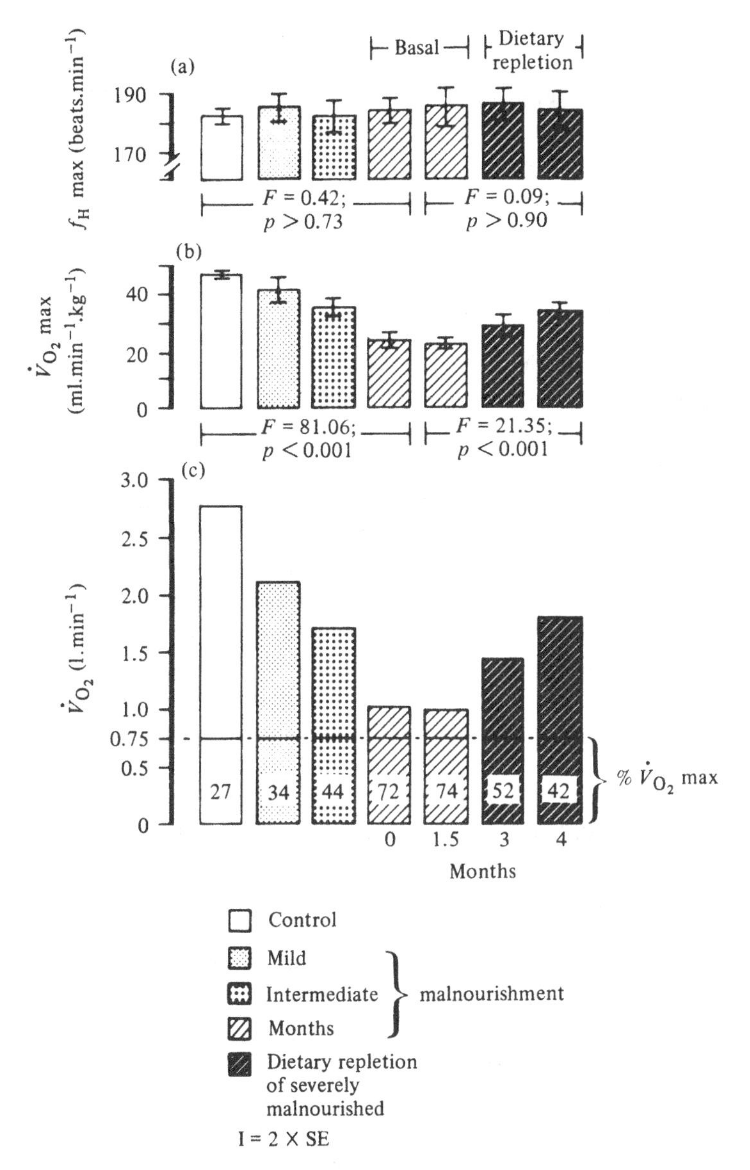

Fig. 3.5 (a) Maximum heart rates (f_H max), (b) aerobic power (see text for details), and (c) $\dot{V}_{O_2}$ max in control, undernourished, and severely malnourished subjects and during dietary repletion of the last group. *F* ratios and significance levels are given below the histograms. Lower panel shows a fixed sub-maximal work load (0.75 l·min^{-1}) in terms of % $\dot{V}_{O_2}$ max (from Spurr, 1983).

subjects, a stepwise multiple regression analysis (Barac-Nieto *et al.*, 1978b) revealed that the $W:H$ ratio ($kg \cdot m^{-1}$), log of the sum of triceps and sub-scapular skinfolds in mm (SK), total body Hb (TotHb) obtained as the product of blood Hb and blood volume ($g \cdot kg$ body weight^{-1}), and daily creatinine (Cr) excretion ($g \cdot day^{-1} \cdot kg^{-1}$) contributed significantly to the variation in $\dot{V}_{O_2}$ max ($l \cdot min^{-1}$);

$$\dot{V}_{O_2}\text{ max} = 0.095\ W:H - 0.152\ \text{SK} + 0.087\ \text{TotHb} + 0.031\ \text{Cr} - 2.550 \quad (1)$$

$$(r = 0.931;\ \text{SEE} = 0.21)$$

All of the variables in the equation are related to nutritional status.

Detailed studies of the body composition of M, I and S subjects demonstrated that over 80% of the difference in $\dot{V}_{O_2}$ max between M and S subjects could be accounted for by difference in muscle cell mass (Barac-Nieto *et al.*, 1980). The remaining difference might be ascribed to reduced capacity for oxygen transport either because of low blood Hb (Fig. 3.4) or reduced maximum cardiac output. There do not seem to be any reports of studies on maximum cardiac output in malnourished subjects. Another possibility is that the skeletal muscle cells have reduced maximum aerobic power because of reduced oxidative enzyme content. Tasker & Tulpule (1964) found a marked decrease in the activities of oxidative enzymes in skeletal muscle of protein deficient rats and Raju (1974) reported that after recovery from 13 weeks of reduced protein intake, rat skeletal muscle had an increase in glycolytic and a decrease in oxidative enzymes and activity. However, there appear to be no studies which have measured similar biochemical changes in humans, although Lopes *et al.* (1982) showed that malnourished patients exhibited marked impairment in muscle function. There were both an increased muscle fatigability in static muscular contraction and a changed pattern of muscle contraction and relaxation which were reversed in patients undergoing nutritional supplementation. Their data would indicate the possibility of a decreased content of ATP and phosphocreatine in the skeletal muscle tissue of malnourished subjects. The data of Heymsfield *et al.* (1982) indicated changes in the biochemical composition of skeletal muscle in both acute and chronic semistarvation, particularly in glycogen and total energy contents. In any event, it should be emphasized that the $\dot{V}_{O_2}$ max not accounted for by differences in muscle cell mass (MCM) is small (<20%; Barac-Nieto *et al.*, 1980). After $2\frac{1}{2}$ months of recovery the $\dot{V}_{O_2}$ max ($l \cdot min^{-1}$) increased significantly (also when it was expressed in terms of body weight and lean body mass (LBM)) but, although mean values were elevated in terms of body cell mass (BCM) and MCM, the increases were not statistically significant. However, at the termination of the experiment, PWC had not returned to values comparable to those seen in mild malnutrition which indicates that the recovery process is a long one, particularly when carried out under the sedentary conditions of the hospital metabolic ward. It is interesting to note that the $\dot{V}_{O_2}$ max increased 45 days after the beginning of the repletion period (after 90 days in hospital), when blood Hb concentration had not yet increased (Fig. 3.4), but MCM was significantly increased over basal values (Barac-Nieto *et al.*, 1980). This also points to a primary dependence of $\dot{V}_{O_2}$ max on MCM. Furthermore, it appears that supplying adequate calories alone was not sufficient to bring about an increase in $\dot{V}_{O_2}$ max or MCM and that only after increasing the protein intake to 100 $g \cdot day^{-1}$ was there improvement in these two variables (Barac-Nieto *et al.*, 1979, 1980). Angeleli *et al.* (1983) reported the results of a lunch supplementation program on physical work performance. The latter was measured by sub-maximal bicycle ergometry before and after three months of supplementation which increased the daily intake by 355 kcal and 20 g protein of mixed quality. The work load required to reach a target sub-maximal heart rate (195 − age (years)) increased significantly, indicating an improvement in overall PWC. It would seem then that it does not require very much supplementation to register an improved PWC in marginally undernourished groups.

Endurance

An endurance test is carried out on a treadmill or bicycle ergometer at a work load ($\dot{V}_{O_2}$) of 70–80% of the subject's maximum until exhaustion supervenes, usually with the f_H within about five beats of f_H max. Because of the difficulty in performing this test, only a few laboratories have attempted measurement of endurance times in normal individuals and, to our knowledge, none except our own, in malnourished subjects.

From a number of sources, it is known that the maximum relative work load that can be sustained for an eight hour working day in physically fit subjects usually does not exceed about 35–40% $\dot{V}_{O_2}$ max (Michael *et al.* (1961); Åstrand (1967); Spurr *et al.* (1975). Sedentary individuals can be expected to have lower upper limits for eight hours of work (Åstrand & Rodahl, 1970).

We have measured maximum endurance times at 80% $\dot{V}_{O_2}$ max (T_{80}) in the groups of malnourished subjects described above (Barac-Nieto *et al.*, 1978b, 1980). We did not find any significant differences between the three groups (M, I and S) of malnourished men; T_{80} averaged 97 ± 12 min (mean ± SE) in all subjects (Barac-Nieto *et al.*, 1978b).

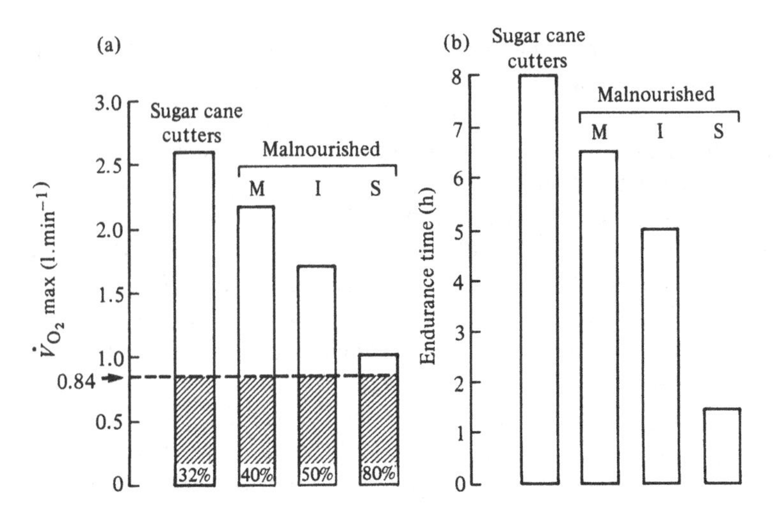

Fig. 3.6 Estimated effect on endurance times at an oxygen consumption of 0.84 $l \cdot min^{-1}$, (b) of mild (M), intermediate (I) and severe (S) malnutrition when working at a % $\dot{V}_{O_2}$ max shown in (a) (data from Barac-Nieto, 1987).

The application of endurance times to daily work and, therefore, to their impact on productivity is illustrated in Fig. 3.6. The average $\dot{V}_{O_2}$ max of all of our sugar cane cutters was 2.6 $l \cdot min^{-1}$ (Maksud *et al.*, 1976). A reasonable value of 0.84 $l \cdot min^{-1}$ for the average $\dot{V}_{O_2}$ that might be sustained during an eight hour work day amounts to 32% $\dot{V}_{O_2}$ max for the nutritionally normal cane cutters, but 40, 50 and 80% of $\dot{V}_{O_2}$ max for the M, I and S malnourished men (Fig. 3.6). Using the two values for endurance times measured in each of our subjects and the negative exponential relationship which exists between % $\dot{V}_{O_2}$ max and endurance time (Åstrand & Rodahl, 1970, Barac-Nieto, 1987), it was possible to estimate (Barac-Nieto *et al.*, 1980) the average endurance of the three groups of malnourished subjects (Fig. 3.6b) if they had been working at the % $\dot{V}_{O_2}$ max levels shown in Fig. 3.6a. The M subjects would have sustained 40% $\dot{V}_{O_2}$ max for 6.5 hours, the I subjects 50% $\dot{V}_{O_2}$ max for about 5 hours and the S men 80% $\dot{V}_{O_2}$ max for 1.5 hours. The implications for productivity in sugar cane cutting and, presumably, for other kinds of heavy physical labor, are obvious.

In the case of Group S during dietary repletion, an interesting change in T_{80} was observed. Endurance times were significantly reduced from 113 min at the first measurement of the basal period to 42 min at the final determination at the end of the dietary repletion (Barac-Nieto *et al.*, 1980). The explanation for this surprising reduction is still not clear. Hansen-Smith, Maksud & Van Horn (1977) reported decreased work endurance times in rats on high protein diets compared to animals ingesting an isocaloric carbohydrate diet, and Bergstrom *et al.* (1967) and Gollnick *et al.* (1972) have shown that diets in which the energy value of carbohydrate has been replaced with fat and/or protein lead to reduced stores of muscle glycogen. Furthermore, Bergstrom *et al.* (1967) demonstrated that the maximum endurance time in humans is directly related to the initial glycogen content of skeletal muscle. During the dietary repletion period of the Group S subjects, carbohydrate intake was reduced from 64% to 50% of calories. In a normal individual this amount of carbohydrate should be sufficient to maintain muscle glycogen stores, but definitive studies seem not to have been done (Durnin, 1982). The rebuilt muscle tissue of Group S subjects may not store glycogen normally and, together with the lack of regular exercise in the protracted sedentary existence in the metabolic ward, may lead to reduced muscle glycogen and shorter endurance times. Heymsfield *et al.* (1982) found reduced muscle glycogen in subjects who had undergone acute or chronic semistarvation prior to death. The areas of muscle nutritive supply, and the metabolic and endocrine responses which regulate them during both short term and prolonged exercise, have not been investigated in malnourished individuals. Even though there is little reason at the moment to suspect abnormal muscle function in acute exercise testing to maximum levels, the responses to prolonged exercise may be worth investigating.

Productivity and physical work capacity

Having established a direct relationship between nutritional status and physical work capacity in undernourished men, attention can now be directed towards the association between $\dot{V}_{O_2}$ max and productivity. The amount of work done in terms of output of a product is usually difficult to measure, particularly in the lighter work tasks where the intellectual component may have as much or more to do with 'productivity' as the physical use of one's body. In moderate and heavy work it has sometimes been possible to estimate productivity by measuring the quantity of product, or income where piece-work is the basis for payment of the worker. Sugar cane cutting and loading are heavy work tasks where the weight of cane cut or loaded is measured carefully since workers are usually paid by the tonnage cut. Because the pay scale in many sugar harvesting operations is very low, one might expect that the motivation factor would be fairly similar in different groups of workers and that they would work close to the limit of their physical capacities. Also, logging is heavy physical work (Durnin & Passmore, 1967) and has been used to relate productivity to worker characteristics. The time required to

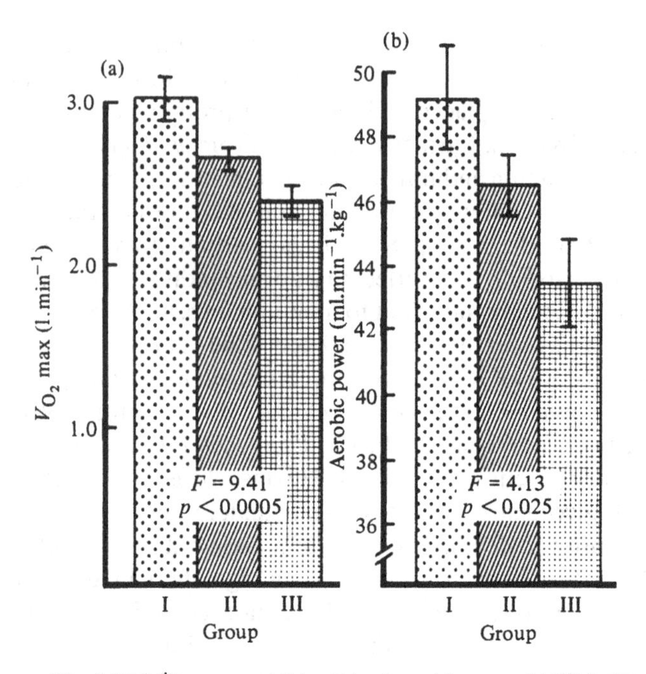

Fig. 3.7 (a) $\dot{V}_{O_2}$ max and (b) maximal aerobic power ($\pm$ SEs) of good (Group I, $n = 8$), average (Group II, $n = 23$) and poor (Group III, $n = 15$) sugar cane cutters aged 18–34 years. *F* ratio values are from one-way analysis of variance. Correlations with productivity: $\dot{V}_{O_2}$ max $r = 0.552$, $p < 0.0002$; aerobic power $r = 0.341$, $p < 0.02$. (From Spurr *et al.*, 1977b.)

accomplish standard work tasks is another method which has been utilized to estimate productivity (Viteri & Torún, 1975).

Hansson (1965) measured sub-maximal work and estimated $\dot{V}_{O_2}$ max in a group of 'top' producing lumberjacks and a group of average producers, and found that the former had a higher estimated $\dot{V}_{O_2}$ max than the latter. Davies (1973b) studied sugar cane cutters in East Africa, dividing them into high, medium and low producers based on the daily tonnage cut. He found no difference in the 3 groups in height, weight, summed skinfolds, LBM, leg volume or the circumferences of biceps and calf, but did encounter a significant correlation between daily productivity and $\dot{V}_{O_2}$ max ($r = 0.46$; $p < 0.001$). Davies *et al.* (1976) also measured productivity in Sudanese cane cutters during a three hour period of continuous cutting and reported a significant correlation between $\dot{V}_{O_2}$ max and rate (kg·min^{-1}) of cane cutting ($r = 0.26$; $p < 0.01$).

We have studied nutritionally normal sugar cane workers in Colombia, where the tasks of cutting and loading the cane are performed by separate gangs of men. The former is a self-paced and continuous task, while the loading of cane is discontinuous, depending on the availability of wagons. The cutters were divided into good (Group I), average (Group II), and poor (Groups III) producers, depending on the daily tonnage cut. The cutters worked at about 35% of their $\dot{V}_{O_2}$ max during the eight hour day (Spurr *et al.*, 1975), which is close to the maximum that can be sustained for this period of time (Spurr *et al.*, 1975; Michael *et al.*, 1961; Åstrand, 1967). In relating various anthropometric measurements and age to productivity, there were statistically significant positive correlations of height, weight and LBM with productivity (Spurr *et al.*, 1977b). The correlations with age and body fat were not significant. Fig. 3.7 summarizes the relation of $\dot{V}_{O_2}$ max and maximal aerobic power with productivity, both of which were significantly correlated. A step-wise multiple regression analysis revealed that $\dot{V}_{O_2}$ max ($l \cdot min^{-1}$), % body fat (F) and height (H; cm) contributed significantly to the variation in productivity ($tons \cdot day^{-1}$) such that

$$\text{Productivity} = 0.81\ \dot{V}_{O_2}\ \text{max} - 0.14\ F + 0.03\ H - 1.962 \qquad (2)$$
$$(r = 0.685;\ p < 0.001)$$

The $\dot{V}_{O_2}$ max and body fat are influenced by present nutritional status (Viteri, 1971; Barac-Nieto *et al.*, 1978a) and adult height by past nutritional status during the period of growth (Martorell, 1985). Equation 2 states simply that those who are presently in poor physical condition or malnourished (low $\dot{V}_{O_2}$ max) or whose height is stunted because of past undernutrition, are at a disadvantage in terms of ability to produce in cutting sugar cane. The negative coefficient for per cent body fat indicates that there is some advantage to low body fat contents. The relatively low correlation coefficients between productivity and $\dot{V}_{O_2}$ max obtained in our studies (Fig. 3.7; Spurr *et al.*, 1977a) and those of others (Davies *et al.*, 1976) preclude the use of regression equations in the prediction of productivity and bring into question the homogeneity of motivation alluded to above. The results shown in Fig. 3.7 indicate that the more physically fit subjects were better producers. Also, since malnutrition reduces $\dot{V}_{O_2}$ max, one can predict that it will have proportional effects on productivity in hard work.

Even in the case of the sugar cane loaders, who do not work continuously, productivity was positively correlated with maximal aerobic power and negatively with resting and working f_H, demonstrating again the relationship of productivity to the physical condition of the worker (Spurr *et al.*, 1977a).

In the case of sugar cane cutting which, at an average expenditure of 5 $kcal \cdot min^{-1}$ per 65 kg of body weight during the eight hour working day

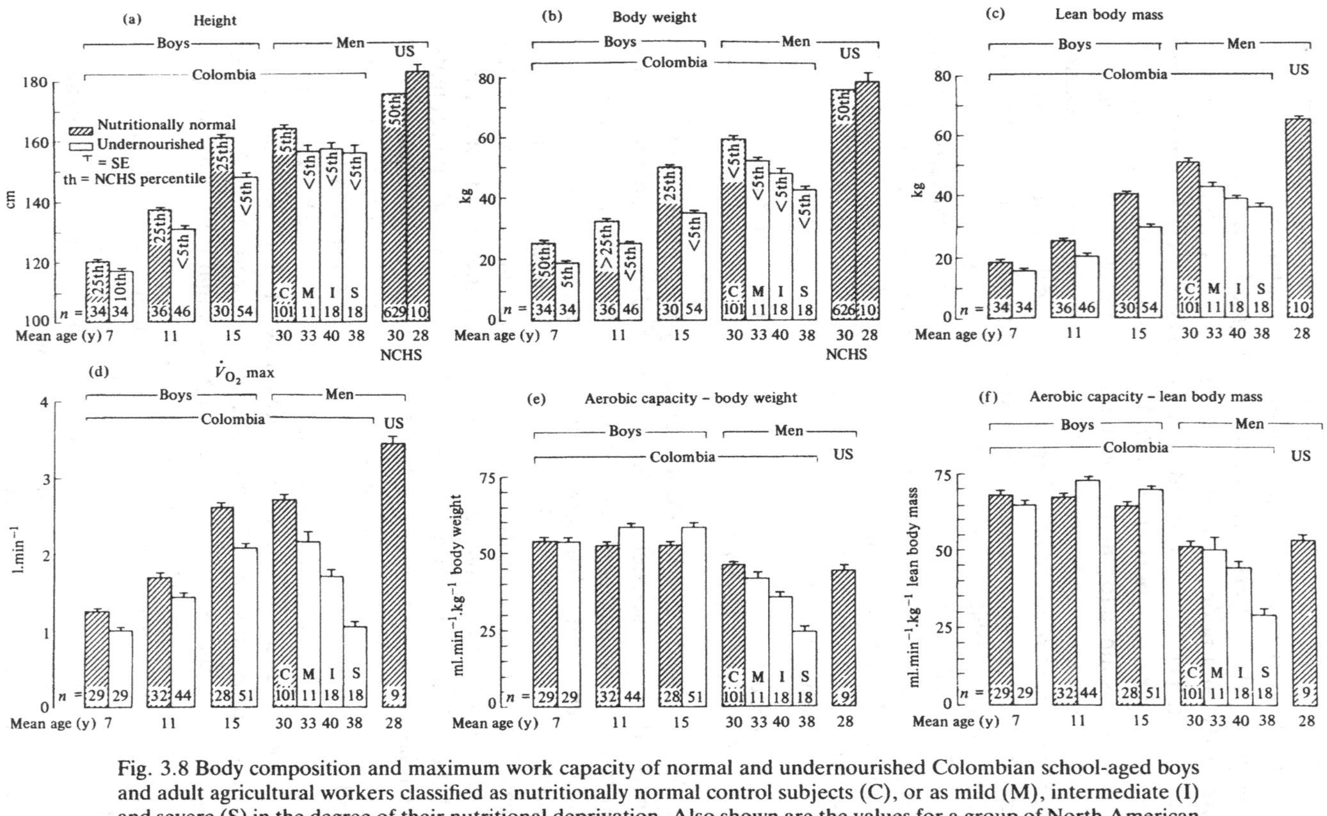

Fig. 3.8 Body composition and maximum work capacity of normal and undernourished Colombian school-aged boys and adult agricultural workers classified as nutritionally normal control subjects (C), or as mild (M), intermediate (I) and severe (S) in the degree of their nutritional deprivation. Also shown are the values for a group of North American men and the US NCHS fiftieth percentile values of weight and height for adult men (from Spurr, 1988).

(Spurr *et al.*, 1975), can be classified as moderate industrial work (Durnin & Passmore, 1967), the worker productivity is related to his body size, height, weight and LBM (Davies *et al.*, 1976). This has also been demonstrated by Satyanarayana *et al.* (1977, 1978) for industrial factory work of presumably less intensity than sugar cane cutting. Their subjects were nutritionally normal workers engaged in the production of detonator fuses which could be measured in terms of the number of fuses produced per day. They found that body weight, height and LBM were significantly correlated with productivity and that after partialling out the effect of height, weight and LBM were still significantly correlated with productivity. That is, the total daily work output was significantly higher in those with higher body weight and LBM.

There also appears to be a relationship between the intensity of a work task and PWC of the worker (Chaffin & De Araujo Cuoto, 1986). Subjects engaged in tasks at various intensities show a positive and significant correlation ($r = 0.84$) of $\dot{V}_{O_2}$ max, as estimated for sub-maximal bicycle ergometry, with the energy cost of the worker's job. One does not know, however, if this is a cause–effect relationship. Does the $\dot{V}_{O_2}$ max increase as a training effect of prolonged intensive work, or are these results a matter of some sort of selection such that individuals with higher natural work capacity choose heavier work (Åstrand & Rodahl, 1970)?

Comparisons of men and boys

Fig. 3.8 is a summary of the anthropometric and work capacity characteristics of our adult and school-aged male subjects studied over a period of 15 years. We have demonstrated (Barac-Nieto *et al.*, 1984) that the empirical equations developed by Pařízková (1961) for estimating body fat from skinfolds in children also apply to these boys. The results of average estimates of LBM derived from these equations are plotted in Fig. 3.8c together with values for height and weight for the 6–8, 10–12 and 14–16 year old boys (Fig. 3.8a & b) to compare with similar values for the four groups of adult Colombian men discussed previously (C, M, I, and S) and a group of ten nutritionally normal North American men. The LBMs of the C group were calculated from the skinfold equations of Pascale *et al.* (1956) and that of the American men from those of Durnin & Womersley (1974). The LBMs of the three groups of malnourished Colombian men (M, I and S) were obtained from measurements of total body water (Barac-Nieto *et al.*, 1978a).

The development of LBM (Fig. 3.8c) of the undernourished boys is significantly attenuated during growth (Barac-Nieto *et al.*, 1984). When expressed as percent of body weight, the undernourished boys had significantly higher values of LBM than well-nourished subjects because of lower fat values in the former (Barac-Nieto *et al.*, 1984).

Growth of work capacity.

The growth of $\dot{V}_{O_2}$ max ($l \cdot min^{-1}$) in the youngest, oldest and middle of the age groups of boys presented in Fig. 3.3 is shown in Fig. 3.8d. The $\dot{V}_{O_2}$ max of the nutritionally deprived boys was significantly lower (<85%) than the normal subjects throughout the age range studied (Spurr *et al.*, 1983b). When expressed as $ml \cdot min^{-1} \cdot kg^{-1}$ body weight, the two oldest groups of undernourished boys had higher aerobic capacities than the normal boys (Fig. 3.8e) which, at first, was thought to be due to differences in body composition (Spurr *et al.*, 1983b). Subsequent studies (Barac-Nieto *et al.*, 1984) demonstrated that even when expressed in terms of LBM, the aerobic capacity of the undernourished subjects, at least in the older age groups, was significantly higher (Fig. 3.8f). That is, the undernourished boys show evidence of better physical condition. These were boys living in the city of Cali. In a similar study on rural Colombian boys, while the difference in aerobic capacity between nutritionally normal and undernourished subjects expressed per kilogram of body weight was similar to that found for urban subjects (Fig. 3.8e) (Spurr *et al.*, 1983b), the difference disappeared when aerobic capacity was calculated in terms of LBM (Fig. 3.8f), i.e. the rural boys did not exhibit a training effect (Barac-Nieto *et al.*, 1984). These differences may be the result of greater access of urban children to sports training facilities (Shephard *et al.*, 1974; Spurr, 1983; Barac-Nieto *et al.*, 1984).

Consequently, it is clear that the lower values of $\dot{V}_{O_2}$ max ($l \cdot min^{-1}$) for the nutritionally deprived children are due to their lower body weights. This is essentially the same conclusion reached by Davies (1973a) and Satyanarayana *et al.* (1979). As with adults, there does not appear to be any basic deficit in muscle function in marginally malnourished children, only in the quantity of muscle available for maximal work. We have deliberately avoided analyzing these data on the basis of so-called 'developmental' age because such an analysis would tend to obscure the differences seen in Fig. 3.8d; the responsibilities of adulthood occur with chronological, not developmental age.

Body size, composition and $\dot{V}_{O_2}$ max

Persons of larger size in general appear to function better than those with smaller stature (Calloway, 1982) in relation to reproduction (Thomson,

Table 3.2[a]. *Correlation coefficients of weight, height, lean body mass (LBM) and maximum oxygen consumption ($\dot{V}_{O_2}$* max; $l \cdot min^{-1}$*) in nutritionally normal boys 6–16 years of age (Spurr* et al., *1983b; Barac-Nieto* et al., *1984) and adult males (Barac-Nieto* et al., *1978a). All are statistically significant ($p<0.01$)*

	Boys ($n=406$)			Men ($n=35$)		
	Weight (kg)	Height (cm)	LBM (kg)	Weight (kg)	Height (cm)	LBM (kg)
Height	0.970	—	—	0.758	—	—
LBM	0.986	0.965	—	0.875	0.702	—
$\dot{V}_{O_2}$ max	0.931	0.911	0.932	0.562	0.489	0.724

[a]From Spurr, 1988.

1980), disease (Reddy *et al.*, 1976), cognition (Klein *et al.*, 1972) and work performance (Spurr, 1983, 1984). Because physical work capacity is a function of body size (Åstrand & Rodahl, 1986), i.e. the mass of muscle tissue involved in the maximum effort, and muscle constitutes about 40% of the body weight and 50% of the LBM (Clarys, Martin & Drinkwater, 1984; Buskirk & Mendez, 1985), it is interesting to note the correlations between three components of body size and $\dot{V}_{O_2}$ max presented in Table 3.2. The correlations in boys are higher than those in men, probably because of a three-fold greater range in values, but in either case it is clear that in non-obese subjects there are significant correlations between parameters of body size and PWC as measured by $\dot{V}_{O_2}$ max. Taller individuals have more LBM and higher $\dot{V}_{O_2}$ max values (Fig. 3.8). Similar relationships exist for adult women, but the correlation coefficients are lower (von Dobeln, 1956).

All of the data presented in Fig. 3.8 are from various studies in our laboratory (Spurr *et al.*, 1975, 1977b, 1983a, b; Barac-Nieto *et al.*, 1978a, b; 1979, 1980) and permit a comparison between Colombian boys and men, and between the latter and a small group of North American adult males. The difference in height between adults of developing and developed countries is well known (ICNND, 1959–63; Spurr, Barac-Nieto & Maksud, 1978). The average value of the C group of men in Fig. 3.8 is very close to that published for low income Colombian men (Spurr *et al.*, 1978) and probably reflects some period(s) of undernutrition during the period of growth. The heights of the three groups of malnourished (M, I, S) men were not significantly different from each other, but were lower than those of the C group. This is probably a result of more severe nutritional deprivation during growth in Groups M, I and

S than occurred in Group C. It is difficult to predict the adult height of the oldest boys, but it is likely that the nutritionally normal children will be taller (Fig. 3.8a) and perhaps have a higher $\dot{V}_{O_2}$ max ($l \cdot min^{-1}$) than Group C (Fig. 3.8d), while undernourished groups of boys will, in adulthood, most likely resemble more closely Group M.

The lower values of aerobic capacity per kilogram of body weight or LBM in adults than in boys is also well known and, at least in part, probably reflects the progressive decline in these measurements with age (Dehn & Bruce, 1972). The differences also may reflect differences in the physical condition of the boys and men.

Physical activity in chronic undernutrition

Deficient energy intake and physical activity

In his early review on the physiological effects of undernutrition, Graham Lusk (1921, p. 538) observed that 'walking gives no pleasure to a seamstress nor golf to a half-starved professor' illustrating the largely anecdotal nature of reports on the relation between physical activity and calorie intake. Benedict *et al.* (1919) reported decreased endurance in chin-up exercise and a tendency towards reduced walking distances in their semistarved subjects. In general there were fairly common complaints of fatigue and weakness and observers noted reductions in endurance, alertness and performance in their athletic activities. Keys *et al.* (1950) reported that most of the subjects in the Minnesota Experiment felt weak and tired easily during semistarvation. However, daily energy expenditure was not measured in either of these early classic studies.

On the other hand, Viteri & Torún (1975), using time and motion studies, demonstrated that unsupplemented agricultural workers in Guatemala slept more and indulged in longer rest periods than similar subjects who were receiving dietary supplementation. In more acute laboratory experiments involving reduced calorie intake, Gorsky & Calloway (1983), using food and activity diary techniques, reported that lower-effort discretionary activities were substituted for higher-effort discretionary activities without significant effect on obligatory activities. The subjects were ingesting about 500 $kcal \cdot d^{-1}$ less during the experimental than during the control period.

Because of evidence that reduced physical activity in young children might be involved in motor development and coordination (Malina, 1984) as well as in cognitive and social development (Sameroff & McDonough, 1984), there has been great interest in the effects of chronic undernutrition on this aspect of daily energy expenditure during growth.

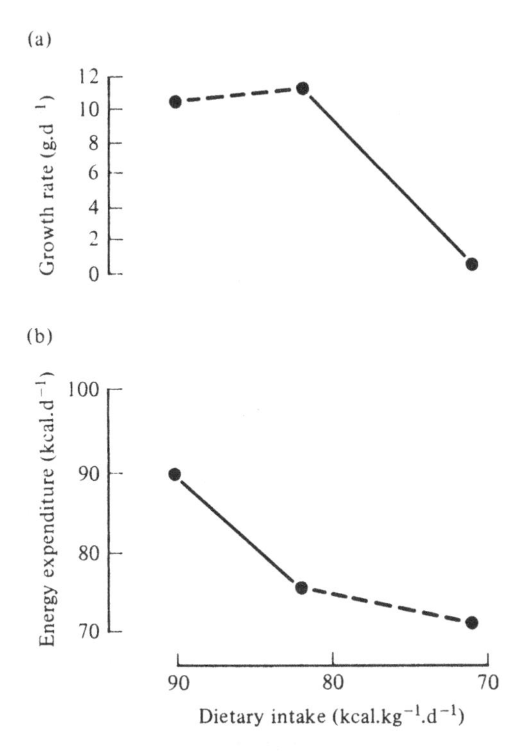

Fig. 3.9 Relationship of (a) rate of growth in body weight and (b) daily energy expenditure to dietary energy intake when the latter is reduced in five pre-school boys, aged 30 ± 8 months. Significant ($p < 0.05$) differences between points joined by solid lines. (Data from Torún & Viteri, 1981.)

Furthermore, Torún *et al.* (1979) reported that physically active children recovering from malnutrition showed enhanced lean body mass and longitudinal growth than comparable children who were more sedentary in their daily activities, indicating a possible role for physical activity in growth itself.

There has been some documentation of a decrease in activity levels in malnourished pre-school children. Chavez & Martinez (1979) reported that unsupplemented children, 0–3 years of age, from poor families were more passive than similar children receiving nutritional supplementation. Rutishauser & Whitehead (1972) found that Ugandan children 1–3 years of age with reduced energy intakes spent significantly less time than European children in activities such as walking and running. Viteri & Torún (1981) also observed decreased spontaneous physical activity in pre-school children associated with reduced energy intake.

In an important study on adaptation to reduced energy intake on activity and growth, Torún & Viteri (1981) found that in pre-school children with an average age of 30 months, a decrease in daily energy intake from 90 to 80 kcal $\cdot$ kg^{-1} body weight resulted in a decrease in total daily energy expenditure without a change in the rate of weight gain, as shown in Fig. 3.9. A further decrease in daily energy intake to 71 kcal $\cdot$ kg^{-1} was associated with a decrease in weight gain without further change in energy expenditure (Fig. 3.9). These results suggest that in pre-school children the first line of defense against low energy intake is reduced activity and only when there is a more severe restriction of ingested energy is there reduced weight gain.

The study of Satyanarayana *et al.* (1979) is the only one of which we are aware that attempted to measure habitual physical activity in undernourished school-aged children. These data were analyzed primarily from the standpoint of their contribution to the physical work capacity of the children studied and it is not possible to decipher the influence of the nutritional status of the subjects on their habitual physical activity although there was a direct and statistically significant relationship between physical activity and body weight (Satyanarayana *et al.* 1979).

Pattern of energy expenditure in undernourished Colombian children

Our own studies on daily energy expenditure in free-ranging marginally malnourished school children have been carried out using the heart rate method in individually calibrated subjects. Initially we employed heart rate accumulation during the awake portion of the day using basal metabolic rate (BMR) for sleep energy expenditure (EE) (Spurr, Reina & Barac-Nieto, 1986; Spurr & Reina, 1987). Our more recent studies have been done with minute-by-minute heart rate recording, also during the awake portion of the day. The latter method has been validated against indirect calorimetry and shown to give excellent estimates of daily EE even in small groups of subjects (Spurr *et al.*, 1988; Ceesay *et al.*, 1989). Because of this and the fact that more information is available from the minute-by-minute heart rate recording, in what follows the discussion will be largely based on the later studies. Furthermore, the basic results of total daily energy expenditure (TDEE) are essentially the same in both types of measurement, i.e. heart rate accumulation and minute-by-minute recording.

The studies were based on measurements in boys and girls 6–8, 10–12 and 14–16 years of age from lower socioeconomic families living in poor neighborhoods of Cali, Colombia. The children were classified as nutritionally normal (control) or marginally malnourished based on Colombian norms of weight-for-age and weight-for-height. There were there-

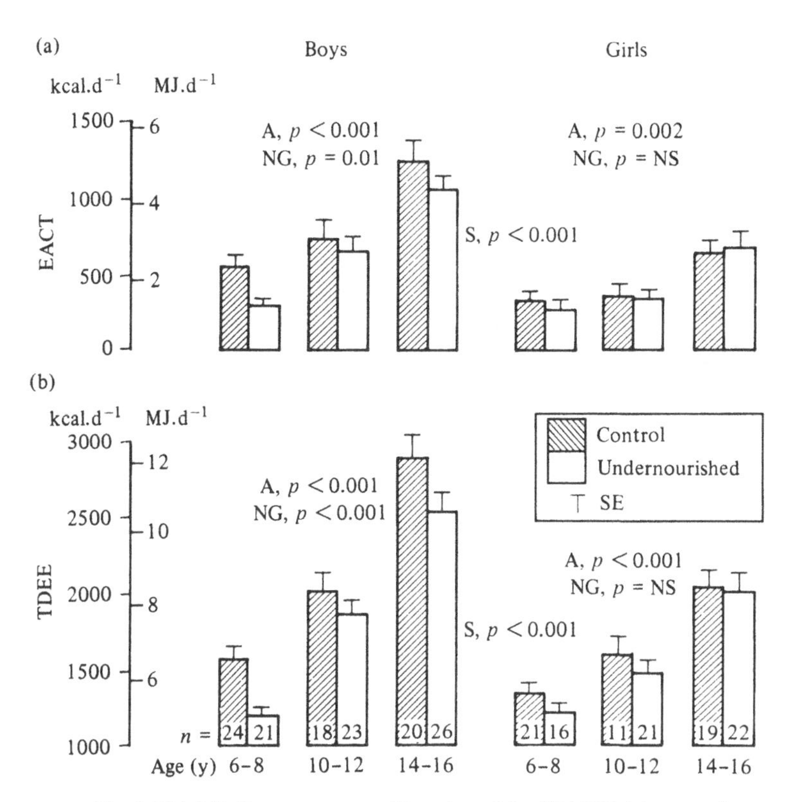

Fig. 3.10 (a) Daily energy expenditure in activity (EACT) (above maintenance energy expenditure) and (b) total daily energy expenditure (TDEE) in nutritionally normal (control) and marginally undernourished boys and girls 6–16 years of age during ordinary school days in Cali, Colombia. Also, shown are standard errors (SE) and probabilities (p) from two-way analyses of variance (ANOVA) for age (A) and nutritional group (NG) effects, for boys and girls separately, and for the sex (S) effects from a three-way ANOVA of all three effects. (Data from Spurr & Reina, 1988a.)

fore 12 groups of subjects; three age groups, and two nutritional groups of boys and girls (Spurr & Reina, 1988a). Since sleep EE was based on BMR and resting EE (RMR) was also obtained as part of the minute-by-minute method of measuring EE (Spurr *et al.*, 1988) it was possible to calculate the maintenance energy expenditure (MEE) as BMR during sleep plus RMR during the remainder of the day. The energy expended in activity (EACT) is then TDEE − MEE.

The means and SEs of TDEE and EACT are presented in Fig. 3.10 together with the number of values for each of the 12 groups of subjects and the results of two-way analysis of variance (ANOVA) for age (A) and nutritional group (NG) effects of boys and girls separately and of the sex (S) effects in a three-way ANOVA. It can be seen that all age effects

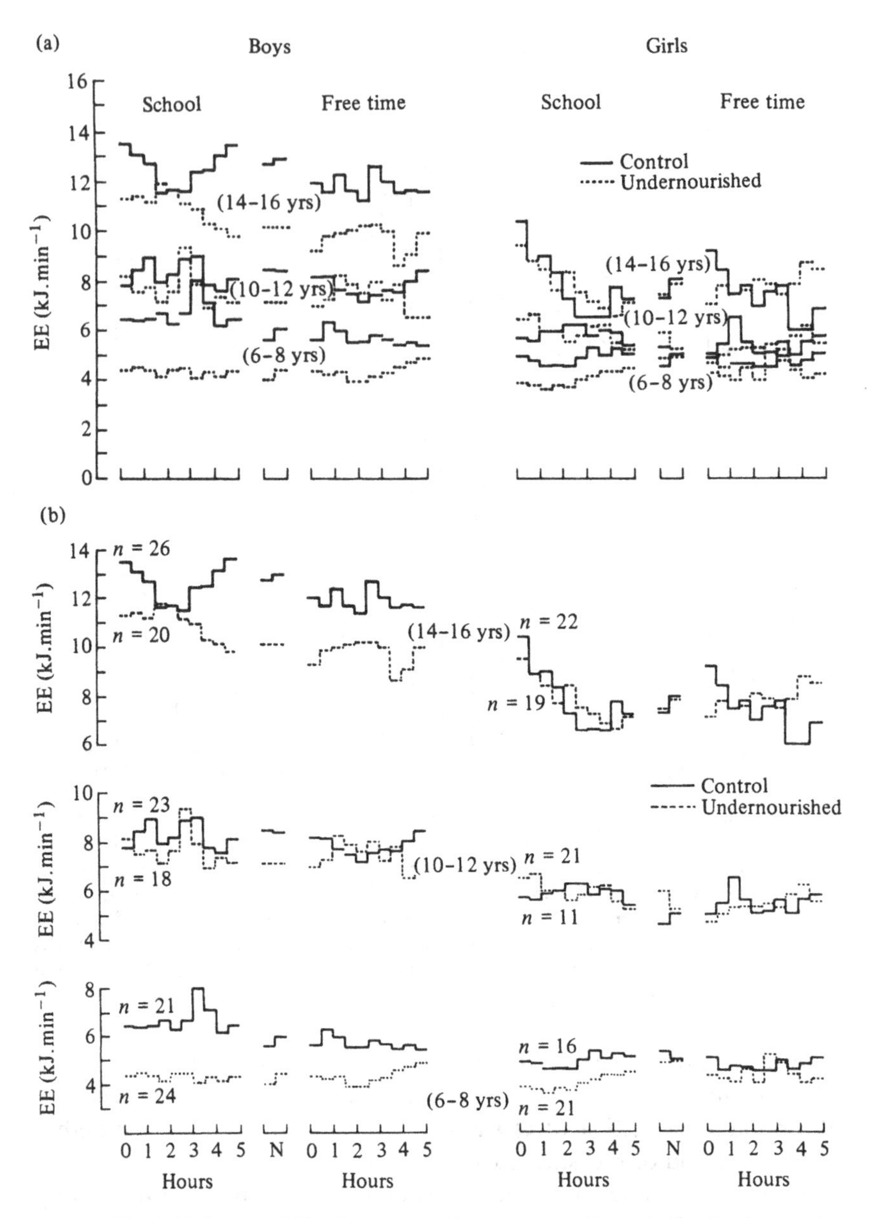

Fig. 3.11 Averaged 30 minute rates of energy expenditure during five hours of school, the noon hour (N) and five hours of free-time in boys and girls 6–8, 10–12 and 14–16 years of age classified as nutritionally normal (control) or marginally undernourished. In (a) the data are plotted on the same y-axis while in (b) the same data are plotted on separate y-axes (from Spurr & Reina, 1988a).

are statistically significant ($p \leq 0.002$) for both TDEE and EACT but only in the case of the boys are the nutritional group effects statistically significant, with the undernourished boys having lower values in both instances. There is a similar tendency in the girls, particularly in the case of TDEE, but the differences are not statistically significant. The values for TDEE and EACT are both significantly less ($p < 0.001$) in girls than in boys.

The schools in Cali operate on two shifts 7 a.m.–12 p.m. and 1 p.m.–6 p.m. In an attempt to standardize the analysis of the pattern of activity of the children, the EEs are expressed as 30 minute averages during five hours of school time, the noon hour, and five hours of free time and presented in Fig. 3.11. In Fig. 3.11a the data are plotted on the same y-axis scale in order to see the age relationships and in Fig. 3.11b on separate y-axes for purposes of separating the three age groups and reveals sex and nutritional group relationships.

It can be seen in Fig. 3.11a that there is an increase in activity levels with age ($p < 0.001$), but that this increase is markedly attenuated in the girls compared to boys. This is believed to be a 'cultural squelch' of the natural exuberance of young girls in this society, perhaps related to the idea that young ladies do not sweat. However, we do not have specific data to confirm this suggestion and believe it to be a phenomenon worthy of some further study.

From Fig. 3.11b it is clear that overall there is a significant reduction in EE in the undernourished children ($p < 0.001$) and that girls are expending less than boys ($p < 0.001$). An exception to the nutritional group effect is seen in the 10–12 year old group where two-way ANOVA reveals that nutritional group effects are not statistically significant.

We have demonstrated that the skinfold equations of Pařízková (1961) can be used to estimate the lean body mass (LBM) of these Colombian children (Barac-Nieto *et al.*, 1984). When MEE is plotted as a function of LBM the average values of all 12 groups fall on the same straight line with a correlation coefficient of 0.98 indicating that the major component of TDEE is related to body size (LBM) alone (Spurr & Reina, 1988a). When the average total EE and EE in activity for five hours of school, five hours of free-time and for the entire day are also plotted on LBM one gets the results shown in Fig. 3.12. The data demonstrate again that the differences in EE, whether due to age or nutritional group are a simple function of body size, i.e. of the LBM. The differences between boys and girls seen on the left of Fig. 3.12 are the result of the voluntary depression of EACT seen on the right of Fig. 3.12, since the involuntary portion of TDEE, the MEE, is also entirely a function of the LBM (Spurr & Reina, 1988a).

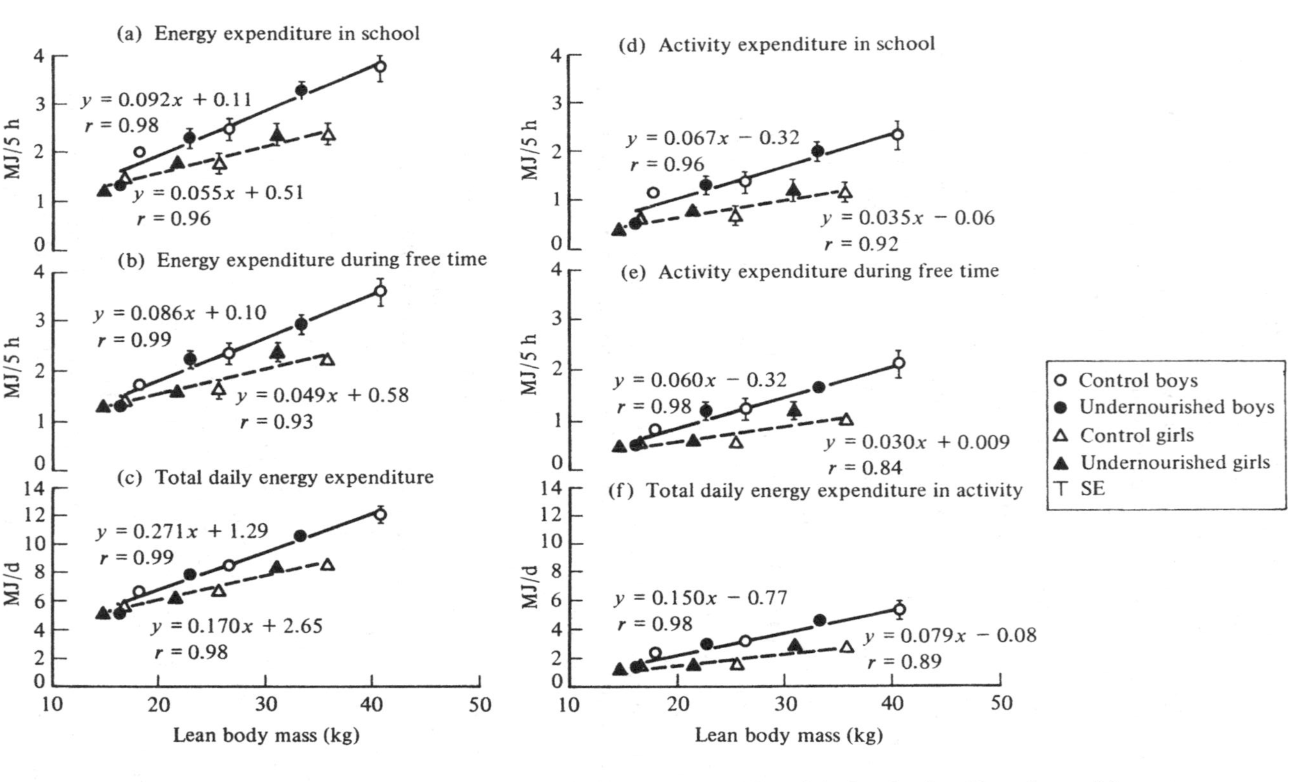

Fig. 3.12 Group averages of total energy expenditure and energy expended in activity in school and free-time and for the entire day as a function of lean body mass in three age groups of boys and girls classified as nutritionally normal (control) and marginally undernourished (from Spurr & Reina, 1988a).

These results indicate then that the slowed growth and smaller body size are the principal adaptation to reduced calorie availability in the diet, resulting in a lower energy expenditure than nutritionally normal children. It also implies that the quality of the activities may not be very different between nutritional groups and that at this level of marginal malnutrition the undernourished children may not be deprived in their cognitive or motor development because of reduced activities associated with school and play.

The next question which naturally occurs is whether these marginally malnourished children can keep up with their nutritionally normal counterparts under conditions where activity levels are artificially increased.

Artificially increased activity and dietary intervention

To answer the question of how undernourished children would fare under conditions in which levels of energy expenditure are artificially increased, we created a summer day-camp in which 14 nutritionally normal and 19 marginally undernourished boys 10–12 yrs of age were invited to participate (Spurr & Reina, 1988b). The subjects were studied in two phases. Phase I occurred during an ordinary school day similar to the measurements described above. In Phase II, approximately six months later during the school vacation in July and August, the boys participated in a summer day-camp which operated Monday through Friday each week. The energy expenditure (EE) measurements during Phase II were made while the boys were under the tutelage of an athletic director who was a university physical education student. The 33 boys were randomly assigned to three groups of eight and one of nine which spent two weeks with the athletic director who did not know the nutritional classification of the individual boys. During this period the director was instructed to encourage the boys to participate in various athletic activities by creating a spirit of friendly competition among the boys without specific pressure being placed on any individual. The boys were picked up at 7 a.m. and taken to a sports facility located in the city of Cali where, under the oversight of the director, they were exposed to various athletic activities specifically designed to increase the level of energy expenditure. These consisted of organized games of soccer, foot races, calisthenics, basketball, etc. At noon they were taken to a location where a hot meal of ~ 760 kcal (3.2 MJ) had been prepared for them consisting of an average of 12.6% protein, 69.7% carbohydrate and 17.8% fat. Following their meal they were returned to the sports complex for an afternoon of activities similar to those of the morning

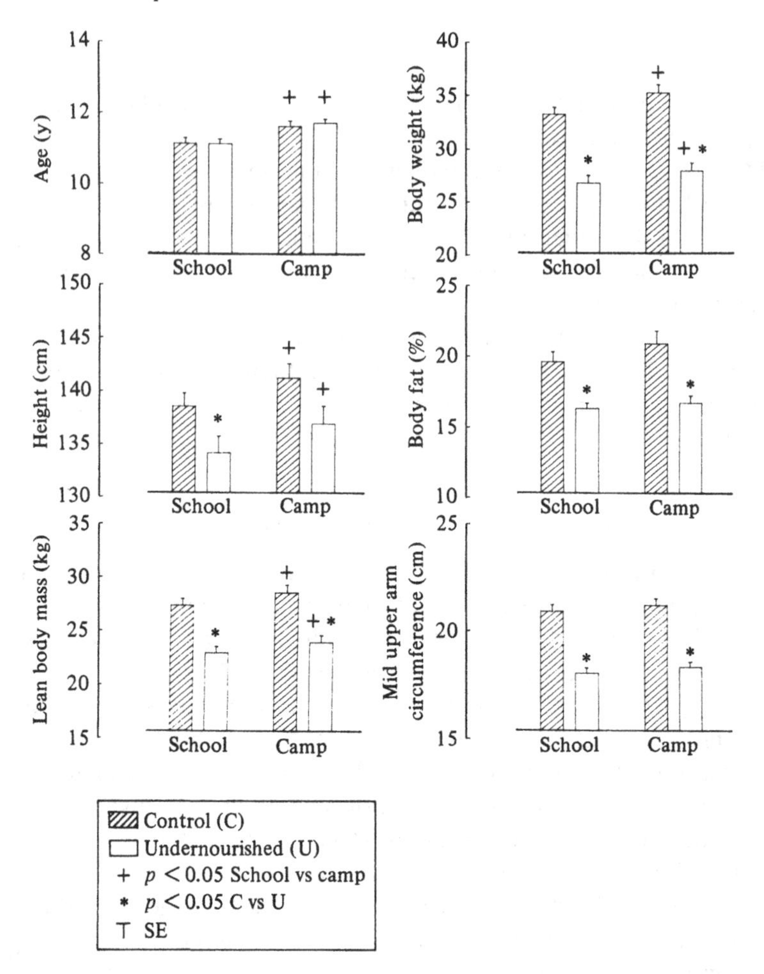

Fig. 3.13 Nutritional anthropometry of a group of nutritionally normal (control, $n = 14$) and marginally undernourished ($n = 19$) 10–12 years old boys during measurements made while attending school and about six months later while attending a summer day-camp. (Data from Spurr & Reina, 1988b.)

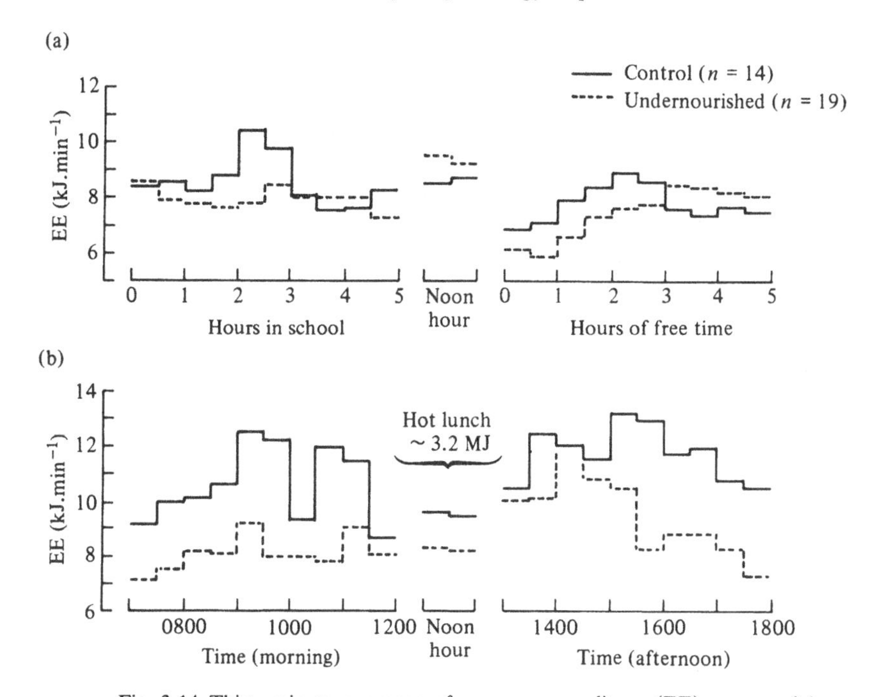

Fig. 3.14 Thirty-minute averages of energy expenditure (EE) measured by minute-by-minute heart rate recording (a) during a school day and (b) while participating in a summer day-camp where an opportunity for increased activity was provided. Also shown is the effect of a hot lunch at mid-day on EE in the afternoon of day-camp (1 kcal = 4.18 kJ). (From Spurr & Reina, 1988b.)

session. All boys were given the hot lunch Monday through Friday throughout July and August (Spurr & Reina, 1988b). In both phases, energy expenditure was measured using the minute-by-minute heart rate recording technique described above. The nutritional anthropometry of the two groups of boys at the time of Phase I (school) and Phase II (camp) studies is presented in Fig. 3.13. It is clear that the undernourished boys were growing, as evidenced by statistically significant increases in height, weight and lean body mass between Phases I and II, and that they were significantly smaller than the control boys.

The patterns of EE during the school day and during the summer day-camp are presented in Fig. 3.14. The data are presented as 30 minute averages during 5 hours of school and 5 hours of free-time in Phase I, as in the study described above. In Phase II, during the summer, EE is presented for five hours of morning and afternoon athletic activities. From a two-way ANOVA it was found that during school time (Phase I) the nutritional group effect was statistically significant ($p < 0.01$), due no doubt to the increased EE of the control boys in mid-morning, but it did not quite achieve statistical significance ($p = 0.06$) in the free-time period.

The most interesting differences between the two nutritional groups were seen during the summer day-camp. In the morning, when the undernourished boys usually arrived without having eaten adequate breakfasts, they were unable or unwilling to increase their EE unlike the control subjects (Fig. 3.14). This difference was statistically significant ($p < 0.01$). Following the noon meal the undernourished boys had significantly higher EE than during the morning, as did the nutritionally normal boys ($p < 0.001$) and for the first two hours increased their EE to levels commensurate with those exerted by the controls. However, this lasted for only about two hours and was followed by a statistically significant ($p < 0.01$) decline in EE as though they tired earlier than their control counterparts.

The demonstration of decreased activity in undernourished pre-school children (Rutishauser & Whitehead, 1972; Chavez & Martinez, 1979; Viteri & Torún, 1981), and the studies of Torún & Viteri (1981) in $2\frac{1}{2}$ year old boys (Fig. 3.9) which showed that activity decreased before the rate of weight gain did when energy intake was reduced, make it appear that the first line of defense against energy deficiency is reduced energy expenditure in this group of subjects. Our own studies have shown that in school-aged children the reduced EE associated with nutritional deficiency is primarily a result of reduced body size (Fig. 3.12; Spurr *et al.*, 1986; Spurr & Reina, 1988a). In addition, we have reported the absence of a positive effect of dietary intervention consisting of a hot lunch program (600 $kcal \cdot d^{-1}$; 2.5 $MJ \cdot d^{-1}$) on the daily EE of marginally malnourished girls 8–11 years of age (Spurr & Reina, 1987). This was despite a significant increase in rate of weight gain and in the sum of three skinfolds (Spurr & Reina, 1987). We suggested that perhaps at the marginal levels of undernutrition seen in our Colombian children, peer pressure in school and play activities to keep up with their nutritionally normal counterparts may interfere with the protective mechanism of decreased activity and result in reduced growth as the first level of protection (Spurr *et al.*, 1986; Spurr & Reina, 1987). The opposite situation in the pre-school subjects of Torún & Viteri (1981), where peer pressure would presumably be absent, may be the more physiological response.

The studies which describe the patterns of energy expenditure of school-aged children, as shown in Fig. 3.11 (Spurr & Reina, 1988a), and those which report the failure of marginally undernourished children to keep up with normal children in situations where activity levels are artificially increased, as presented in Fig. 3.14 (Spurr & Reina, 1988b), make it clear that the minute-by-minute heart rate method is a valuable tool for physiological and anthropological investigations. Furthermore, the data reported offer the possibility of new avenues of enquiry into the relationship between energy expenditure and behavior. For example, the undernourished children 10–12 years of age in Fig. 3.11b were not significantly lower in their energy expenditure than their control counterparts while the other age groups were (Spurr & Reina, 1988a). Is there a difference in this group in peer pressure to keep up? What are the sociological factors which pressure young girls to put a damper on their youthful exuberance (Fig. 3.11) and how can they be reversed? Indeed, would reversal be a positive factor for the individual? Does the apparent inability of marginally malnourished children to increase their energy expenditure in the face of a challenge contribute in a negative way to their cognitive or motor development? These questions may form the basis for fascinating future studies. Furthermore, the methodology employed in these studies has wide applicability in the study of energetics in normal and obese subjects as well as in the undernourished.

Summary

The physical work capacity is measured by the maximum oxygen consumption ($\dot{V}_{O_2}$ max) which is related to work and productivity when expressed in terms of $l \cdot min^{-1}$. The normative expression of $\dot{V}_{O_2}$ max in terms of $ml \cdot min^{-1} \cdot kg^{-1}$ of body weight or lean body mass is an expression of the individual's physical condition (training), not necessarily of capacity to carry out heavy work.

Chronic nutritional deficiency in childhood leads to reduced growth with the result that adults in disadvantaged societies have smaller body size than those from more advantaged countries. The reduced growth and smaller adult size results in decreased physical work capacity of the affected groups such that there is a negative affect on the productivity of those who have to engage in moderate to heavy physical work as the expression of their economic participation in their society. Poor nutritional status also results in reduced physical work capacity with similar connotations for the economic viability of the population.

The reduced energy expenditure of marginally malnourished school-aged children, which occurs during ordinary activities, is primarily the result of decreased body size and lean body mass. Their response to peer pressure in school may prevent the more physiological response, decreased energy expenditure, from being the first line of defense against deficit of energy availability as it appears to be in pre-school children where peer pressure is presumably absent or at least might be expected to have less effect on behavior.

On the other hand, when activity levels are artificially increased, marginally malnourished boys are unable to keep up with their nutritionally normal counterparts until they receive a hot meal. Even then they appear to tire more easily since they return to lower levels of energy expenditure earlier than controls.

Acknowledgements

Various aspects of the work described from our laboratory have been supported by USAID contracts AID/CSD 2943 and AID/TA-C 1424, NIH Grant HD10814, Nestlé Research Grants Programme, United Nations University, Zablocki VA Medical Center and the Fundación para la Educación Superior, Cali, Colombia.

References

Angeleli, W.A., Vichi, F.L., Vannucchi, H., Desai, I.D. & Dutra de Olivera, J.E. (1983). Dietary supplementation and improvement in physical work performance of agricultural migrant workers of Southern Brazil. *Archives of Latinamerican Nutrition*, **23**, 158–69.

Åreskog, N.H., Selinus, R. & Vahlquist, B. (1969). Physical work capacity and nutritional status in Ethiopian male children and young adults. *American Journal of Clinical Nutrition*, **22**, 471–9.

Åstrand, I. (1967). Degree of strain during building work as related to individual aerobic capacity. *Ergonomics*, **10**, 293–303.

Åstrand, P.O. & Rodahl, K. (1970). *Textbook of work physiology*. McGraw-Hill, New York, p. 292.

Åstrand, P.O. & Rodahl, K. (1986). *Textbook of work physiology*. McGraw-Hill, New York.

Barac-Nieto, M. (1987). Physical work determinants and undernutrition. *World Review of Nutrition and Dietetics*, **49**, 22–65.

Barac-Nieto, M., Spurr, G.B., Dahners, H.W. & Maksud, M.G. (1980). Aerobic work capacity and endurance during nutritional repletion of severely undernourished men. *American Journal of Clinical Nutrition*, **33**, 2268–75.

Barac-Nieto, M., Spurr, G.B., Lotero, H. & Maksud, M.G. (1978a). Body composition in chronic undernutrition. *American Journal of Clinical Nutrition*, **31**, 23–40.

Barac-Nieto, M., Spurr, G.B., Lotero, H., Maksud, M.G. & Dahners, H.W. (1979). Body composition during nutritional repletion of severely undernourished men. *American Journal of Clinical Nutrition*, **32**, 981–91.

Barac-Nieto, M., Spurr, G.B., Maksud, M.G. & Lotero, H. (1978b). Aerobic work capacity in chronically undernourished adult males. *Journal of Applied Physiology*, **44**, 209–15.

Barac-Nieto, M., Spurr, G.B. & Reina, J.C. (1984). Marginal malnutrition in

school-aged Colombian boys: body composition and maximal O_2 consumption. *American Journal of Clinical Nutrition*, **39**, 830–9.
Benedict, F.G., Miles, W.R., Roth, P. & Smith, H.M. (1919). *Human vitality and efficiency under prolonged restricted diet*. Washington DC. (Publication No. 280), Carnegie Institute.
Bergstrom, J., Hermansen, L., Hultman, E. & Saltin, B. (1967). Diet, muscle glycogen and physical performance. *Acta Medica Scandinavica*, **71**, 140–50.
Buskirk, E.R. & Mendez, J. (1985). Lean-body-tissue assessment, with emphasis on skeletal-muscle mass. In: *Body Composition Assessments in Youth and Adults*, ed. A.F. Roche, pp. 59–65. Sixth Ross Conference on Medical Research, Ross Labs, Columbus OH.
Calloway, D.G. (1982). Functional consequences of malnutrition. *Review of Infectious Diseases*, **4**, 736–45.
Ceesay, S.M., Prentice, A.M., Day, K.C., Murgatroyd, P.R., Goldberg, G.R., Scott, W. & Spurr, G.B. (1989). The use of heart rate monitoring in the estimation of energy expenditure using whole body calorimetry: a validation study. *British Journal of Nutrition*, **61**, 175–86.
Chaffin, D.B. & De Araujo Couto, H. (1986). Correlation of aerobic capacity of Brazilian workers and their physiologic work requirements. *Journal of Occupational Medicine*, **28**, 509–13.
Chavez, A. & Martinez, C. (1979). Consequences of insufficient nutrition on child character and behavior. In: *Malnutrition, Environment and Behavior*, ed. D.A. Levitzky, pp. 238–55. Cornell University Press, Ithaca, NY.
Clarys, J.B., Martin, A.D. & Drinkwater, D.T. (1984). Gross tissue weights in the human body by cadaver dissection. *Human Biology*, **56**, 459–73.
CVC (1984). *Corporación Autonoma Regional del Cauca*. División de Estudios Tecnicos – Sección de Hidroclimatología. Rio Melendez, Cali, Colombia.
Davies, C.T.M. (1973a). Physiological responses to exercise in East African children. II. The effects of shistosomiasis, anaemia and malnutrition. *Journal of Tropical Pediatrics*, **19**, 115–19.
Davies, C.T.M. (1973b). Relationship of maximum aerobic power output to productivity and absenteeism of East African sugar cane workers. *British Journal of Industrial Medicine*, **30**, 146–54.
Davies, C.T.M., Brotherhood, J.R., Collins, K.J., Dore, C., Imms, F., Musgrove, J., Weiner, J.S., Amin, M.A., Ismail, H.M., El Karim, M., Omer, A.H.S. & Sukkar, M.Y. (1976). Energy expenditure and physiological performance of Sudanese cane cutters. *British Journal of Industrial Medicine*, **30**, 146–54.
Dehn, M.M. & Bruce, R.A. (1972). Longitudinal variations in maximal oxygen intake with age and activity. *Journal of Applied Physiology*, **33**, 805–7.
Durnin, J.V.G.A. (1982). Muscle in sports and medicine – nutrition and muscular performance. *International Journal of Sports Medicine*, **3**, 52–7.
Durnin, J.V.G.A. & Passmore, R. (1967). *Energy, work and leisure*. Heinemann, London.
Durnin, J.V.G.A. & Womersley, J. (1974). Body fat assessed from total body density and its estimation from skinfold thickness: measurements on 481 men and women aged from 16 to 72 years. *British Journal of Nutrition*, **32**, 77–97.
Gollnick, P.D., Piehl, K., Saubert, C.W., Armstrong, R.B. & Saltin, B. (1972). Diet, exercise and glycogen changes in human muscle fibers. *Journal of Applied Physiology*, **33**, 421–5.

Gorsky, R.D. & Calloway, D.H. (1983). Activity pattern changes with decreases in food energy intake. *Human Biology*, **55**, 577–86.

Hamill, P.V.V., Drizd, T.A., Johnson, C.L., Reed, R.B., Roche, A.F. & Moore, W.M. (1979). Physical growth: National Center for Health Statistics percentiles. *American Journal of Clinical Nutrition*, **32**, 607–29.

Hansen-Smith, F.M., Maksud, M.G. & van Horn, D.L. (1977). Influence of chronic undernutrition on oxygen consumption of rats during exercise. *Growth*, **41**, 115–21.

Hansson, J.E. (1965). The relationship between individual characteristics of the worker and output of logging operations. *Studia Forestalia Suecia*, **29**, 68–77. Skogshogstolan, Stockholm.

Heymsfield, S.B., Stevens, V., Noel, R., McManus, C., Smith, J. & Nixon, D. (1982). Biochemical composition of muscle in normal and semistarved human subjects: relevance to anthropometric measurements. *American Journal of Clinical Nutrition*, **36**, 131–42.

ICNND (1959–63). *Interdepartmental Committee on Nutrition for National Defense Nutritional Surveys*. US Govt. Printing Office, Washington, DC. (a) Ecuador: July, 1959; (b) Chile: March, 1960; (c) Colombia: May, 1960; (d) Uruguay: March–April, 1962; (e) North East Brazil: May, 1963; (f) Venezuela: May, 1963.

Keller, W. (1988). The epidemiology of stunting. In: *Linear Growth Retardation in Less Developed Countries*, ed. J.C. Waterlow, pp. 17–339. Raven Press, New York.

Keys, A., Brožek, J., Henschel, A., Mickelsen, O. & Taylor, H.L. (1950). *The biology of human starvation*. University of Minnesota Press, Minneapolis.

Klein, R.E., Breeman, H.E., Kagan, J. & Yarbrough, C. (1972). Is big smart? The relation of growth to cognition. *Journal of Health and Social Behavior*, **13**, 219–25.

Krahenbuhl, G.J., Skinner, J.S. & Kohrt, W.M. (1985). Developmental aspects of maximal aerobic power in children. *Exercise and Sport Science Review*, **13**, 503–38.

Lopes, J., Russell, D.M., Whitewell, J. & Jeejeebhoy, K.N. (1982). Skeletal muscle function in malnutrition. *American Journal of Clinical Nutrition*, **36**, 602–10.

Lusk, G. (1921). The physiologic effects of undernutrition. *Physiological Reviews*, **1**, 523–52.

Maksud, M.G., Spurr, G.B. & Barac-Nieto, M. (1976). The aerobic power of several groups of laborers in Colombia and the United States. *European Journal of Applied Physiology*, **35**, 173–82.

Malina, R. (1984). Physical activity and motor development/performance in populations nutritionally at risk. In: *Energy Intake and Activity*, ed. E. Pollitt & P. Amante, pp. 285–302. A.R. Liss, New York.

Martorell, R. (1985). Child growth retardation: a discussion of its causes and its relationship to health. In: *Nutritional Adaptation in Man*, ed. K.L. Blaxter & J.C. Waterlow, pp. 13–30. John Libbey, London.

Michael, E.D., Hutton, K.E. & Horvath, S.M. (1961). Cardiorespiratory responses during prolonged exercise. *Journal of Applied Physiology*, **16**, 997–1000.

Mirwald, R.L. & Bailey, D.A. (1981). Longitudinal comparison of aerobic power in active and inactive boys aged 7.0 to 17.0 years. *Annals of Human Biology*, **8**, 405–14.

Pařízková, J. (1961). Total body fat and skinfolds in children. *Metabolism*, **10**, 794–807.

Pascale, L.R., Grossman, M.I., Sloan, H.J. & Frankel, T. (1956). Correlations between thickness of skinfolds and body density in 88 soldiers. *Human Biology*, **28**, 165–75.

Raju, N.V. (1974). Effect of early malnutrition on muscle function and metabolism in rats. *Life Science*, **15**, 949–60.

Reddy, V., Jagadeesan, V., Ragharamulu, N., Bharkaram, C. & Srikantia, S.G. (1976). Functional significance of growth retardation in malnutrition. *American Journal of Clinical Nutrition*, **29**, 3–7.

Rueda-Williamson, R., Luna-Jaspe, H., Ariza, J., Pardo, F. & Mora, J.O. (1969). Estudio seccional de crecimiento, desarrollo y nutrición en 12, 138 niños de Bogotá, Colombia. *Pediatría*, **10**, 337–49.

Rutishauser, I.H.E. & Whitehead, R.G. (1972). Energy intake and expenditure in 1–3 year old Ugandan children living in a rural environment. *British Journal of Nutrition*, **28**, 145–52.

Sameroff, A.J. & McDonough, S.C. (1984). The role of motor activity in human cognitive and social development. In: *Energy Intake and Activity*, ed. E. Pollitt & P. Amante, pp. 331–53. A.R. Liss, New York.

Satyanarayana, K., Nadamuni Naidu, A., Chatterjee, B. & Narasinga Rao, B.S. (1977). Body size and work output. *American Journal of Clinical Nutrition*, **30**, 322–5.

Satyanarayana, K., Nadamuni Naidu, A. & Narasinga Rao, B.S. (1978). Nutrition, physical work capacity and work output. *Indian Journal of Medical Research*, **68** (suppl.), 88–93.

Satyanarayana, K., Nadamuni Naidu, A. & Narasinga Rao, B.S. (1979). Nutritional deprivation in childhood and the body size, activity and physical work capacity of young boys. *American Journal of Clinical Nutrition*, **32**, 1769–75.

Shephard, R.J., Lavallée, H., Larivière, G., Rajic, M., Brisson, G.R., Beucage, C., Jequier, J-C. & La Barre, R. (1974). La capacité physique des enfants canadiens: une comparison entre les enfants canadiens-français, canadien-anglais et esquimaux: I. Consommation maximale d'oxygène et débit cardiaque. *Union Medica*, **103**, 1767–77.

Smil, V. (1979). Energy flow in the developing world. *American Scientist*, **67**, 522–31.

Spurr, G.B. (1983). Nutritional status and physical work capacity. *Yearbook of Physical Anthropology*, **26**, 1–35.

Spurr, G.B. (1984). Physical activity, nutritional status and physical work capacity in relation to agricultural productivity. In: *Energy Intake and Activity*, ed. E. Pollitt & P. Amante, pp. 207–61. A.R. Liss, New York.

Spurr, G.B. (1988). Body size, physical work capacity, and productivity in hard work: is bigger better? In: *Linear Growth Retardation in Less Developed Countries*, ed. J.C. Waterlow, pp. 215–43. Raven Press, New York.

Spurr, G.B., Barac-Nieto, M. & Maksud, M.G. (1975). Energy expenditure cutting sugar cane. *Journal of Applied Physiology*, **39**, 990–6.

Spurr, G.B., Barac-Nieto, M. & Maksud, M.G. (1977b). Productivity and maximal oxygen consumption in sugar cane cutters. *American Journal of Clinical Nutrition*, **30**, 316–21.

Spurr, G.B., Barac-Nieto, M. & Maksud, M.G. (1978). Childhood undernutrition: Implications for adult work capacity and productivity. In: *Environmental Stress: Individual Human Adaptions*, ed. L.J. Folinsbee, J.A. Wagner, J.F. Borgia, B.L. Drinkwater, J.A. Gliner & J.F. Bedi, pp. 165–81. Academic Press, New York.

Spurr, G.B., Barac-Nieto, M., Reina, J.C. & Ramirez, R. (1984). Marginal malnutrition in school-aged Colombian boys: efficiency of treadmill walking in submaximal exercise. *American Journal of Clinical Nutrition*, **39**, 452–9.

Spurr, G.B., Maksud, M.G. & Barac-Nieto, M. (1977a). Energy expenditure, productivity, and physical work capacity of sugar cane loaders. *American Journal of Clinical Nutrition*, **30**, 1740–6.

Spurr, G.B., Prentice, A.M., Murgatroyd, P.R., Goldberg, G.R., Reina, J.C. & Christman, N.T. (1988). Energy expenditure using minute-by-minute heart rate recording: comparison with indirect calorimetry. *American Journal of Clinical Nutrition*, **48**, 552–9.

Spurr, G.B. & Reina, J.C. (1987). Marginal malnutrition in school-aged Colombian girls: dietary intervention and daily energy expenditure. *Human Nutrition: Clinical Nutrition*, **41C**, 93–104.

Spurr, G.B. & Reina, J.C. (1988a). Patterns of daily energy expenditure in normal and marginally undernourished school-aged Colombian children. *European Journal of Clinical Nutrition*, **42**, 819–34.

Spurr, G.B. & Reina, J.C. (1988b). Influence of dietary intervention on artificially increased activity in marginally undernourished Colombian boys. *European Journal of Clinical Nutrition*, **42**, 835–46.

Spurr, G.B. & Reina, J.C. (1989). Maximum oxygen consumption in marginally malnourished Colombian boys and girls 6–16 years of age. *American Journal of Human Biology*, **1**, 11–19.

Spurr, G.B., Reina, J.C. & Barac-Nieto, M. (1983a). Marginal malnutrition in school-aged Colombian boys: anthropometry and maturation. *American Journal of Clinical Nutrition*, **37**, 119–32.

Spurr, G.B., Reina, J.C. & Barac-Nieto, M. (1986). Marginal malnutrition in school-aged Colombian boys: metabolic rate and estimated daily energy expenditure. *American Journal of Clinical Nutrition*, **44**, 113–26.

Spurr, G.B., Reina, J.C., Dahners, H.W. & Barac-Nieto, M. (1983b). Marginal malnutrition in school-aged Colombian boys: functional consequences in maximum exercise. *American Journal of Clinical Nutrition*, **37**, 83–47.

Tasker, K. & Tulpule, P.G. (1964). Influence of protein and calorie deficiencies in the rat on the energy transfer reactions of the striated muscle. *Biochemical Journal*, **92**, 391–8.

Thomson, A.M. (1980). The importance of being tall. *Human Ecology Forum*, **10**, 4–10.

Torún, B., Schultz, Y., Viteri, F. & Bradfield, R.B. (1979). Growth, body composition and heart rate/V_{O_2} relationship changes during the nutritional recovery of children with two different physical activity levels. *Bibliotheca Nutrition & Dieta*, **27**, 55–6.

Torún, B. & Viteri, F.D. (1981). Energy requirements of pre-school children and effects of varying energy intakes on protein metabolism. In: *Protein–Energy*

Requirements of Developing Countries: Evaluation of New Data, (*Supplement 5*), ed. B. Torún, V.R. Young & W.M. Rand, pp. 229–41. United Nations University World Hunger Program, Food and Nutrition Bulletin, Tokyo.

United Nations (1980). *Demographic Year Book (1979)*. Department of International Economy and Social Affairs. UN, New York.

Viteri, F.E. (1971). Considerations on the effect of nutrition on the body composition and physical working capacity of young Guatemalan adults. In: *Amino Acid Fortification of Protein Foods*, ed. N.S. Schrimshaw & A.M. Altshull, pp. 350–75. MIT Press, Cambridge, MA.

Viteri, F.E. & Torún, B. (1975). Ingestión calórica y trabajo fisico de obreros agrícolas en Guatemala. Efecto de la suplementación alimentaria y su luger en los programas de salud. *Bolitin de la Oficina Sanitaria Panamericana*, **78**, 58–74.

Viteri, F.E. & Torún, B. (1981). Nutrition, physical activity, and growth. In: *The Biology of Normal Human Growth*, ed. M. Ritzen, A. Aperia, K. Hall, A. Larsson, A. Zetterberg & R. Zetterstrom, pp. 265–73. Raven Press, New York.

von Dobeln, W. (1956). Human standard and maximal metabolic rate in relation to fat free mass. *Acta Physiologica Scandinavica*, **37** (Suppl. 126), 1–79.

Waterlow, J.C. (ed.) (1988). *Linear Growth Retardation in Less Developed Countries*. Raven Press, New York.

Zavaleta, A.N. & Malina, R.M. (1980). Growth, fatness and leanness in Mexican-American children. *American Journal of Clinical Nutrition*, **33**, 2008–20.

4 *Morbidity, nutritional deficiencies and child development in developing countries*

SALLY GRANTHAM-MCGREGOR

Children's development is intimately related to both socio-cultural and biological factors in their environment. In developing countries it is probable that biological factors play a more important role than they do in developed countries. On the one hand, nutrient deficiencies and morbidity are more prevalent and more severe, while on the other hand, health care is poorer. Although a large amount of research has been carried out on the effects of certain nutrient deficiencies and child development, there have been very few studies on morbidity. In this paper I will give a brief overview of the relationship between nutrient deficiencies, morbidity and child development focussing on conditions of special importance to the third world.

Modifying factors

Both diseases and nutrient deficiencies occur more frequently in circumstances of poverty, which include poor parental care and educational levels, poor housing, sanitation and water supply and overcrowding. Some of these conditions themselves have a detrimental effect on child development. It is extremely difficult to separate these effects from those of morbidity and nutrient deficits. Many studies have failed to take social background into account, particularly those focussing on morbidity. Further, it is important to determine whether the effects are additive or interactive. There is some suggestion of an interaction with social background from the studies of small birth weight, and those of postnatal protein–energy malnutrition to be discussed later.

In addition to the social context in which the conditions occur, other factors may be important. The severity and duration of the condition, the presence of other diseases or nutrient deficiencies, and the stage of development of the child, all probably affect the behavioural outcome.

Table 4.1. *Prenatal conditions important in lesser developed regions and affecting child development*

Protein–energy malnutrition
Iodine deficiency
Infections
Specific infections, e.g. rubella, syphillis
Lead contamination

Prenatal

The relationship between morbidity, nutrient deficiencies and child development begins from conception (Table 4.1). Maternal infections and protein–energy malnutrition (PEM) are common and lead to small birth weights and prematurity.

The prevalence of small birth weight in developing countries ranges from 50% in Bangladesh, through 30% in India, 25% in Nigeria to 12% in Jamaica. This contrasts with rates between 4 and 10% in developed countries (Grant, 1987). Small birth weight is associated with poor psychomotor performance in childhood (Lasky *et al.*, 1975). Furthermore, there is an interaction between the quality of the environment and the effects of small birth weight, with poverty increasing the deficit attributed to any perinatal stress. In contrast, better socio-economic situations minimise the detrimental effects (Werner, Bierman & French, 1971). It is therefore probable that not only is small birth weight more prevalent but it is also associated with more problems in developing than in developed countries. There is a need for more research into this question.

Foetal iodine deficiency is the other prenatal condition which must be mentioned. It is still prevalent in Asia, Africa and Latin America and has a devastating effect on brain development, causing cretinism. There is also some evidence of neurological and intellectual deficits occurring with lesser degrees of deficiency (Stanbury, 1987). Other diseases which cause specific damage to the foetus, such as syphillis and rubella, may not be particularly important in developing countries where health care is adequate. However, where appropriate treatment or immunisation is not available the prevalence and effects are unnecessarily great.

Postnatal

In postnatal life infections and nutrient deficiencies are particular problems (Table 4.2). Toxins may not be of special concern to the third world at present. However, recent prospective studies have produced strong

Table 4.2. *Postnatal conditions important in lesser developed regions and possibly affecting child development*

Nutrient deficiencies
Iodine
Vitamin A
Iron
Short term food deprivation
Protein–energy malnutrition
Infections of sense organs
Otitis media
Onchocerciasis
General infections
Repeated respiratory infections & gastroenteritis
Specific diseases, e.g. malaria, schistosomiasis
Gastrointestinal helminths
Toxins
Lead

evidence, which suggests that even low levels of lead contamination affect infant mental development (Davis & Svendsgaard, 1987). It is to be expected that developed countries will initiate stricter controls, and undeveloped countries will probably fall behind in this. The possible effects of contamination with pesticides also need to be investigated.

Infections

Some of the more important childhood infections, in terms of severity and prevalence, are gastroenteritis, respiratory infections, gastrointestinal parasites, malaria and schistosomiasis. In addition there are those affecting the sense organs such as onchocerciasis (river blindness) and otitis media. In general there has been extremely little attempt to determine whether these conditions affect child development.

Pollitt (1983) is one of the few investigators to examine the possible effects of repeated gastroenteritis and respiratory infections on infant development. In a Taiwan study, he demonstrated a significant relationship in the first six months of life. Several studies have been conducted on the effects of schistosomiasis infection but they have yielded inconsistent results. This may be attributed to the difficulties in separating prevalence and burden of the parasite, and the presence of multiple infestations. In addition there is some evidence that the more gregarious, active children are the most likely to be infected (Kvalsvig, 1988).

Mechanisms

There are several plausible mechanisms whereby infections could affect child development. Diseases cause apathy secondary to chronic inflammation, pain or fever. This in turn would reduce young children's ability to explore their environment and thus acquire skills. In addition the child's attention and concentration might be affected.

The disease itself might directly affect the senses, such as sight with onchocerciasis, and hearing with otitis media. Several studies have shown that children who suffer from recurrent otitis media in early childhood have poor language development and school performance, although this has not been found in all studies (Rapin, 1979; Horowitz & Leake, 1980). It appears that those with chronic disease, those that are not well treated medically, and possibly children from the poorest homes, are most at risk of poor development.

Infections may also cause iron deficiency, especially intestinal helminths and malaria, and this in turn may affect behaviour. Similarly they may cause PEM which may then affect behaviour.

Chronic diseases in children in developed countries are associated with emotional problems (Pless & Roghmann, 1971). Repeated infections in underdeveloped countries may have a similar effect and this could in turn affect other areas of development. Repeated illness would contribute to school absence and subsequent poor school performance. This also occurs in children with chronic disease in developed countries (Fowler, Johnson & Atkinson, 1985).

Conclusions

There are good reasons to believe that the repeated and chronic infections suffered by many third world children would affect their development. However, there are remarkably few studies on the subject, and an urgent need to begin investigations.

Nutrient deficiencies

Nutrient deficiencies likely to affect child development are shown in Table 4.2. Both iodine and vitamin A deficiency are very prevalent. Vitamin A deficiency affects vision and would be expected to affect development and school attendance. Iodine deficiency is the subject of Chapter 6, so will not be discussed here.

Some nutrient effects on behaviour appear to be transient, while others are more long term. Transient ones are readily reversed whereas long term ones are more difficult or impossible to reverse.

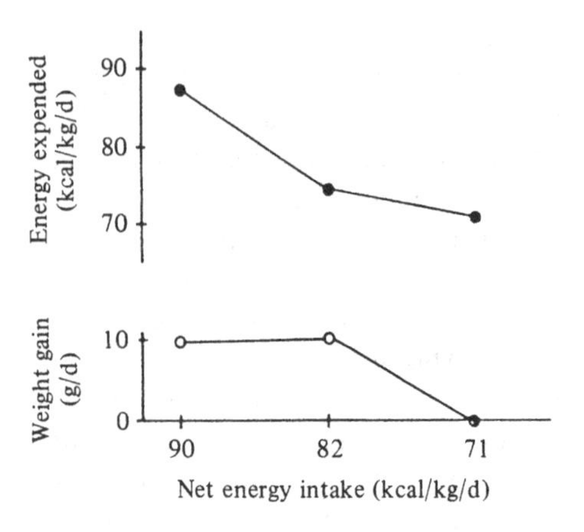

Fig. 4.1 Energy expenditure and weight gains of children fed 1.73 g of corn and bean proteins/kg body weight, with changes in energy intake at 40 day intervals (Torun & Viteri, 1981).

Transient effects

Torun & Viteri (1981) showed that when energy intake was reduced in children, their activity levels rapidly declined (Fig. 4.1). They did not report the return of normal activity levels with increasing intakes. However, in Jamaica we compared severely malnourished children with adequately nourished children who were sick with other diseases (Waterlow, Grantham-McGregor & Tompkins, 1989). Both groups were admitted to hospital. The malnourished group was less active and exploring than the adequately nourished children on admission, but this behaviour improved readily with nutritional rehabilitation. It is not clear which nutrients were responsible for these changes.

In experimental situations, missing breakfast has also been shown to have transient effects on cognitive function, in adequately nourished children (Pollitt, Leibel & Greenfield, 1981; Conners & Blouin, 1983; Pollitt *et al.*, 1985). It is not known whether missing breakfast on a regular basis also affects behaviour. In a recent study children who were undernourished, in that they had low weights-for-height or low heights-for-age, were compared with adequately nourished children. The undernourished children's cognitive functions were detrimentally affected when they missed breakfast. In contrast, the adequately nourished children's were not (Simeon & Grantham-McGregor, 1987). This suggests that the transient effects of food deprivation vary with the children's underlying nutritional status.

Iron deficiency may also have transient behavioural effects. It is the commonest cause of anaemia, and estimates of the prevalence of anaemia range from 51% in less developed regions of the world to 10% in developed regions (DeMaeyer & Adiels-Tegman, 1985). An association has been demonstrated in many studies between iron deficiency and poor levels of mental development. However, this is frequently confounded by poor social background.

In the last decade, several trials of the effect of iron therapy on the cognitive function of iron-deficient, anaemic children have been reported. Four studies have been conducted with children under 24 months of age, using short term iron therapy of two weeks or less. In two of the studies therapy produced improvements in psychomotor function (Oski & Honig, 1978; Walters, Kovalskys & Stekel, 1983). In the other two no benefit was found (Lozoff *et al.*, 1982; 1987). When treatment was continued for three months in children in the last study, only those with complete haematological recovery showed a significant benefit.

Two studies of longer treatment were conducted with children between 18 months and six years of age, in one study no benefit was found in the children's developmental quotients or IQ scores (Deinard *et al.*, 1986). In the other the children showed a significant benefit in discrimination learning (Pollitt *et al.*, 1986).

Two studies have been conducted with school children who received three to four months of treatment. In one, the treated children showed a significant benefit in their efficiency at problem solving (Pollitt *et al.*, 1986). In the other, a significant benefit in school achievement was produced by treatment (Soemantri, Pollitt & Kim, 1985).

There are some obvious inconsistencies in these findings; however, some of the studies were not true clinical trials. In some there was no random assignment to treatment group (e.g. Walters *et al.*, 1983), others had very small groups (e.g. Oski & Honig, 1978). The two studies with school children (Soemantri *et al.*, 1985; Pollitt *et al.*, 1983) are perhaps the most convincing in terms of having robust study designs and significant improvements with treatment. These indicate a definite and transient effect on behaviour. However, after three months of treatment the treated anaemic group still had scores below the non-anaemic group which suggests that some of the behavioural effects may be more difficult to reverse.

Long term effects

The precise mechanism of how long term effects are produced is unknown. It is possible that they result if transient effects continue for long periods of time. For instance if a child is apathetic and inactive, or has limited ability to pay attention, he may develop skills more slowly than normal. Over time, a sizeable developmental lag could accumulate which would be difficult to reverse. In addition the apathetic behaviour of the children could secondarily produce less responsiveness in the caretakers.

An alternative hypothesis is that permanent changes may occur in the central nervous system (CNS). This certainly happens in foetal iodine deficiency and cretinism. Severe PEM in rats produces permanent changes to the brain (Bedi, 1987). However, there is no evidence of CNS changes in children with PEM being directly linked to cognitive deficits.

I would hypothesise that severe PEM is the most likely type of PEM to produce long term deficits in mental development. The prevalence of severe PEM is usually between 1 and 3% in developing countries (Grant, 1987).

Many studies have been conducted in which school aged survivors of severe PEM, have been compared with their siblings or controls carefully matched for social background. These have been extensively reviewed previously (Pollitt & Thomson, 1977; Grantham-McGregor, 1989a, b).

We reviewed eight well documented studies using siblings as controls (Birch *et al.*, 1971; Evans, Moodie & Hansen, 1971; Hertzig *et al.*, 1972; Nwuga, 1977; Bartel *et al.*, 1978; Graham & Adrianzen, 1979; Pereira, Sundararaj & Begum, 1979; Moodie *et al.*, 1980). In four of these, the index children performed worse on tests of cognitive function or school achievement (Birch *et al.*, 1971; Hertzig *et al.*, 1972; Nwuga, 1977; Pereira *et al.*, 1979). In none were they better. We found six studies in which reasonably well matched controls were used (Champakam, Srikantia & Gopalen, 1968; Hertzig *et al.*, 1972; Hoorweg & Stanfield, 1976; Nwuga, 1977; Bartel *et al.*, 1978; Galler *et al.*, 1983). In all but one (Bartel *et al.*, 1978), the index children performed significantly worse on tests of cognitive function, school achievement or motor function.

Neither siblings nor children matched for social background make perfect controls. Siblings are often malnourished themselves, thus they minimise the disadvantage of the index group. It is almost impossible to match control groups exactly for all factors which may affect mental development (Richardson, 1974).

The other problem with the above studies is that they were all retrospective, beginning after the children became malnourished. It is therefore possible that the groups were not equivalent preceding the onset of malnutrition. Only one study, conducted in a Mexican village, was prospective (Cravioto & Delicardie, 1972). In this, the children who became severely malnourished had normal levels of development preceding the onset of malnutrition.

In general, these studies indicate that children who are severely malnourished in early childhood are likely to have poor levels of mental development, at least through school age. However, because of limitations in the study design a causal association cannot be established unequivocally. In addition, poor mental development follows only when children return to poor environments. An important finding is that children who are malnourished secondary to other diseases in developed countries are much less likely to suffer poor mental development (Rush, 1984). However, most studies of these children have no measures of development immediately following the acute episode. It is therefore unclear whether they are initially affected in development, then recover with a supportive environment, or whether they are never deterimentally affected.

It is possible that the effect of malnutrition interacts with the social background, and that the disadvantages are only apparent in poor social circumstances. Richardson demonstrated that survivors of severe malnutrition in Jamaica, who came from better home backgrounds had higher levels of intelligence at school age than survivors living in worse homes (Richardson, 1976). In addition, it has been shown in studies with malnourished rats that increased stimulation at the time of malnutrition reduced the behavioural effects of malnutrition (Levitsky, 1979). However, in other studies no sign of an interaction between malnutrition and stimulation was found in the neuroanatomy of rat brains (Bedi & Bhide, 1988).

Mild to moderate malnutrition is more prevalent than severe malnutrition, affecting around 30% of the population in less developed regions (Grant, 1987). Consequently it is extremely important to determine its effects on mental development. Several studies of nutritional intervention have been carried out in an attempt to answer this question. These have been reviewed in detail previously (Rush, 1984; Grantham-McGregor, 1987).

Briefly, the studies have been conducted in communities where malnutrition is endemic. Four major studies have involved pregnant women and/or their subsequent offspring. In one study in Taiwan, only the mothers were supplemented in pregnancy and lactation (Joos *et al.*, 1983). In three others the children were also supplemented for three years or more. These were in Bogota, Columbia (Waber *et al.*, 1981), in Guatemala (Freeman *et al.*, 1980) and in Mexico (Chavez & Martinez, 1982). All of these studies suffered from some problems in study design (Grantham-McGregor, 1987). One of the main problems was the difficulty in ensuring a substantial increase in the subjects' dietary intakes. However, in all four studies at least a small benefit was found in the children's psychomotor development or cognitive function. This was so while the supplement was being given. There is less evidence

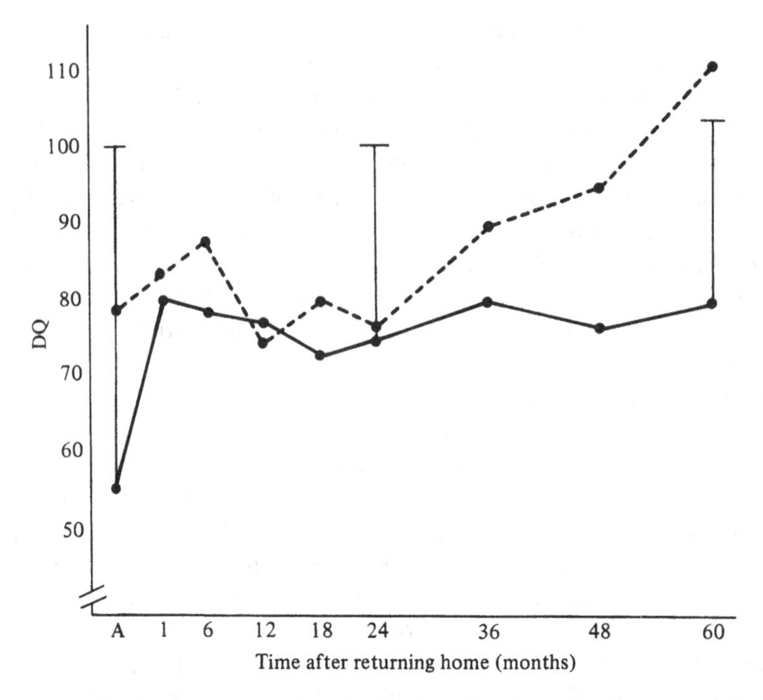

Fig. 4.2 Comparison of the developmental quotients (DQ) on the Griffiths Test of children with severe malnutrition, (A) on admission to hospital and up to 60 months after returning home. Solid lines: mean (and two SDs) of 16 children; dotted line: DQ of child adopted 24 months after leaving hospital.

for the benefits being sustained following the cessation of supplementation.

The effect of supplementation in children who are already undernourished is not well established. In a study in Cali, Columbia (McKay *et al.*, 1978), no benefit was found, while a small benefit was found in another study in Bogota (Mora *et al.*, 1974).

In summary, long term effects on mental development are likely to occur following severe malnutrition in children living in circumstances of poverty. Mild to moderate malnutrition is associated with concurrent deficits, but it is unknown whether there are associated long term effects.

Rehabilitation

Studies of adopted children suggest that vast improvements in development can take place in enriched environments (Winick, Meyer & Harris, 1975). We followed one boy who suffered from severe malnutrition and was subsequently adopted two years later by a middle class family (Grantham-McGregor & Buchanan, 1982). This child was one of a group of 17 severely malnourished children, who were being studied. Fig. 4.2 shows the dramatic improvement which the child experienced in his developmental level, compared with the rest of the group.

We also conducted an intervention programme of increased stimulation with a group of severely malnourished children in their own homes (Grantham-McGregor, Schofield & Powell, 1987). The children were played with in hospital, then visited at home weekly for two years, then every two weeks for a third year. During the visits the mothers were shown how to play with their children and were given home made toys. They were compared with two other groups who had also been in hospital; a second severely malnourished group and an adequately nourished group who had been ill with other diseases. Initially, both malnourished groups had developmental levels seriously lower than the adequately nourished children. The intervened group showed marked improvements compared with the non-intervened malnourished group. Their developmental level caught up with that of the adequately nourished children. Following the cessation of intervention they showed a small decline.

It remains unclear whether children can completely recover with rehabilitation.

Conclusion

The development of third world children is in obvious jeopardy. This is not only from an array of socio-cultural disadvantages but also from repeated and chronic infections and nutrient deficiencies. There is some evidence of an interaction between social background and undernutrition. The possibility of an interaction with infection should also be considered. There is extremely little data available on the role of infection and certain specific nutrient deficiencies on the developmental process. There is therefore an urgent need for further investigation.

Acknowledgements

I acknowledge the support of the Wellcome Trust, UK, and the Ford Foundation, USA. I thank M. Lewis for typing assistance.

References

Bartel, P.R., Griesel, R.D., Burnett, L.S., Freiman, I., Rosen, E.V. & Geefhuysen, J. (1978). Long term effects of kwashiorkor on psychomotor development. *South African Medical Journal*, **53**, 360–2.

Bedi, K.S. (1987). Lasting neuro anatomical changes following undernutrition. In: *Early Nutrition and Later Achievement*, ed. J. Dobbing, pp. 1–49. Academic Press, London.
Bedi, K.S. & Bhide, P.G. (1988). Effects of environmental diversity on brain morphology. In: *Current Topics in Research on Synopses*, ed. D.G. Jones. A.R. Liss, New York.
Birch, H.G., Pineiro, C., Alcalde, E., Toca, T. & Cravioto, J. (1971). Relation of kwashiorkor in early childhood and intelligence at school age. *Pediatric Research*, **5**, 579–85.
Champakam, M.A., Srikantia, S.G. & Gopalan, G. (1968). Kwashiorkor and mental development. *American Journal of Clinical Nutrition*, **21**, 844–50.
Chavez, A. & Martinez, C. (1982). Neurological maturation and performance on mental tests. Chapter in: *Growing Up in a Developing Community*. INCAP, Guatemala City.
Conners, K.C. & Blouin, A.G. (1983). Nutritional effects on behavior of children. *Journal of Psychiatric Research*, **17**, 193–201.
Cravioto, J. & Delicardie, E. (1972). Environmental correlates of severe clinical malnutrition and language development in survivors from kwashiorkor or marasmus. In: *Nutrition. The Nervous System and Behaviour*, pp. 73–94. PAHO Sci. Publ. No. 251, Washington.
Davis, J.M. & Svendsgaard, D.J. (1987). Lead and child development. *Nature*, **329**, 297–300.
Deinard, A.S., List, A., Lindgren, B., Hunt, J.V. & Chang, P. (1986). Cognitive deficits in iron-deficient and iron-deficient anaemic children. *Journal of Pediatrics*, **108**, 681–9.
DeMaeyer, E. & Adiels-Tegman, M. (1985). The prevalence of anaemia in the world. *World Health Statistics Quarterly*, **38**, 302–16.
Evans, D.E., Moodie, A.D. & Hansen, J.D.L. (1971). Kwashiorkor and intellectual development. *South African Medical Journal*, **25**, 1413–26.
Fowler, M.G., Johnson, M.P. & Atkinson, S.S. (1985). School achievement and absence in children with chronic health conditions. *Journal of Pediatrics*, **106**, 683–7.
Freeman, H.E., Klein, R.E., Townsend, J.W. & Lechtig, A. (1980). Nutrition and cognitive development among rural Guatemalan children. *American Journal of Public Health*, **70**, 1277–85.
Galler, J., Ramsey, F., Solimano, G., Lowell, W.E. & Mason, F. (1983). The influence of early malnutrition on subsequent behavioural development. 1. Degree of impairment in intellectual performance. *Journal of the American Academy of Child Psychiatry*, **22**, 8–15.
Graham, G.G. & Adrianzen, B.T. (1979). Status in school of Peruvian children severely malnourished in infancy. In: *Behavioural Effects of Energy and Protein Deficits*, ed. J. Brozek, pp. 185–94. DHEW (NIH) Publ. No. 79–1906, Washington.
Grant, J.P. (1987). The state of the World's Children 1987. UNICEF, New York.
Grantham-McGregor, S.M. (1987). Field studies in early nutrition and later achievement. In: *Early Nutrition and Later Achievement*, ed. J. Dobbing, pp. 128–74. Academic Press, London.
Grantham-McGregor, S.M. (1989a). The effects of undernutrition on mental development. In: *Psychobiology of Human Eating and Nutritional*

Behaviour. Wiley Psychophysiology Series, ed. R. Shepherd, Wiley, Chichester, GB.

Grantham-McGregor, S.M. (1989b). Malnutrition and mental function. In: *The Malnourished Child*, 19th Nestlé Nutrition Workshop, ed. R. Suskind. Raven Press, New York.

Grantham-McGregor, S.M. & Buchanan, E. (1982). The development of an adopted child recovering from severe malnutrition – a case report. *Human Nutrition: Clinical Nutrition*, **36C**, 251–6.

Grantham-McGregor, S.M., Schofield, W. & Powell, C. (1987). Development of severely malnourished children who received psychosocial stimulation: six year follow-up. *Pediatrics*, **79**, 247–54.

Hertzig, M.E., Birch, H.G., Richardson, S.A. & Tizard, J. (1972). Intellectual levels of school children severely malnourished during the first two years of life. *Pediatrics*, **49**, 814–24.

Hoorweg, J. & Stanfield, J.P. (1976). The effects of protein energy malnutrition in early childhood and intellectual and motor abilities in later childhood and adolescence. *Developmental Medicine and Child Neurology*, **18**, 330–50.

Horowitz, F.D. & Leake, H. (1980). Effects of otitis media on cognitive development. *Annals of Otology, Rhinology and Laryngology* (Suppl. 68), **89**, 264–8.

Joos, S.K., Pollitt, E., Mueller, N.H. & Albright, D.L. (1983). The Bacon Chow Study: Maternal nutritional supplementation and infant behavioural development. *Child Development*, **54**, 669–76.

Kvalsvig, J.D. (1988). The effects of parasitic infection on cognitive performance. *Parasitology Today*, **4**, 206–8.

Lasky, R.E., Lechtig, A., Delgado, H., Klein, R.E., Engle, P., Yarbrough, C. & Martorell, R. (1975). Birth weight and psychomotor performance in rural Guatemala. *American Journal of Diseases in Children*, **129**, 566–70.

Levitsky, D.A. (1979). Malnutrition and hunger to learn. In: *Malnutrition, Environment and Behaviour*, ed. D.A. Levitsky, pp. 161–79. Cornell University Press, Ithaca.

Lozoff, B., Brittenham, G.M., Viteri, F.E., Wolf, A. & Urrutia, J.J. (1982). The effects of short-term oral iron therapy on developmental deficits in iron-deficient anaemic infants. *Journal of Pediatrics*, **100**, 351–7.

Lozoff, B., Brittenham, G.M., Wolf, A.W., McClish, D.K., Kuhnert, P.M., Jimenez, E., Jimenez, R., Mora, L.A., Gomez, I. & Krasukoph, D. (1987). Iron deficiency anaemia and iron therapy effects on infant developmental test performance. *Pediatrics*, **79**, 981–95.

McKay, H., Sinisterra, L., McKay, A., Gomez, H. & Lloreda, P. (1978). Improving cognitive ability in chronically deprived children. *Science*, **200**, 270–8.

Moodie, A.D., Bowie, M.D., Mann, M.D. & Hansen, J.D.L. (1980). A prospective 15-year follow-up study of kwashiorkor patients. *South African Medical Journal*, **58**, 677–81.

Mora, J.O., Amezquita, A., Castro, L., Christiansen, J., Clement-Murphy, J., Cobbs, L.F., Cremer, H.O., Dragastin, S., Elias, M.F., Franklin, D., Herrera, M.G., Ortiz, N., Pardo, F., de Paraedes, B., Ramos, C., Riley, R., Rodriguez, H., Vuori-Christiansen, L., Wagner, M. & Stare, F.J. (1974). Nutrition, health and social factors related to intellectual performance. *World Review of Nutrition and Dietetics*, **19**, 205–36.

Nwuga, U.C.B. (1977). Effect of severe kwashiorkor on intellectual development among Nigerian children. *American Journal of Clinical Nutrition*, **30**, 1423–30.
Oski, F.A. & Honig, A.S. (1978). The effects of therapy on the developmental scores of iron-deficient infants. *Journal of Pediatrics*, **92**, 21–5.
Pereira, S.M., Sundararaj, R. & Begum, A. (1979). Physical growth and neurointegrative performance of survivors of protein–energy malnutrition. *British Journal of Nutrition*, **42**, 165–71.
Pless, I.B. & Roghmann, K.J. (1971). Chronic illness and its consequences: Observations based on three epidemiologic surveys. *Journal of Pediatrics*, **79**, 351–9.
Pollitt, E. (1983). Morbidity and infant development: A hypothesis, *International Journal of Behavioural Development*, **6**, 461–73.
Pollitt, E., Leibel, R. & Greenfield, D. (1981). Brief fasting, stress and cognition in children. *American Journal of Clinical Nutrition*, **34**, 1526–33.
Pollitt, E., Lewis, N., Garcia, C. & Shulman, R. (1983). Fasting and cognitive functioning. *Journal of Psychiatric Research*, **17**, 169–74.
Pollitt, E., Saco-Pollitt, C., Leibel, R.L. & Viteri, E. (1986). Iron deficiency and behavioural development in infants and preschool children. *American Journal of Clinical Nutrition*, **43**, 555–65.
Pollitt, E., Soemantri, A.G., Yunis, F. & Scrimshaw, N.S. (1985). Cognitive effects of iron-deficiency anaemia. *Lancet*, i, 158.
Pollitt, E. & Thomson, C. (1977). Protein–calorie malnutrition and behaviour: A view from psychology. In: *Nutrition and the Brain*, Vol 2, ed. R. Wurtmann & J. Wurtmann, pp. 261–306. Raven Press, New York.
Rapin, I. (1979). Conductive hearing loss effects on children's language and scholastic skills. *Annals of Otology, Rhinology and Laryngology* (Suppl. 60), **88**, 3–12.
Richardson, S.A. (1974). The background history of school children severely malnourished in infancy. In: *Advances in Pediatrics*, Vol 21, ed. I. Schulman, pp. 167–95. Yearbook Medical Publishers, Chicago.
Richardson, S.A. (1976). The relation of severe malnutrition in infancy to the intelligence of school children with differing life histories. *Pediatric Research*, **10**, 57–61.
Rush, D. (1984). The behavioural consequences of protein–energy deprivation and supplementation early life. An epidemiological perspective. In: *Human Nutrition. A Comprehensive Treatise*, ed. J. Galler, pp. 119–54. Plenum Press, London.
Simeon, D. & Grantham-McGregor, S.M. (1987). Cognitive function, undernutrition and missed breakfast. *Lancet*, ii, 737–8.
Soemantri, A.G., Pollitt, E. & Kim, I. (1985). Iron deficiency anaemia and educational achievement. *American Journal of Clinical Nutrition*, **42**, 1221–8.
Stanbury, J.B. (1987). The iodine deficiency disorders: Introduction and general aspects. In: *The Prevention and Control of Iodine Deficiency Disorders*, ed. B.S. Hetzel, J.T. Dunn & J.B. Stanbury, pp. 35–47. Elsevier, Amsterdam.
Torun, B. & Viteri, F.E. (1981). Energy requirements of preschool children and effects of varying energy intakes on protein metabolism. In: *Protein–Energy Requirements of Developing Countries: Evaluation of New Data*, ed. B. Torun, V.R. Young & W.M. Rand, Suppl. 5, pp. 229–41. United Nations University Food and Nutrition Bulletin, Tokyo.

Waber, D.P., Vuori-Christiansen, L., Ortiz, N., Clement, J.R., Christiansen, N.E., Mora, J.O., Reed, R.B. & Herrera, M.G. (1981). Nutritional supplementation, maternal education, and cognitive development of infants at risk of malnutrition. *American Journal of Clinical Nutrition*, **34**, 807–13.

Walters, T., Kovalskys, J. & Stekel, A. (1983). Effect of mild iron deficiency on infant mental development scores. *Journal of Pediatrics*, **102**, 519–22.

Waterlow, J.C., Grantham-McGregor, S.M. & Tomkins, H. (1989). *Protein–Energy Malnutrition in Third World Children*. E. Arnold of Hodder & Stoughton.

Werner, E.E., Bierman, J.M. & French, F.E. (1971). *The Children of Kauai: A Longitudinal Study from the Prenatal Period to Age Ten*. University of Hawaii Press, Honolulu.

Winick, M., Meyer, K.K. & Harris, R. (1975). Malnutrition and environmental enrichment by early adoption. *Science*, **190**, 1173–5.

5 *The aetiology of kwashiorkor*

A.A. JACKSON

Introduction

The kwashiorkor syndrome

The term kwashiorkor was introduced into the medical lexicon by Williams in 1933. The term was that used by the local people in the area of West Africa where she was working to describe a clinical syndrome which afflicted children during the second year of life. In her first description of the condition, Williams attempted to differentiate the disorder from pellagra and took great care to describe the progression of the skin lesions. Over the next few years a controversy developed as to the specific aetiology of the condition: the workers in East Africa believed it to be primarily a vitamin deficiency (Trowell, Davies & Dean, 1954), the current fashion in nutrition at that time. Williams was in favour of a more general dietary causation. The demonstration that the children did poorly on supplements of niacin, but responded to a general improvement in care and diet appeared to resolve the issue. Over the next few years the idea that kwashiorkor was a disease associated with an inadequate intake of high quality protein took increasing hold, until kwashiorkor was seen as the archetype of a primary protein deficiency disorder. The evidence for this association was never strong and at all times there were dissenting voices (Landman & Jackson, 1980).

Nutrient deficiency syndromes

For most of this century the concept that clinical nutrition is primarily a discipline that defines and treats deficiency diseases has held sway. The perception has been that there is a clinical syndrome associated with a dietary deficiency of a specific nutrient and the approach to treatment is the adequate provision of the missing nutrient. In the first instance generous supplements are provided to make good depleted stores. On this basis the logical way in which to treat a disease that is identified as being caused by a deficiency of protein is to give abundant protein of good quality. In practice the effect of this approach to treatment resulted in children being given massive amounts of protein without any obvious benefit, and frequently with fatal consequence (Waterlow, Cravioto & Stephen, 1960).

In 1961 Waterlow was the first to suggest that the rate of recovery might be determined more by the level of energy intake, than that of protein. In 1973, the expert Committee of the FAO/WHO reported on their review of the recommendations for dietary energy and protein intakes to maintain health in healthy populations (FAO/WHO, 1973). On the basis of the evidence available the Committee considered that the recommendation for protein in childhood should be reduced compared with earlier recommendations. One consequence of this reduction in the recommended intake was that protein deficiency was virtually eliminated at a stroke as a specific nutritional problem of global significance.

There were two important comments upon this series of events. On the one hand McLaren noted that as a public health problem, in terms of the world as a whole, kwashiorkor was in fact less common, and by implication less important than marasmus, or wasting disease (McLaren, 1974). Although this may have been true as an observation of fact, given the specific clinical problems associated with kwashiorkor, the elucidation of its aetiology represents an important step in our appreciation of a disease process that is unlikely to be uncovered by following individuals with simple marasmus. On the other hand Waterlow and Payne in commening upon 'The Great Protein Fiasco' acknowledged that there was a need to reconsider with care the specific aetiology of kwashiorkor if it was not to be considered as a primary disorder of protein deficiency (Waterlow & Payne, 1975). One of the biggest problems in removing the idea of protein deficiency as the cause of kwashiorkor is that it left a vacuum of thought, with no realistic alternative suggestion to fill its place.

Proposed aetiology of kwashiorkor

In order to be able to think constructively about a disease it is necessary to create a concept of the disease process. The difficulty in having no framework within which to think about the aetiology of kwashiorkor, meant that it was almost impossible to develop further ideas in terms of the causation and treatment. Therefore it was imperative that alternative suggestions as to the possible aetiology be developed. This is the situation that obtained at the end of the 1970s and the past decade has seen the exploration of possible alternative models for the disease process.

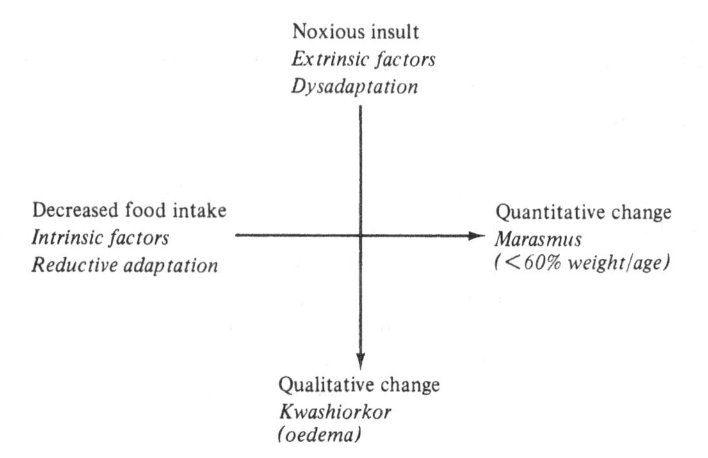

Fig. 5.1 Malnutrition is a consequence of two independent, but related processes taking place, a decrease in food intake and the impact of a noxious insult. A decreased food intake might be the consequence of a number of factors, including simple unavailability. Noxious insults might be infectious, chemical or physical. Kwashiorkor is now defined by the presence of oedema, and marasmus by a reduction in weight to less than 60% of that expected for age.

One approach has been to look for a specific agent or set of circumstances that can be associated with the onset of the disease. In this model the cause is seen as being primarily extrinsic, and not necessarily of nutritional origin. The most recent example of this approach is the suggestion by Hendrickse that aflatoxins might play a major determinant role in the development of kwashiorkor (Hendrickse, 1984). He has presented convincing evidence to support his hypothesis as it applies to the situation found in parts of West Africa, but the limitation of the idea is that it does not appear to be generalisable to all situations in which the kwashiorkor syndrome develops.

Waterlow, on the other hand, argues in favour of dietary factors playing a dominant role (Waterlow, 1984). Without strong evidence to support his position he suggests that dietary protein deficiency has an important part to play in the causation of the disorder. This approach places nutritional factors and an inadequate diet at the centre of the causation.

A third framework considers that the primary pathology is intrinsic to the individual who, because of dysadaptation or maladaptation, succumbs to disease (Gopalan, 1968; Fig. 5.1). This implies the primacy of an intrinsic factor, or the host's ability to withstand the stresses imposed by the external environment. The concept of dysadaptation carries with it a fine ringing tone which implies understanding, but here we reach our first major difficulty. Our appreciation of the normal adaptive processes is so limited that we are not well placed to understand either the nature or the implications of dysadaptation. The tendency therefore has been for investigators to characterise one component of the dysadaptive response and then to raise that to the level of a major causal factor. The nature and cause of oedema provides a useful example of this exercise and is discussed more fully below.

For many investigators the interaction between intrinsic and extrinsic factors has been represented by mutual interaction of nutrition and infection. This interaction has been placed in a wider framework by Golden who has attempted to place the problem within an holistic framework (Golden, 1985; Golden & Ramdath, 1987). He has taken all the observations that he can find on children with oedematous malnutrition and tried to weave them into a single unifying hypothesis, embracing all the observations. This approach has two main advantages; (i) that it seeks to create a new paradigm of thought which gives some weight to all the information available, and (ii) that it creates an hypothesis that should potentially be testable. For these reasons Golden's hypothesis has been welcomed. It has filled the potential vacuum and thereby offered a way forward. Sufficient time should have passed since this model was first enunciated to allow us to assess the extent to which the hypothesis can stand up to critical experimentation, and this is discussed further below. The model itself allows for an interaction between intrinsic and extrinsic factors with the major emphasis being upon the capability of intrinsic factors to withstand a range of adverse environmental influences. The nub of the theory is that cell and tissue damage is produced primarily by the toxic effect of oxidative free radical species, and that in essence the disorder is an inability of the body to maintain an adequate defence against these damaging species: the so-called 'free radical theory of kwashiorkor'.

The nature of kwashiorkor

The clinical syndromes of infantile or childhood malnutrition cover a wide range and variety of physical signs and pathophysiological changes, but most authorities agree that there are two processes taking place together, either concurrently or consecutively. A decrease in food intake leads to wasting disease, with the associated adaptive responses which characterise the process of reductive adaptation: weight loss and increased susceptibility to environmental perturbations or stress, see Fig. 5.2. The second process is more difficult to characterise specifically, but is recognised as being qualitative in nature; a response to some

Table 5.1. *Relative mortality rates for children admitted with severe malnutrition to the Tropical Metabolism Research Unit, University of the West Indies, Jamaica, over a 25-year period, coming from adjacent lowland (South Clarendon) or upland (St Catherine Hills) areas, compared with overall mortality (Jackson & Golden, 1986)*

Area of origin	Admissions	Deaths	% Mortality
South Clarendon	169	28	17
St Catherine Hills	108	2	2
Overall	2032	224	11

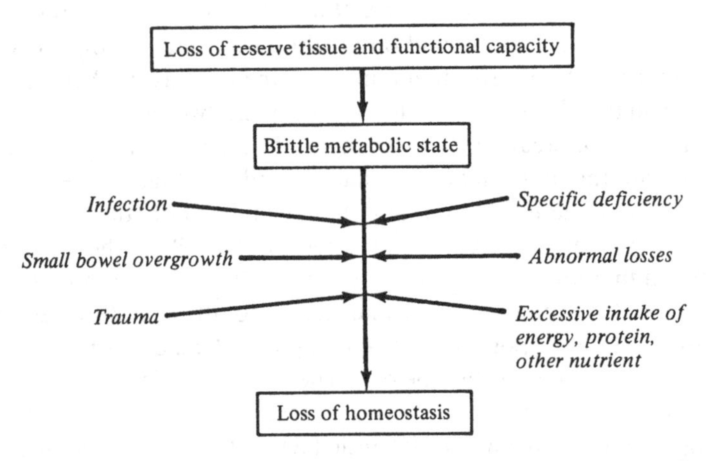

Fig. 5.2 Loss of weight and loss of tissue leads to a loss of the reserve functional capacity known as reductive adaptation. This leads to a brittle metabolic state which is easily perturbed by a number of influences or stresses.

damaging influence and giving rise to derangement in function of a range of systems or organs. In their extreme forms the two processes have been characterised as marasmus and kwashiorkor (Fig. 5.1). The hallmark of kwashiorkor has, by convention, been recognised as the presence of pitting oedema, which may characterise the disorder, but does little in itself to add to our appreciation of the cause.

The clinical presentation of children with severe malnutrition varies in time and place (Jackson & Golden, 1986; Table 5.1). Therefore it is never entirely clear the extent to which differences reported by different workers represent a part of the natural variation of the disease process or real differences in the aspects of the disease itself from place to place. An important step towards clarification was made when the terms kwashiorkor, marasmus and marasmic-kwashiorkor were specifically defined by the Wellcome working party (1970). These terms had previously been used to identify clinical syndromes, but now were to be used to define qualitative and quantitative aspects of severe malnutrition. The quantitative expression, marasmus, was to be used to identify a significant weight deficit: less than 60% of that expected for the child's age. Kwashiorkor was to be used to identify a qualitative change: the presence of nutritional oedema. Although this emphasis upon oedema as being of clinical importance was appropriate, it had the disadvantage of detracting attention from other important aspects of the disease that were not necessarily associated with oedema, but nevertheless represent an important part of the clinical syndrome, for example skin lesions and hepatic changes (Williams, 1933; Waterlow, 1948).

Inevitably within this framework of definition the aetiology of kwashiorkor became synonymous with the aetiology of nutritional oedema. The overwhelming weight of opinion considered the cause of oedema to be primarily a manifestation of hypoalbuminaemia, which in turn was considered to be a manifestation of a dietary protein deficiency. The circle was complete and the conundrum insoluble. If the disease was primarily one of dietary protein deficiency why was it not treatable by generous presentation of protein? There are three propositions which have to be explored:

(i) the evidence in favour of a primary dietary protein deficiency,
(ii) the evidence which relates dietary protein to plasma albumin concentration,
(iii) the evidence that relates plasma albumin to the presence of oedema.

Protein, hypoalbuminaemia and oedema

Dietary protein deficiency

In 1980 we reviewed the information available in the literature at that time on the relationship between the dietary intake of protein and energy and the development of oedematous malnutrition (Landman & Jackson, 1980). We were unable to show for any diet, reported from any part of the world, that there was a shortfall in protein in the diet without a greater shortfall in energy. This interpretation was based upon the 1973 recommended intakes for protein and energy, but persisted even if the earlier recommendations for protein that had been set at a higher level were used. In other words, for all the diets available the children would have been deficient in energy before they became deficient in protein, and any diet that was deficient in protein was even more severely restricted in energy. This interpretation could be incorrect if the recommended intake of protein were set at too low a value, but there would have to be a substantial increase in the recommendations to alter the interpretation, and there is no evidence to justify this change. Alternatively the energy recommendation may have been set too high, but again the error would have to have been considerable if it were to alter the interpretation. Although there may be situations in which there is a specific inadequacy of dietary protein, there is no good evidence in the literature to support this proposition.

Dietary protein and plasma albumin

Much of the evidence which draws an association between dietary protein intake and the concentration of plasma albumin, its rate of synthesis and degradation was reviewed by Waterlow & Alleyne (1971). They concluded that in both animal and human experiments the effect of reduction in dietary protein would be to reduce the rate of both synthesis and degradation of plasma albumin without having any marked effect upon circulating concentrations, which supports the contention that a low protein intake in itself is unlikely to give rise to a low plasma albumin concentration. For many animals a diet low in protein causes a reduction in voluntary food intake. In rats, low protein diets are associated with a fall in albumin concentration when the energy intake is maintained at a level sufficient to cover energy expenditure (Lunn & Austin, 1983). This implies that if the metabolic demand for protein is maintained at a high level in the face of a limited intake then albumin synthesis is reduced and the plasma concentration falls (Jackson, 1985). This mechanism appears to act at the level of the genome (Sakuma *et al.*, 1987).

There are other situations in which the metabolic demand for protein is maintained at a high level in the face of a low or inadequate protein intake. These are most commonly seen during the acute phase response to infection, stress or trauma (Jackson & Grimble, 1989; Fig. 5.3). This response is thought to be a specific effect of cytokines acting at the level of the genome, switching protein synthesis in the liver away from albumin synthesis to the production of acute phase reactants (Perlmutter *et al.*, 1986). Plasma albumin concentrations can fall dramatically as part of the acute phase response, and the magnitude of the change cannot simply be ascribed to a change in the rate of production of albumin (Fleck *et al.*, 1985; Hamilton *et al.*, 1986; Liao, Jefferson & Taylor, 1986). The evidence would suggest that the albumin moves from the plasma into an ill defined 'third space' (Fleck, 1989). Some studies show that the gastro-intestinal tract may represent an important component of the

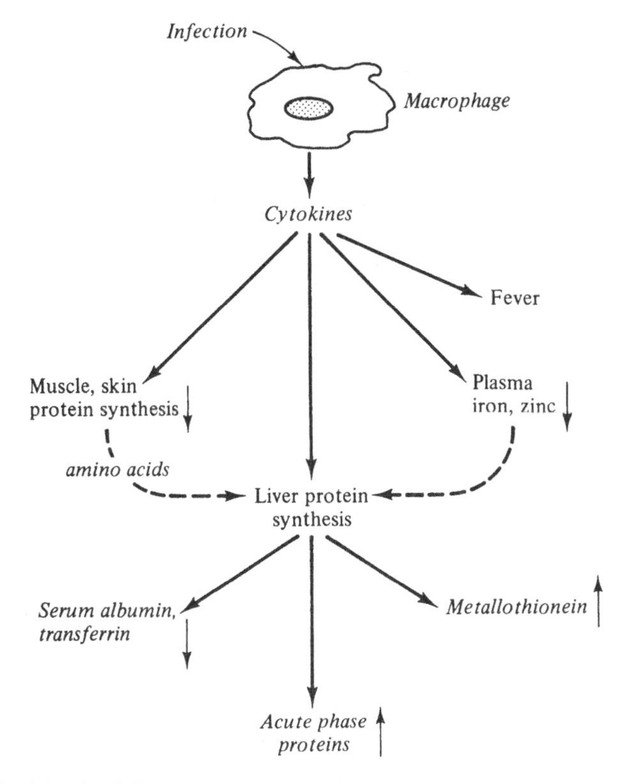

Fig. 5.3 The inflammatory response is a series of coordinated changes in metabolism mediated by chemical messengers, cytokines, which are produced by macrophages in response to infection, trauma or other stress (see text for explanation).

'third space', although the magnitude of the movement of albumin into the bowel is not clear at this time (Lunn, Northrop & Wainwright, 1988).

Increasingly, the consensus of opinion is beginning to favour the idea that a low plasma albumin concentration is not a useful indication of protein intake, and is more appropriately perceived as a negative acute phase reactant, marking the acute phase response (Golden, 1982a). If this is so then it means that children with malnutrition who have a low plasma albumin level are more likely to be demonstrating an acute phase response than those with a normal level.

Plasma albumin and oedema

The concept that hypoalbuminaemia is the primary determinant of nutritional oedema is based upon Starling's (1896) hypothesis which relates the movement of water at the arterial and venous ends of a capillary to the balance between hydrostatic and plasma oncotic forces acting across the capillary membrane. Children presenting with oedematous malnutrition are best treated in the first instance by the carefully controlled presentation of nutrients at a level sufficient to satisfy their maintenance requirements for energy and protein (Jackson & Golden, 1987). Thus on an energy intake of 100 kcal/kg/d and a protein intake of 0.6 g/kg/d, children are able to re-establish homeostatic control, correct specific nutrient deficiencies, clear infection, and lose oedema. This pattern of nutrient intake does not allow for plasma albumin concentrations to move into the normal range, and so the observation that oedema fluid can be cleared without any change in plasma albumin concentration argues in favour of, but does not prove, that albumin concentration is not the primary determinant of nutritional oedema (Golden, Golden & Jackson, 1980).

Children with a congenital deficiency of albumin (analbuminaemia) do not have oedema as a feature of their disease condition (Bearn & Litwin, 1978). The intravenous infusion of albumin in patients with nutritional oedema may produce a transient increase in plasma albumin concentration, but has little effect upon the mobilisation of oedema fluid. In a retrospective study of the mobilisation of oedema in 103 children taking a range of protein and energy intakes, a close association was found between an adequate intake of energy and the mobilisation of oedema (Golden, 1982b). No particular relationship could be found with the intake of protein. Attention has been drawn to the difficulty in formulating a diet that reproducibly gives rise to oedematous malnutrition in experimental animals (Golden, 1985). Part of the reason might be ascribed to the variable composition of the minerals and micronutrients in the dietary formulations that have been used. In exploring this suggestion McGuire & Young (1986) have shown that the relative content of sodium and potassium in the experimental diets may be of particular importance.

During the 1950s it was shown that specific potassium deficiency results in sodium retention and expansion of the extracellular fluid volume (Black & Milne, 1952). The diet on which oedematus malnutrition could be most reliably reproduced in rats, was low in protein and potassium and relatively high in sodium (McGuire & Young, 1986). There are reports of children with marasmus developing pitting oedema when they are treated with oral rehydration solutions that contain a relatively high sodium content.

Therefore oedema can be produced by nutritional manipulation, but a low protein intake by itself is unlikely to be sufficient cause. Appropriate manipulation of dietary energy, potassium and sodium at least are likely to be necessary if reproducible results are to be obtained.

Severely undernourished individuals of all ages appear to have a specific deficit and deficiency of total body potassium (Garrow, 1965; Shizgal, 1980). In childhood, one of the most important contributing factors is thought to be loss of potassium in diarrhoeal fluids, although this is difficult to quantify (Garrow, Smith & Ward, 1968). Certainly, during management the provision of generous supplements of potassium is one of the important components of the successful rehabilitation regime (Waterlow & Alleyne, 1971).

Cellular function and oedema

Impaired liver function

Although hepatic enlargement was noted by Williams (1933) in her first descriptions of kwashiorkor, it was Waterlow (1948) who drew particular attention to this feature during his investigation of the occurrence of malnutrition in the Carribean during the 1940s. He noted that many of the children he saw had a liver which was increased in size, mainly because of an intense infiltration with fat. At the time the fatty liver specific to malnutrition had to be differentiated from other causes of liver disease, childhood cirrhosis, the toxic effect of Senecio and Crotolaria taken as bush teas, and vomiting sickness with hypoglycaemia due to ackee poisoning (Rhodes, 1957). It was noted that the fat accumulation was not particularly responsive to the usual lipotropic agents, but appeared to clear on an adequate diet without any notable after effects. It was remarkable that despite massive fatty infiltration and the impairment of hepatic function there was very little evidence of cellular necrosis or an inflammatory response. Recovery from the insult appeared to be complete (Waterlow, 1975). There did not seem to be any specific impairment of hepatic albumin synthesis over and above that associated with the general condition of the children. It was not possible to show a specific relationship between impairment of hepatic function and either the presence or the mobilisation of oedema.

Renal disease

Under normal circumstances fluid and electrolyte balance is maintained by the kidney, and it is reasonable to consider the extent to which the salt and water retention seen in kwashiorkor is a consequence of deranged renal function. The changes found in renal function have been well characterised and by themselves are insufficient to account for the accumulation of oedema (Alleyne *et al.*, 1977). The specific role played by complex mineral deficiencies, such as phosphate and magnesium, have not been fully explored, nor the potential impact of trace mineral deficiencies such as vanadium (Golden & Golden, 1980), or disturbances in acid base balance.

Impaired membrane function

It is clear that the mechanisms that normally control the distribution of fluid between the intracellular, interstitial and vascular spaces are deranged in kwashiorkor. There is evidence to show that there are alterations in the function (Patrick, 1979) and composition (Coward, 1971; Brown *et al.*, 1978) of cell membranes in kwashiorkor. Patrick has shown that the handling of potassium and sodium across the cell membrane of leucocytes is altered in oedematous malnutrition (Patrick, 1978), but it is not clear the extent to which this represents an appropriate response to an unusual environment, or an inappropriate response (Patrick, 1977). Patrick, Golden & Golden (1980) have presented evidence that specific nutrient deficiencies, such as zinc may play a part in influencing the behaviour of the Na/K-ATPase which has a primary function in maintaining the normal gradient of sodium and potassium between the intracellular and extracellular compartments. Golden has argued in favour of membranes being damaged by the unlimited action of free radicals (Golden & Ramdath, 1987). It is not clear how these short lived species, with effects that are strictly localised in space, can exert such wide ranging effects on many systems.

Therefore it is reasonable to assert that we are not clear as to the general or the specific aetiology of oedema. It has not been possible to demonstrate whether oedema is simply one aspect of a condition with protein manifestations, or whether oedema represents the final common pathway for a wide range of disorders or insults.

Adaptation to diets low in energy and protein

Waterlow has recently discussed the evidence for adaptation to changes, or reductions, in dietary energy and protein (Waterlow, 1986). In this conference he has discussed the importance of social adaptation, compared with biochemical or metabolic adaptation, as an effective mechanism for enhancing the efficiency with which energy can be conserved. These conservations may be of considerable importance in normal individuals.

Malnourished individuals exist in a state of reductive adaptation

(Jackson & Golden, 1987), in which the reserve function of all organs and tissues is decreased as a part of the mechanism whereby the body economises in energy expenditure. There is a cost to the body of reductive adaptation, which means that it is not possible to support a full range of homeostatic control. As a result many aspects of metabolic control are brittle, with only a limited ability to withstand any perturbation.

Against the background of reductive adaptation, any stress arising from infection or other insults, will exert its effects through a limited range of final common pathways. Although having certain general features in common, the detailed expression of the metabolic stress of dysadaptation will be varied depending upon which organ or tissue carries the brunt of the deranged function. As a nutritionist I wish to argue that food has a primary determinant effect upon setting the background conditions against which a range of environmental factors will exert a modulating influence. In order to be able to do this I first have to try to develop a clearer idea of what I consider might be some of the important aspects of the process of adaptation to changes in nutritional intake. Then I must try to explore the ways in which disturbances, either in the process, or as a consequence of the process, might account for the disease condition.

I should like to consider the process of adaptation by exploring some of the effects of changes in dietary energy and protein and the interaction between energy and protein. There are a number of identifiable, specific factors that will either interfere with or modulate the adaptive response. I wish to focus upon the influence of specific nutrient deficiencies, infection in general and gastro-intestinal infection in particular.

Energy

The amount of food eaten is determined primarily by the need to ingest sufficient energy to satisfy the expenditure of energy. For any given diet the pattern of energy intake may vary in the relative proportions of fat, carbohydrate and protein. Although it is clear that individuals may take widely differing amounts of energy, it would seem that for any individual there is only a limited range over which they can accommodate a change in energy intake without any associated change in either function or body composition. The mechanisms whereby individuals accommodate a reduction in energy intake is not clear. Many studies have looked at rather extreme situations in which there has been a severe reduction in energy intake or total fasting, neither situation is a good model for exploring subtle responses.

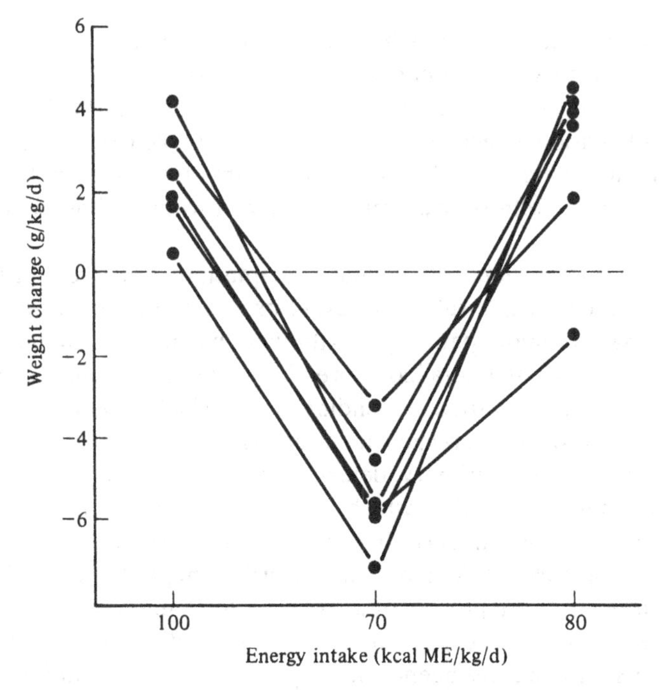

Fig. 5.4 A group of six young children were maintained for a week on each of three diets which provided 1.7 g protein/kg/d and 100, 70 and 80 kcal metabolisable energy (ME)/kg/d. All children gained weight during the first week (linear regression of weight on time for seven days) on 100 kcal/kg/d, and lost weight on 70 kcal/kg/d. During the third week on 80 kcal/kg/d five of the six gained weight, and on average the weight gain was similar to that during the first week of 100 kcal/kg/d. These data imply an accommodation to a low energy intake during the second week that was carried through into the third week (Kennedy *et al.*, 1987).

There are only a few studies in which the responses to relatively small alterations in total energy intake have been examined. This problem was tackled as part of an international study which sought to identify safe levels of protein intake and adequate levels of energy intake in infants and young children (see Rand, Uauy & Scrimshaw, 1984). In these studies a net energy intake of about 90 kcal/kg/d appeared to be adequate to sustain normal growth and activity in young children. However, at a net intake of 80 kcal/kg/d either energy expenditure or nitrogen balance was reduced.

In a subsequent study we investigated the response to a further lowering of energy intake to 70 kcal/kg/d. We were very interested in the time course of the response to the lower intake, which at first seemed to be clearly inadequate. Six children received an adequate intake of energy, 100 kcal metabolisable energy/kg/d, and abundant protein,

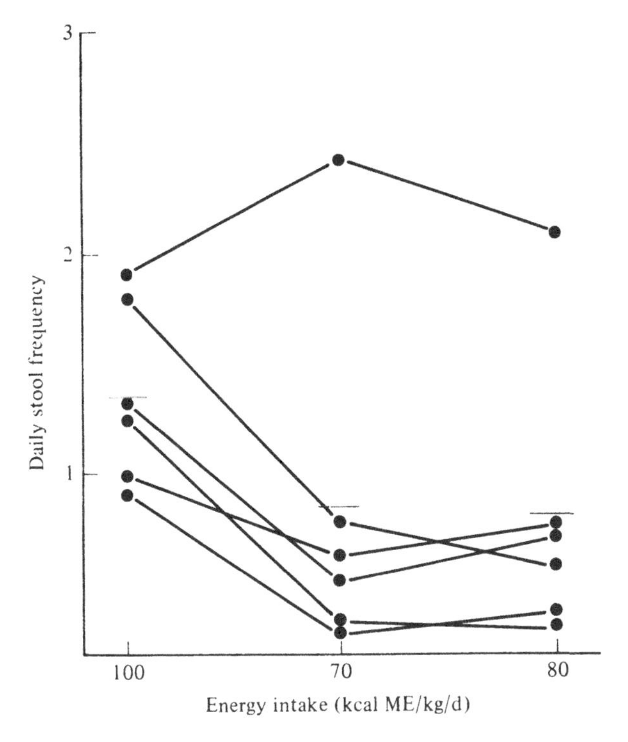

Fig. 5.5 The daily stool frequency was recorded for the children shown in Fig. 5.4. On reducing the energy intake from 100 kcal ME/kg/d to 70 kcal/kg/d, stool frequency was significantly reduced in 5 of the 6 children (bars represent mean values). This reduction was maintained during the third week on 80 kcal/kg/d (bar = mean value) (Kennedy *et al.*, 1987).

1.7 g/kg/d, for one week. During this time they gained weight at an average of 2.3 g/kg/d and were in positive nitrogen balance (Kennedy, Badaloo & Jackson, 1987). During the second week they received the same amount of protein, but the energy intake was reduced to 70 kcal metabolisable energy/kg/d. All of the children lost weight rapidly for the first day or so but the rate of weight loss diminished after the third day, and by the end of the week the weight was tending towards an asymptote. Nitrogen balance was significantly positive at this time. In the third week the children were given 80 kcal/kg/d. The children gained weight on this intake at a rate, 2.7 g/kg/d, that was indistinguishable from the weight gain achieved on 100 kcal/kg/d (Fig. 5.4). In a previous study when the children had received 80 kcal/kg/d after a period on either 90 or 100 kcal/kg/d, they had either lost weight or had no change in weight (Jackson *et al.*, 1983). Hence, the observation that weight could be gained following a period on 70 kcal/kg/d implied a measure of adaptation on the low energy intake which lasted through into the period on 80 kcal/kg/d. The magnitude of the changes was far greater than could be simply accounted for by water shifts, or changes in glycogen content of the body. It is not possible to say the nature or the mechanism of this accommodation to a low energy intake, but it is worth observing that during the period on 70 kcal/kg/d there was a decrease in stool frequency from an average of 1.4 to 0.8 stools per day; a reduction which was maintained into the period on 80 kcal/kg/d, see Fig. 5.5.

From these studies we may conclude that:

(i) in young children a reduction of 10% in energy intake was seen to result in a change in the pattern of behaviour which led to a reduction in energy expenditure and may be seen as being purposive.
(ii) there is a defence of nitrogen balance, if necessary at the cost of body weight, that can be demonstrated down to a reduction in energy intake to 70% of normal; hence there is an extent to which there is important protection of physiological function.
(iii) one of the changes associated with the energy intake of 70% of normal is a reduction in the stool frequency, indicating that a part of the adaptive response to a low energy intake takes place at the level of gastro-intestinal function.

Protein

The mechanisms whereby the body accommodates to changes in protein intake have been the subject of extensive investigation over many years (Waterlow & Stephen, 1981; Blaxter & Waterlow, 1985), and I only wish to consider some of the broad general principles.

Over any extended period of time the weight of an individual remains reasonably constant, with a fairly constant body composition, and nitrogen balance is maintained. For a given intake of protein an equivalent amount of nitrogen is lost from the body on a daily basis in urine, stools, skin, sweat and other routes to maintain nitrogen balance. In addition to this external balance there is also an internal exchange of nitrogen through the processes of protein turnover whereby the proteins of the body are constantly being synthesised and degraded at different, but characteristic, rates. The reutilisation of amino acids from protein degradation for protein synthesis can be very efficient (Stephen, 1981; Fig. 5.6).

Down to a certain level a reduction in protein intake is followed by a series of changes which result in a decrease in the excretion of nitrogen and the restoration of nitrogen balance. Quantitatively the most important aspect of these changes relates to the production and handling of urea in the body, as a reduction in urinary nitrogen is the major factor

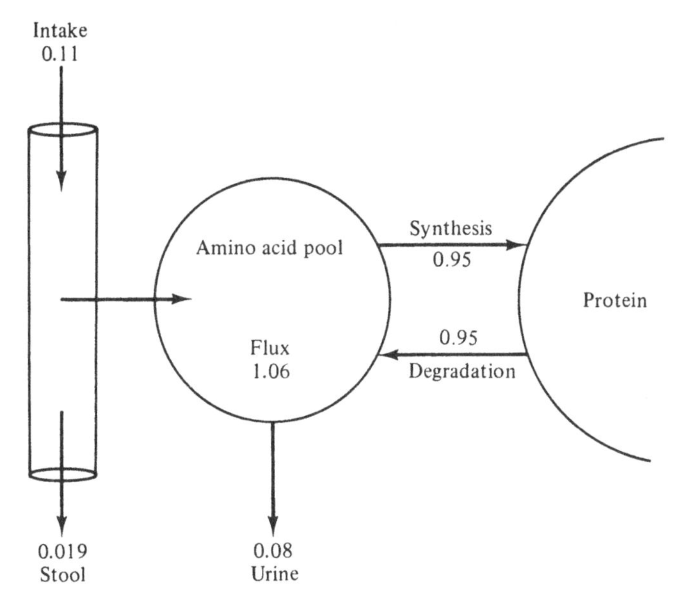

Fig. 5.6 Protein turnover (g protein N/kg/d) was measured in two children who had received an adequate intake of energy (100 kcal ME/kg/d) and a marginal intake of nitrogen (0.11 gN/kg/d) for one week. Amino acids from protein degradation were efficiently utilised for protein synthesis (90%) (Jackson *et al.*, 1983).

which contributes to the restoration of nitrogen balance. As the intake of protein is reduced, amino acids are shunted preferentially into synthetic pathways (Stephen, 1981). There is a decrease in the activity of the enzymes of urea production in the liver, with a consequent reduction in the amount of urea formed (Picou & Phillips, 1972). Even on a generous intake of protein, only about two thirds of the urea produced is excreted in the urine, with the remaining third being passed into the bowel, where the colonic microflora hydrolyse the urea, making the nitrogen available for further metabolic interaction (Jackson, Picou & Landman, 1984). As the dietary protein is reduced, urea production falls; a smaller proportion is excreted in the urine, with a greater proportion being retained by the body (Jackson, Landman & Picou, 1981).

The factors which determine the nature of the fate of the urea produced can be identified in part, see Table 5.2. They show a complex interrelationship, but it is possible to demonstrate that both dietary protein and dietary energy exert an influence (Doherty *et al.*, 1989). The microflora are dependant for their source of energy upon endogenous secretions, such as mucus (van Soest, 1988) and non-digestible carbohydrate taken in the diet (Cummings, 1988). We have been able to show

Table 5.2. *Factors that have been shown to exert an influence upon the rate of utilisation of urea nitrogen following hydrolysis of urea in the gastro-intestinal tract*

Factors
Energy intake
Protein intake
Metabolic demand
Quality of dietary energy
Gastroenteritis
Antiobiotic therapy

that in children showing rapid catch up growth during recovery from severe malnutrition, the most efficient use is made of the endogenous urea production on a diet which is relatively low in protein and high in carbohydrate (Doherty *et al.*, 1989). The available evidence indicates that urea nitrogen can be used for the synthesis of essential and non-essential amino acids (Tanaka *et al.*, 1980). Previously it had been thought that the hydrolysis of urea was unlikely to play any important nutritional role as the colon had been considered to be impermeable to urea and amino acids. We now know that the colon is effectively permeable to both urea and amino acids (Heine *et al.*, 1987; Moran, Persaud & Jackson, 1989b), and there is a rapid and extensive exchange of nitrogen between the host and his colonic flora. By measuring urea kinetics it is possible to quantify the magnitude of this exchange (Jackson, 1986a), which in the normal adult might amount to as much as 230 mg nitrogen/kg/d, equivalent to 100 g protein/d, a significant contribution to the daily intake.

Many observers have shown that the movement of urea into the large bowel can be altered or reduced by the use of broad spectrum antibiotics. Although there are no good measurements of nitrogen losses in the stools during episodes of diarrhoea, the evidence would suggest that they are increased and that the normal activity of the colonic microflora is profoundly disturbed. Under these circumstances one could anticipate derangements of functional significance in the movement and handling of nitrogen in the colon. As urea represents only one component, about 10 to 15%, of a much larger movement of nitrogen through the large bowel (Jackson *et al.*, 1984; Wrong, Vince & Waterlow, 1985), one might anticipate the overall impact to be even greater. Hence, the mechanism whereby the body accommodates to a reduction in dietary protein involves the metabolic activity of the gastro-intestinal tract.

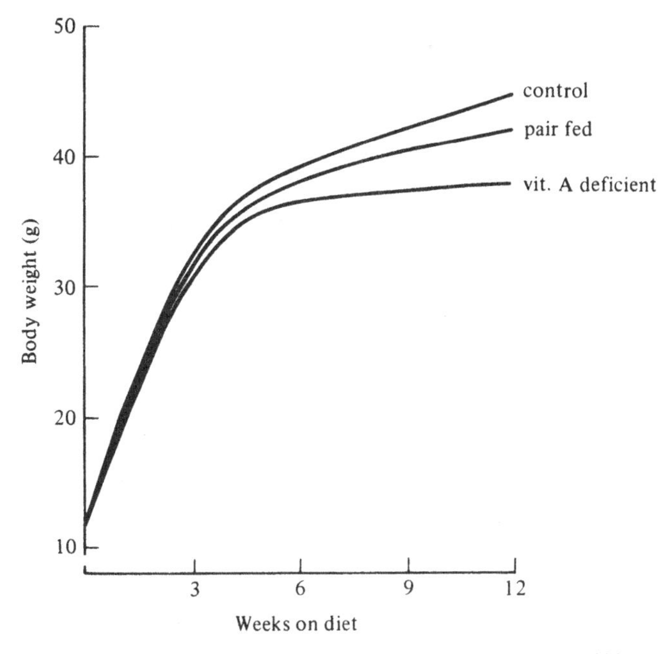

Fig. 5.7 Groups of mice were reared on one of three regimes; *ad libitum* access to a control diet, *ad libitum* access to a vitamin A deficient diet, and pair fed the control diet to the intake of the mice fed the vitamin A deficient diet. Weight gain virtually ceased in mice on the vitamin A deficient diet after 5 or 6 weeks. The same amount of a complete diet allowed for nearly normal weight gain in the pair fed group. This shows a greater efficiency of utilisation of a complete diet compared to a deficient diet (Ahmed, 1989).

Interaction of energy and protein

There are well characterised processes whereby dietary energy and protein interact. An increased energy intake improves nitrogen balance and makes it possible to achieve nitrogen balance at a lower protein intake. Energy is utilised more efficiently on an adequate protein intake, and specific deficiencies in individual amino acids may cause a significant increase in energy expenditure, not associated with any obvious metabolic advantage. The different forms of dietary energy fulfill specific functions, and dietary fat, simple and complex carbohydrates, and protein are not readily or easily interchangeable. The details of these interactions are dealt with in greater detail elsewhere (see Garrow & Halliday, 1985; Rand *et al.*, 1984; FAO/WHO/UNU, 1985).

Factors affecting adaptation

Specific nutrient deficiencies

Specific nutrient deficiencies may arise by a combination of any of three factors: a marginal or inadequate intake; increased demands for growth or associated with pathological processes; increased losses in urine, stool, skin, sweat or other routes. In the face of a specific deficiency a diet that would have been adequate under normal circumstances may become inadequate. Specific deficiencies have been associated with characteristic pathological features, but they can also have more general effects upon metabolism that would tend to interfere with the development of an effective adaptive state. The most important of these general effects may be the tendency of specific deficiencies to provoke an inefficient utilisation of dietary energy, combined with a later, secondary effect upon appetite depression (Kleiber, 1945).

Fig. 5.7 shows the body weight of three groups of mice. The animals fed a control, complete diet *ad libitum* showed a normal growth curve. Those fed a vitamin A deficient diet *ad libitum* showed a slowing of weight gain after about six weeks with a weight plateau from seven to eight weeks. The animals that were pair fed a complete diet to the intake of the vitamin A deficient group, continued to gain weight for the duration of the study. Evidently the pair fed animals are using the dietary energy much more efficiently than the vitamin A deficient animals (Ahmed, 1989). This effect of altered efficiency of utilisation of dietary energy may be more obvious for some mineral deficiencies than for vitamin deficiencies, but where the appropriate experiments have been carried out, it would appear to be a general phenomenon. The effect may also be seen with specific deficiencies of essential and 'non-essential' amino acids (Harper, Benevenga & Wohlhueter, 1970).

Although a dietary inadequacy is the most usual way to produce a specific nutritional deficiency in the experimental situation, in practice specific deficiencies are more likely to be a consequence of increased losses, especially in the stool. Thus, for example, losses of zinc and copper in diarrhoeal fluid (Castillo-Duran, Vial & Uauy, 1988) probably make an important contribution to the promotion of a deficiency state. Similar consequences may be associated with the losses of fat soluble vitamins in association with fat malabsorption.

General infections

High levels of infection and infestation are characteristic of environments where malnutrition is prevalent. The mutual interaction between nutritional status and infection has been the subject of extensive investigation for many years. Scrimshaw has emphasised that the nature of the interaction may vary with the type of infection and either the overall status of the host organism or deficiencies of specific nutrients (Scrimshaw, Taylor & Gordon, 1968, Morehead *et al.*, 1974, Neumann *et al.*, 1975). In most cases the interaction is one of synergy. However, in some cases, especially with viral and protozoal infections, it may as equally well be antagonistic, with a nutrient deficiency affording some measure of protection. In severely malnourished children the infections are often multiple, with one study showing that on average each child had five different foci of infection (Christie, Heikens & MacFarlane, 1985). Organisms that are usually commensals, or that are relatively innocuous in a well nourished host, can become pathogenic in a malnourished individual. The most frequent sites for infection appear to be the gastro-intestinal tract and the upper respiratory tract (surface infections), although invasive infections are also common.

In the well nourished host there is a characteristic response to infection, with a local inflammatory reaction and a more generalised systemic response, involving an integrated series of changes throughout the whole body. These responses are often diminished or absent in malnutrition. Infection and infestation bring about an activation of the immune system during which macrophage populations around the body are stimulated to produce a range of cytokines (monokines and lymphokines) (Grimble, 1989). The circulating levels of cytokines may be reduced in malnutrition (Kauffman, Jones & Kluger, 1986; Hoffman-Goetz, 1988). Cytokines, such as IL1, IL6 and tumour necrosis factor/cachectin (TNF), bring about an enhancement of the immune response. In animal models this involves a radical shift in the intensity and direction of protein metabolism, resulting in a negative nitrogen balance (Fig. 5.3, Grimble, 1989). There is a loss of protein from muscle and skin, and the amino acids released as a consequence are made available for the synthesis of acute phase proteins in the liver, or for the immune system as substrates for gluconeogenesis. The profile of the secretory proteins produced by the liver changes. Serum albumin synthesis is curtailed and there is an increase in the synthesis of proteins which are acute phase reactants such as C-reactive protein, fibrinogen and the zinc binding protein metallothionein. In the mouse TNF has been shown to have a specific effect upon the gastro-intestinal tract. Following large doses of

TNF mice developed a watery diarrhoea, which was associated with necrosis of the tips of the villi in the small bowel, and a vascular leak syndrome with preferential loss of fluid into the small bowel (Remick *et al.*, 1987). This type of response may well underlie the non-specific diarrhoea that paediatricians have associated with infection for many years.

Gastro-intestinal infection with parasites is a common finding in children coming from areas in which malnutrition is prevalent. Ascariasis has been directly implicated in causing a deterioration in nutritional status (Crompton, Nesheim & Pawlowski, 1985), and giardiasis is well recognised as being an important cause of malabsorption syndrome (Farthing, 1988). The recent observations by Cooper and Bundy of the particular risks that can be ascribed to a heavy infection with trichuriasis are not as widely appreciated (Cooper & Bundy, 1988). These workers have been able to show a specific association between severe undernutrition, stunting and wasting and the dysenteric features of a heavy infection with trichuris. Appropriate treatment of the parasites can effect a dramatic response with catch up in both weight and height. The mechanisms whereby the organisms are able to breach or compromise the host's immune system are a matter of active investigation. However, one of the few situations in which it has been possible to measure high circulating levels of TNF is in individuals with intestinal parasitosis (Scuderi *et al.*, 1986).

The detailed mechanisms as to how the host's responses to cytokines are brought into play are not completely clear at this time, but there are a number of points in the cascade reactions associated with the responses that might be especially sensitive to nutritional state. Eicosanoids (prostanoids, leukotrienes and thromboxanes) play an important role in mediating the inflammatory response. As the membrane precursors for these metabolites are themselves formed from dietary essential fatty acids, one might expect the response to be influenced or modulated by the pattern of dietary fat. Grimble and his group have shown that aspects of the inflammatory response can be modulated by alterations in the dietary fats (Fig. 5.8); (Wan & Grimble, 1987; Grimble, 1989). Reducing the intake of essential fatty acids, by feeding rats a diet rich in coconut oil, results in a marked suppression of many aspects of the acute phase response (Table 5.3).

There is an extensive literature on the responses of the specific immune system to severe malnutrition. Most observers have found that there is a measurable depression in the cell mediated immune response, with relative preservation of the humoral response, although the particular situation in any individual is complex. As the normal specific

Table 5.3. *The effect of TNFα on muscle and liver protein and serum copper concentration (mean ± SEM) in rats fed corn oil or coconut oil in the diet. (Data of Dr R. Grimble, pers. comm.)*

	TNFα (30 μg/kg)	Corn oil	Coconut oil
Tibialis anterior, total protein* (mg)	–	150 ± 3^{a}	154 ± 3^{a}
	+	136 ± 3^{b}	$158 + 3^{a}$
Liver, total protein* (mg)	–	2290 ± 60^{a}	2340 ± 50^{a}
	+	2450 ± 60^{b}	2180 ± 100^{a}
Serum copper† (μg/l)	–	1512 ± 103^{a}	1570 ± 50^{a}
	+	2080 ± 60^{b}	1717 ± 80^{a}

Values bearing different superscripts are significantly different by ANOVA.
*24 hours after injection
†8 hours after injection

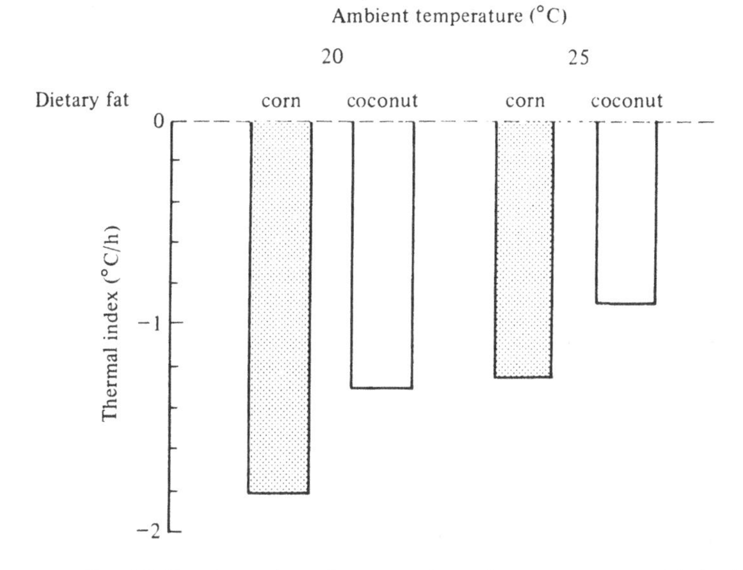

Fig. 5.8 The change in body temperature of rats was measured following the injection of TNFα, and the results were expressed as a thermal index. Rats that had been fed a diet based on corn oil showed a more profound hypothermia than rats on coconut oil, at both 20 and 25 °C ambient temperature. Coconut oil obtunds the response to TNFα. (Data of R. Grimble, pers. comm.)

response to an antigenic challenge requires the multiplication of cells and the active synthesis of protein and other macromolecules, it is not surprising that it is possible to demonstrate impairment of the process with a specific deficiency of almost any nutrient (Gross & Newberne, 1980). However, the one nutrient that might be limiting most frequently for the cell mediated immune response is probably zinc. In both animal studies and human investigation zinc has been shown to confer a benefit or improvement in the cell mediated immune responsiveness of malnourished individuals (Keusch, 1989). The maintenance of a normal zinc status is dependent upon an adequate dietary intake. However, diarrhoeal disease is probably an important factor in the promotion of clinical zinc deficiency (Castillo-Duran *et al.*, 1988). Furthermore, one of the clinical features of zinc deficiency itself is diarrhoea. There is evidence to suggest that a large proportion of malnourished children have a degree of zinc deficiency (Golden, Golden & Bennett, 1985). This is likely to be particularly true in children with oedematous malnutrition and is almost invariably true for those children who have ulcerated skin lesions (Golden *et al.*, 1985).

Diarrhoeal disease

Diarrhoeal disease and varying degrees of malabsorption are an almost invariable accompaniment to severe malnutrition. Although it is impossible to determine the extent to which this represents cause and effect, it is quite clear that the loss of specific nutrients in stool can contribute to compromised function. Acute infective diarrhoea is a self limiting condition and the threat it presents to health is a result of acute disturbances in fluid and electrolyte balance. However, the losses of specific nutrients may create the potential for more chronic problems associated with nutrient imbalance. The example of zinc deficiency as a consequence and cause of diarrhoea is a good example.

We have recently had the opportunity to look at the interaction of rotavirus infection with vitamin A deficiency in a mouse model of malnutrition (Ahmed, 1989). Vitamin A deficiency can be produced after 12 weeks on a deficient diet. In control and pair fed animals rotavirus infection did not produce any marked changes in the histology of the gastro-intestinal tract, but in the face of vitamin A deficiency there was a profound disorganisation of the small bowel mucosa, with extensive damage to the tips of the villi, see Fig. 5.9. The metabolic basis of this interaction requires further investigation, but it is interesting to note that one effect of zinc deficiency in malnourished children is to impair the normal metabolic handling of vitamin A (Mathias, 1982). There is clearly

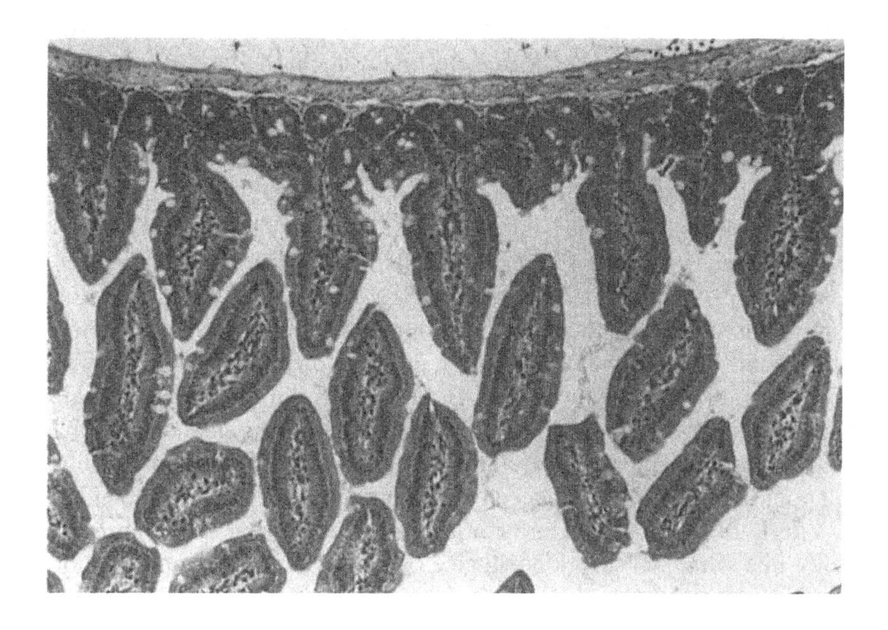

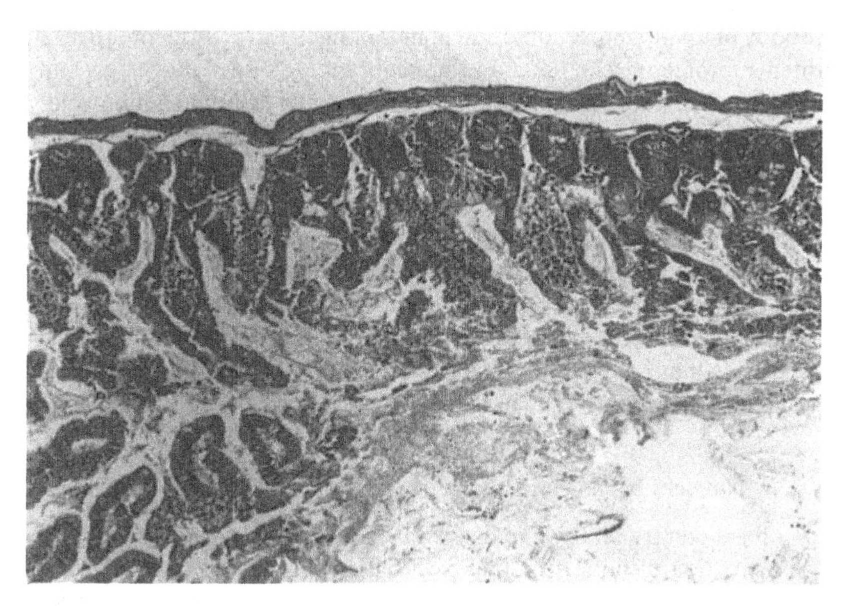

Fig. 5.9 Histological sections of the small intestine of mice are shown under low power magnification. When mice that have been fed a vitamin A deficient diet for 12 weeks are challenged with rotavirus the mucosal architecture is extensively damaged (a). This effect of rotavirus infection is not seen in animals that have been pair fed a complete diet (b) (Ahmed, 1989).

the potential for complex nutrient–nutrient interactions in maintaining the normal integrity of the gastro-intestinal tract in the face of infective disease.

Although specific pathogens may be the cause of chronic diarrhoea, it is unusual to be able to draw an association with a particular organism. In communities where malnutrition and chronic diarrhoea are prevalent, it has been shown that the feeds given to young children are heavily contaminated with faecal and pathogenic micro-organisms (Barrell & Rowland, 1979; Hibbert & Golden, 1981). Against the background of compromised host defences, this is thought to be one important factor which contributes to the development of overgrowth of the small bowel with faecal organisms (Jackson & Golden, 1978; Gracey, 1981). Small bowel overgrowth gives rise to bile salt deconjugation and bile salt malabsorption (Gracey, 1981). There is a marked reduction in the bile salt pool which is further reduced in the presence of diarrhoea (Schneider & Viteri, 1974). Bile salts are conjugates of either glycine or taurine, hence bile salt malabsorption will give rise to a drain on the body's pool of sulfur amino acids and glycine/serine.

The sulfur amino acids are limiting on many diets, including milk based diets. In normal metabolism they play an important role in clearing the body of toxins and xenobiotics, maintaining the structural integrity of proteins, protecting the function of membrane proteins, antibodies and hormones. We know that glycine is a conditionally essential amino acid, which appears to play an important role in the acute phase response (Jackson, 1982; Grimble, 1989). The evidence would suggest that in malnutrition the endogenous production of glycine is inadequate to satisfy the demand. The carbon skeleton of glycine, glyoxyllic acid, is a toxic compound which may be present in increased concentrations in the blood of the most severely malnourished children (Jackson, 1986c).

The chronic diarrhoea which is associated with severe gastro-intestinal dysfunction may lead to increased losses of total nitrogen, fat, fat soluble vitamins and trace elements in faeces.

Nutrient balance/imbalance

For a fixed dietary intake unbalanced faecal losses of nutrients will lead to the creation of nutrient imbalance. These imbalances cannot necessarily be corrected on a normal intake, and there will be the need for specific supplements to make good the deficiency. Increased losses of potassium have long been recognised as a particular problem in watery diarrhoea, and this is the justification for the use of potassium salts in oral rehydration fluids. Diarrhoeal losses of zinc and copper can be held to account for part of the deficiency seen in some malnourished children. We have little knowledge of the extent to which diarrhoea may increase the requirements for water soluble vitamins, and although it is appreciated that fat malabsorption may lead to increased losses of fat soluble vitamins such as A and E we are unable to make any quantitative estimate as to how important this might be. Macronutrient losses in the form of energy and protein have not been extensively documented. However, if as suggested above, an important part of the mechanism of adaptation to a low protein and energy intake is related to the activity of the gastro-intestinal flora, then inevitably this function is going to be affected, most probably compromised, in both acute and chronic diarrhoeal states.

Malnutrition represents a situation of imbalance, and there is imbalance at two levels. On the one hand the mismatch between dietary intake and unusual losses represents an external imbalance. However, there is also the need to match the available nutrients to the metabolic demands for tissue synthesis and repair, and the mounting of an effective inflammatory response. If the relative availability of individual nutrients to meet these requirements can not be adequately satisfied then there will be an internal imbalance. The consequence of this is that instead of synthetic activities being coordinated in an appropriate pattern towards satisfying the needs of the body, there will be a pattern of synthetic activity determined by the available pattern of nutrients.

The evidence for a primary deficiency of protein as the main causal factor in oedematous malnutrition is not strong. However, a dietary deficiency is but one way in which protein metabolism can be disturbed (Table 5.4). An inability to utilise the dietary protein, increased losses of protein, or an increase in the metabolic demand for protein may be expected to induce a response that has some similarities to a dietary deficiency of protein (Jackson, 1986a). Indeed as the metabolic demands for protein change the pattern of amino acids that did support normal maintenance and growth may no longer be appropriate (Grimble, 1989). The safe intake of protein, or the 'recommended daily allowance' as defined by expert committees, are designed to maintain the majority of a healthy population in good health. We do not have any reference point to define unusual needs for protein or the specific needs for individual amino acids for the particular conditions of illness, or other situations where the demands for protein synthesis are altered.

The general conclusion that can be drawn is that adaptation to reduced intakes of food and nutrients involves a number of systems, but in particular the gastro-intestinal tract, and the normal relationship

Table 5.4. *A number of different factors may contribute to a disturbance in protein balance and metabolism*

Insufficient dietary protein
Impaired digestion or absorption of protein and amino acids
Inability to utilise protein
Increased demand for protein or amino acids
Increased losses of protein or nitrogen
Inappropriate quality of protein:
(a) absolutely
(b) relative to an increased demand

between the host and his gastro-intestinal flora. The environments in which malnutrition occurs are hostile with respect to the infective load of pathogens presented to a child. These pathogens have direct effects upon the host, but also have important indirect effects, some of which involve interference with gastro-intestinal function. This leads to an unbalanced loss of nutrients. The consequence is that in the face of infection it is not possible for the normal adaptive processes to maintain their functional integrity.

Free radicals

Golden has presented a theory which proposes that the primary cause of oedematous malnutrition is the inability of the host to protect his functional and cellular integrity against the excessive production of oxidative free radical induced damage (Golden, 1985). He argues that on the one hand there is an increased generation of free radicals and on the other impairment of the normal protective mechanisms. As we do not have any reliable methods of measuring either the rate of production or the rate of consumption of free radicals, we have no direct way of assessing the balance between the two processes. The indirect methods that have been used for assessing this balance are not considered to be reliable. Golden's own observations have been based upon measurements of the activity, concentration or function of the free radical protective mechanisms. He has presented data to show that there is impairment of free radical protection. It is reasonable to expect that if his theory has substance then there should be a major difference in the complement of the protective mechanisms between oedematous and non-oedematous malnutrition. Although there may be a tendency for clustering of abnormal findings in the sickest children, or those with oedematous malnutrition, there is no clear cut differentiation between the two groups for any of the specific nutrients, minerals or vitamins that are central to the protective mechanisms (Golden *et al.*, 1985; Golden & Ramdath, 1987).

Millward's group have tried to produce an animal model of kwashiorkor in rats by exposing vitamin E deficient animals to the effects of endotoxin (Omer, Bates & Millward, 1986). To date they have had difficulty in producing a convincing model. Therefore, although free radical induced damage might make an important contribution to the morbidity and mortality of severe malnutrition, there is no evidence which suggests that it is specifically involved in the aetiology of oedema. Free radicals are generated as a normal part of cellular metabolism. It has been suggested that uncompensated free radical production is most likely to be associated with gastro-intestinal infection. Animal experiments suggest that the intact gastro-intestinal tract has a particularly enhanced capability to provide protection against lipid peroxidation produced by free radicals compared to other tissues (Balasubramanian, Manohar & Mathan, 1988).

Glutathione and the aetiology of kwashiorkor

We can identify factors that have to be considered as making an important contribution to the aetiology of kwashiorkor, but we are as yet unable to explicitly characterise or specify the aetiology or the aetiological process.

The most severe prognostic indicators of adverse outcome in severe malnutrition have been identified as: indices associated with impaired or failing hepatic function; factors associated with disturbances of fluid and electrolyte balance, including the sick cell syndrome; and factors associated with an imbalance between the generation and consumption of oxidative free radicals (Jackson, 1986b). There are a number of factors which might be considered to be common to all three processes, but there is only one which has been found to approach an absolute distinction between oedematous malnutrition (kwashiorkor and marasmic-kwashiorkor) and the wasting syndrome marasmus. This is the blood level of glutathione (Jackson, 1986b). The maintenance of the cellular level of reduced glutathione plays a central role in assuring the integrity of normal function as well as defending the cell against a range of toxic metabolites, including oxygen derived free radicals (Kosower & Kosower, 1978; Reed & Fariss, 1984; Table 5.5). The cyclical oxidation and reduction of glutathione carries all the features of a control point in metabolism (Newsholme, 1970): a reversible reaction with independent catalytic control of the forward and backward reactions. To maintain the cellular level of glutathione in an effective, reduced state requires

Table 5.5. *A number of cellular functions are either associated with or dependent upon a normal glutathione status*

Protein metabolism:
amino acid transport
protein synthesis
protein structure
protein function: enzyme reactions
Cellular regulation:
second messenger
prostaglandins/leukotrienes
calcium homeostasis
cell membrane thiol status
hormone receptor activity
oxido-reductive state
Cell structure and function:
microtubules and cytoskeleton
phagocytosis
free radical protection and radioprotection
removal of lipid peroxide
conjugation and detoxification

interaction with cofactors and metabolites whose activity depends upon a range of specific minerals and vitamins (Fig. 5.10). Because the redox cycling of glutathione, and the maintenance of glutathione in the reduced state, is fundamental to normal cellular function the active form of the tripeptide is protected by a series of checks and balances.

We were first attracted to the measurement of glutathione when we were looking at differences in the mechanism of production of fatty liver by either a choline deficient or a protein deficient diet (Jahoor & Jackson, 1982). Whereas the choline deficient diet acted through mechanisms related to the provision of methyl groups and produced an intense centrilobular infiltration of fat, protein deficiency acted through a completely different mechanism. On a low protein diet the fatty infiltration is predominantly periportal, and was associated with impaired uptake and clearance of indocyanine green. As indocyanine green is metabolised by the glutathione-s transferase series of enzymes, there is an obligate requirement for glutathione as a cofactor for its effective clearance. Therefore impaired clearance of indocyanine green was considered to reflect limited availability of glutathione. The degree of fatty infiltration on the low protein diet was inversely related to the hepatic level of glutathione (Davis, Hibbert & Jackson, 1988). The hepatic level of glutathione could vary directly with changes in the dietary intake of cysteine.

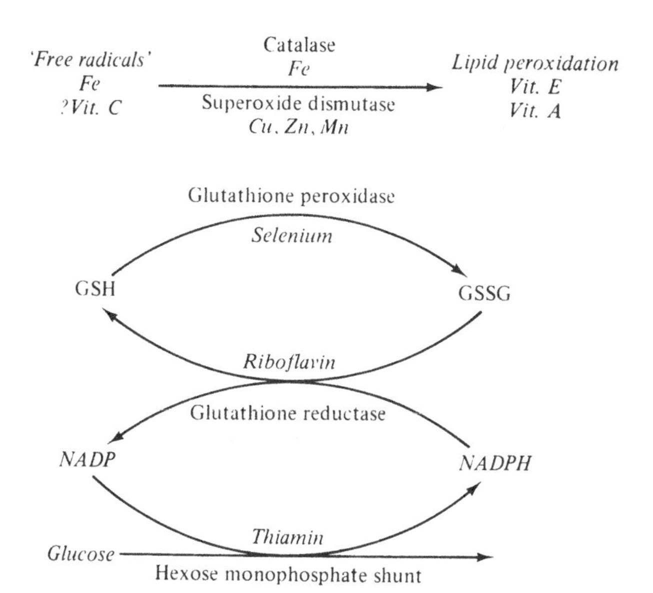

Fig. 5.10 The cyclical oxidation and reduction of glutathione (GSH) is central to the mechanisms whereby the cell protects itself from the potential damaging effects of free radicals. The processes associated with the protection from the adverse effects of free radicals involve many specific nutrients that have been found to be deficient in severe malnutrition. (GSSG = oxidised GSH.)

Glutathione is a tripeptide of glutamic acid, cysteine and glycine, all of which have been considered to be non-essential. Moreover, glycine and cysteine should now be considered to be conditionally essential (Jackson, 1989). As a fall in glutathione concentration must imply a limited ability for its synthesis, this argues in favour of a primary limitation in amino acid metabolism. The implication of these data is that periportal fatty infiltration on a low protein diet is caused by the limited availability of specific amino acids, rather than the usual lipotrophic factors. A low level of glutathione would increase the susceptibility of the cell to any other potentially damaging influence. As sulfur amino acids tend to be limiting on milk based diets (and as the demands for sulfur amino acids can be increased by the demands for conjugation, detoxification, and bile salt malabsorption), nitrogen, amino acid or protein deficiency in a specific sense, may contribute to the aetiology of oedematous malnutrition. Thus an imbalance between the amino acid availability and the metabolic demands of the body effectively creates a deficient state, which is not simply responsive to the provision of abundant quantities of all amino acids as abundant protein. A similar concept has been developed in relation to the demands and requirements for arginine and glutamine for tissue repair, the maintenace of immune function and gastro-intestinal integrity (Crabtree & Ardawl, 1985; Barbul, 1986).

Under normal circumstances there is a net output from the gastro-intestinal tract of glutathione, glycine and probably sulfur amino acids (Elwyn, 1970; Rerat *et al.*, 1988). It is not known the extent to which this is perturbed during diarrhoeal disease, but we have found low levels of blood glutathione in adults with compromised gastro-intestinal function, in both fasted patients and patients on total parenteral nutrition (Sutton, Taylor & Jackson, 1987). It required an adequate enteral intake to bring the level into the normal range. Similarly, in adults who require parenteral nutritional support for gastro-intestinal failure associated with a range of inflammatory conditions, the body appears to be unable to synthesise sufficient glycine to satisfy its requirements (Moran, Persaud & Jackson, 1989a).

Conclusion

Do we have a useful conceptual framework within which we can consider the aetiology and pathophysiology of kwashiorkor? Each of the models that has been proposed has served to enhance our understanding of aspects of the disease. There is evidence to support the suggestion that both extrinsic and intrinsic factors make a major contribution to the disease process. Although there is little evidence for a protein deficient diet playing a primary role in itself, there is abundant evidence that protein, nitrogen and amino acid metabolism are disturbed. Furthermore, the combination of unbalanced losses and unusual demands for amino acids makes it unlikely that requirements can be satisfied by the habitual dietary intake. It is unlikely that the normal mechanisms that enable the body to accommodate changes in energy and protein intake are maintained intact.

At the present time we would argue in favour of deranged gastro-intestinal function as having an important, if not primary, role in the aetiology of oedematous malnutrition. Diarrhoea provokes unbalanced losses of nutrients, in particular potassium and zinc; it also disturbs one of the important mechanisms whereby adaptation to low energy and protein intakes takes place. One of the major sources of unusual free radical loads to the body is gastro-intestinal infection. Glutathione, the central control mechanism designed to defend the body against these potentially toxic species, appears to be dependent upon the integrity of the gastro-intestinal tract for its normal functioning. The common finding of a decreased blood concentration of glutathione in oedematous malnutrition is a reflection of a major disturbance in protein and amino

Table 5.6. *A fall in intracellular glutathione (GSH) concentration is likely to be the consequence of an interaction between a number of factors*

A limited production or availability of the precursor amino acids; glutamate, cysteine and glycine
An ability to synthesise GSH: (a) decreased availability of ATP (b) impaired enzyme function (c) limited availability of cofactors
Increased utilisation, consumption or loss of GSH, glutamate, cysteine or glycine

acid balance in the body (Table 5.6). Anecdotal reports record that famine oedema can be cleared by the use of glutathione or cysteine, and suggest the possibility of a direct causal association (Winick, 1979).

The varied clinical presentation of malnutrition is probably determined by the individual complex of interactions between specific nutrient deficiencies, infections and trauma. A full elucidation of these processes will require the effective application of modern techniques and understanding to a condition which characterises poverty and deprivation.

References

Ahmed, F. (1989). The interaction of vitamin A deficiency and rotavirus infection. PhD Thesis, University of Southampton.

Alleyne, G.A.O., Hay, R.W., Picou, D.I., Stanfield, J.P. & Whitehead, R.G. (1977). *Protein–energy malnutrition*. Edward Arnold, London.

Balasubramanian, K.A., Manohar, M. & Mathan, V.I. (1988). An unidentified inhibitor of lipid peroxidation in the intestinal mucosa. *Biochimica et Biophysica Acta*, **962**, 51–8.

Barbul, A. (1986). Arginine: biochemistry, physiology and therapeutic implications. *Journal of Parenteral and Enteral Nutrition*, **10**, 227–37.

Barrell, R.A. & Rowland, M.G. (1979). Infant foods as a potential source of diarrhoeal illness in rural West Africa. *Transactions of the Royal Society of Tropical Medicine and Hygiene*, **73**, 85–90.

Bearn, A.G. & Litwin, S.D. (1978). Deficiencies of circulating enzymes and plasma proteins. In: *The Metabolic Basis of Inherited Disease*, ed. J.B. Stanbury, J.B. Wyngaarden & D.S. Fredrickson, pp. 1712–25. McGraw-Hill, New York.

Black, D.A.I. & Milne, M.D. (1952). Experimental potassium deficiency in man. *Clinical Science*, **11**, 397–415.

Blaxter, K.L. & Waterlow, J.C. (ed.) (1985). *Nutritional Adaptation in Man*. John Libbey, London.

Brown, K.H., Suskind, R.M., Lubin, B., Kulapongs, P., Leitzman, C. & Olson, R.E. (1978). Changes in the red blood cell membrane in protein–calorie malnutrition. *American Journal of Clinical Nutrition*, **31**, 574–8.

Castillo-Duran, C., Vial, P. & Uauy, R. (1988). Trace mineral balance during acute diarrhoea in infants. *Journal of Pediatrics*, **113**, 452–7.

Christie, C., Heikens, G.T. & MacFarlane, D.E. (1985). Infections in malnourished children. *West Indian Medical Journal*, **34** (Suppl.), 47–8.

Cooper, E.S. & Bundy, D.A.P. (1988). Trichuris is not trivial. *Parasitology Today*, **4**, 301–6.

Coward, W.A. (1971). The erythrocyte membrane in kwashiorkor. *British Journal of Nutrition*, **25**, 145–51.

Crompton, D.W.T., Nesheim, M.C. & Pawlowski, Z.S. (1985). Ascariasis and its public health significance. Taylor & Francis, London.

Cummings, J.H. (1988). The colon: physiology. In: *Recent Advances in Gastroenterology*, vol. 7, ed. R. Pounder, pp. 313–36. Churchill Livingstone, Edinburgh.

Davis, E.E., Hibbert, J.M. & Jackson, A.A. (1988). Relationship between hepatic glutathione and fatty liver in weanling rats on a low protein diet. *Nutrition Reports International*, **37**, 847–56.

Doherty, T., de Benoist, M-H., Hibbert, J., Persaud, C. & Jackson, A.A. (1989). The effect of the level of dietary protein and the quality of dietary energy on urea kinetics in young children recovering from severe malnutrition. *Proceedings of the Nutrition Society*, **48**, 55A.

Elwyn, D.H. (1970). The role of the liver in the regulation of amino acid and protein metabolism. In: *Mammalian Protein Metabolism IV*, ed. H.N. Munro, pp. 523–57. Academic Press, London.

Farthing, M.J.G. (1988). Tropical gastroenterology. In: *Recent Advances in Gastroenterology*, ed. R. Pounder, pp. 177–96. Churchill Livingstone, Edinburgh.

FAO/WHO (1973). *Energy and Protein Requirements*. FAO Nutrition Meetings Report Series, No. 52. WHO, Geneva.

FAO/WHO/UNU (1985). *Energy and Protein Requirements: report of an Expert Consultation*. Technical Report Series 724, WHO, Geneva.

Fleck, A. (1989). *Proceedings of the Nutrition Society*.

Fleck, A., Raines, G., Hawke, F., Trotter, J., Wallace, P.I., Ledingham, I.McA. & Calman, K.C. (1985). Increased vascular permeability: a major cause of hypoalbuminaemia in disease and injury. *Lancet*, i, 781–4.

Garrow, J.S. (1965). Total body potassium in kwashiorkor and marasmus. *Lancet*, ii, 455–8.

Garrow, J.S. & Halliday, D. (1985). *Substrate and Energy Metabolism in Man*. John Libbey, London.

Garrow, J.S., Smith, R. & Ward, E.E. (1968). *Electrolyte Metabolism in Severe Infantile Malnutrition*. Pergamon Press, Oxford.

Golden, M.H.N. (1982a). Transport proteins as indices of protein status. *American Journal of Clinical Nutrition*, **35**, 1159–65.

Golden, M.H.N. (1982b). Protein deficiency, energy deficiency, and the oedema of malnutrition. *Lancet*, i, 1261–5.

Golden, M.H.N. (1985). The consequence of protein deficiency in man and its relationship to the features of kwashiorkor. In: *Nutritional Adaptation in Man*, ed. K.L. Blaxter & J.C. Waterlow, pp. 169–87. John Libbey, London.

Golden, M.H.N. & Golden, B.E. (1980). Trace elements: potential importance in human nutrition with particular reference to zinc and vanadium. *British Medical Bulletin*, **37**, 31–6.

Golden, M.H.N., Golden, B.E. & Bennett, F.I. (1985). Relationship of trace element deficiencies to malnutrition. In: *Trace Elements in Nutrition of Children*, ed. R.K. Chandra, pp. 185–207. Nestlé Nutrition/Raven Press, New York.

Golden, M.H.N., Golden, B.E. & Jackson, A.A. (1980). Albumin and nutritional oedema. *Lancet*, i, 114–16.

Golden, M.H.N. & Ramdath, D. (1987). Free radicals in the pathogenesis of kwashiorkor. *Proceedings of the Nutrition Society*, **46**, 53–68.

Gopalan, G. (1968). Kwashiorkor and masarmus; evolution and distinguishing features. In: *Calorie Deficiencies and Protein Deficiencies*, ed. R.A. McCance & E.M. Widdowson, pp. 49–58. Churchill, London.

Gracey, M.S. (1981). Nutrition, bacteria and the gut. *British Medical Bulletin*, **37**, 71–5.

Grimble, R.F. (1989). Cytokines: their relevance to nutrition. *European Journal of Clinical Nutrition*, **43**, 217–30.

Gross, R.L. & Newberne, P.M. (1980). Role of nutrition in immunologic function. *Physiological Reviews*, **60**, 188–302.

Hamilton, S.M., Johnston, M.G., Fong, A., Pepevnak, C., Semple, J.L. & Movat, H.Z. (1986). Relationship between increased vascular permeability and extravascular albumin clearance in rabbit inflammatory responses induced with Escherichia coli. *Laboratory Investigation*, **55**, 580–7.

Harper, A.E., Benevenga, N.J. & Wohlhueter, R.M. (1970). Effects of ingestion of disproportionate amounts of amino acids. *Physiological Reviews*, **50**, 428–558.

Heine, W., Wutzke, K.D., Richter, I., Walther, F. & Plath, C. (1987). Evidence for colonic absorption of protein nitrogen in infants. *Acta Paediatrica Scandinavica*, **76**, 741–4.

Hendrickse, R.G. (1984). The influence of aflatoxins on child health in the tropics with particular reference to kwashiorkor. *Transactions of the Royal Society of Tropical Medicine and Hygiene*, **78**, 427–35.

Hibbert, J.M. & Golden, M.H.N. (1981). What is the weanling's dilemma? Dietary faecal bacteria ingestion of normal children in Jamaica. *Journal of Tropical Paediatrics*, **27**, 255–8.

Hoffman-Goetz, L. (1988). Lymphokines and monokines in protein–energy malnutrition. In: *Nutrition and Immunology*, ed. R.K. Chandra, pp. 9–23. A.R. Liss, New York.

Jackson, A.A. (1982). Amino acids: essential and non-essential? *Lancet*, i, 1034–7.

Jackson, A.A. (1985). Nutritional adaptation in disease and recovery. In: *Nutritional Adaptation in Man*, ed. K.L. Blaxter & J.C. Waterlow, pp. 111–26. John Libbey, London.

Jackson, A.A. (1986a). Dynamics of protein metabolism and their relationship to adaptation. In: *Proceedings of the XIII International Congress of Nutrition, 1985*, ed. T.G. Taylor & N.K. Jenkins, pp. 403–9. John Libbey, London.

Jackson, A.A. (1986b). Blood glutathione in severe malnutrition in childhood. *Transactions of the Royal Society of Tropical Medicine and Hygiene*, **80**, 911–13.

Jackson, A.A. (1986c). Severe undernutrition in Jamaica. *Acta Pediatrica Scandinavica*, Suppl. 323, 43–51.

Jackson, A.A. (1989). Optimizing amino acid and protein supply and utilization in the newborn. *Proceedings of the Nutrition Society*, **48**, 293–301.

Jackson, A.A. & Golden, M.H.N. (1978). The human rumen. *Lancet*, ii, 764–7.

Jackson, A.A. & Golden, M.H.N. (1986). Protein energy malnutrition: kwashiorkor and marasmic-kwashiorkor, physiopathology. In: *Clinical Nutrition of the Young Child*, Nestlé Nutrition, pp. 133–42. Vevey/Raven Press, New York.

Jackson, A.A. & Golden, M.H.N. (1987). Severe malnutrition. In: *Oxford Textbook of Medicine*, ed. D.J. Weatherall, J.G.G. Ledingham & D.A. Warrell, pp. 8.12–8.28. Oxford Medical Publications.

Jackson, A.A., Golden, M.H.N., Byfield, R., Jahoor, F., Royes, J. & Soutter, L. (1983). Whole-body protein turnover and nitrogen balance in young children at intakes of protein and energy in the region of maintenance. *Human Nutrition: Clinical Nutrition*, **37C**, 433–46.

Jackson, A.A. & Grimble, R.F. (1989). Malnutrition and amino acid metabolism. In: *The Malnourished Child*, ed. R. Suskind. Nestlé Nutrition Workshop Series, Vol. 19. Rowen Press, New York.

Jackson, A.A., Landman, J. & Picou, D. (1981). Urea production rates in man in relation to the dietary intake of nitrogen and the metabolic state of the individual. In: *Nitrogen Metabolism in Man*, ed. J.C. Waterlow & J.M.L. Stephen, pp. 247–51. Applied Science Publishers, London.

Jackson, A.A., Picou, D. & Landman, J. (1984). The non-invasive measurement of urea kinetics in normal man using a constant infusion of $^{15}N^{15}N$-urea. *Human Nutrition: Clinical Nutrition*, **38C**, 339–54.

Jahoor, F. & Jackson, A.A. (1982). Hepatic function in rats with dietary induced fatty liver, as measured by the uptake of indocyanine green. *British Journal of Nutrition*, **47**, 391–9.

Kauffman, C.A., Jones, M.D. & Kluger, M.J. (1986). Fever and malnutrition: endogenous pyrogen/interleukin-1 in malnourished patients. *American Journal of Clinical Nutrition*, **44**, 449–52.

Kennedy, N., Badaloo, V. & Jackson, A. (1987). Metabolic adaptation to a marginal energy intake. *Proceedings of the Nutrition Society*, **46**, 85A.

Keusch, G.T. (1989). Malnutrition infection and immune function. In *The Malnourished Child*, ed. R. Suskind. Nestlé Nutrition Workshop Series, Vol. 19. Raven Press, New York.

Kleiber, M. (1945). Dietary deficiencies and energy metabolism. *Nutrition Abstracts and Reviews*, **15**, 207–22.

Kosower, N.S. & Kosower, E. (1978). Glutathione status of cells. *International Review of Cytology*, **54**, 109–60.

Landman, J. & Jackson, A.A. (1980). The role of protein deficiency in the aetiology of kwashiorkor. *West Indian Medical Journal*, **29**, 229–38.

Liao, W.S.L., Jefferson, L.S. & Taylor, J.M. (1986). Changes in plasma albumin concentration, synthesis rate and mRNA level during acute inflammation *American Journal of Physiology*, **251**, C928–C934.

Lunn, P.G. & Austin, S. (1983). Dietary manipulation of plasma albumin concentration. *Journal of Nutrition*, **113**, 1791–802.

Lunn, P.G., Northrop, C.A. & Wainwright, M. (1988). Hypoalbuminemia in energy-malnourished rats infected with *Nippostrongylus brasiliensis* (Nematoda). *Journal of Nutrition*, **118**, 121–7.

Mathias, P. (1982). The effect of zinc supplementation on plasma levels of vitamin A and retinol binding protein (RBP) in children recovering from protein–energy malnutrition. *Proceedings of the Nutrition Society*, **41**, 52A.

McGuire, E.A. & Young, V.R. (1986). Nutritional edema in rat model of protein deficiency; significance of the dietary potassium and sodium content. *Journal of Nutrition*, **116**, 1209–24.

McLaren, D.S. (1974). The great protein fiasco. *Lancet*, ii, 93–6.

Moran, B., Persaud, C. & Jackson, A.A. (1989a). Urinary excretion of 5-oxoproline in severe inflammatory illness. *Proceedings of the Nutrition Society*, **48**, 75A.

Moran, B., Persaud, C. & Jackson, A.A. (1989b). Urea absorption by the functioning human colon. *Proceedings of the Nutrition Society*, **48**.

Morehead, D.C., Morehead, M., Allen, D.M. & Olson, R.E. (1974). Bacterial infections in malnourished children. *Environmental Child Health*, **00**, 141–7.

Neumann, C.G., Lawlor, G.J., Stiehm, E.R., Swendseid, M.E., Newton, C., Herbert, J., Ammann, A.J. & Jacob, M. (1975). Immunologic responses in malnourished children. *American Journal of Clinical Nutrition*, **28**, 89–104.

Newsholme, E.A. (1970). Theoretical and experimental considerations on the control of glycolysis in muscle. In: *Essays in Metabolism*, ed. W. Bartley, H.L. Kornberg & J.R. Quayle, pp. 189–223. Wiley, London.

Newsholme, E.A., Crabtree, B. & Ardawi, M.S.M. (1985). Glutamine metabolism in lymphocytes: its biological, physiological and clinical importance. *Quaterly Journal of Experimental Physiology*, **70**, 473–89.

Omer, A.B., Bates, P.C. & Millward, D.J. (1986). Response of the vitamin-E-deficient rat to severe protein deficiency and the *Escherichia coli* endotoxin. *Proceedings of the Nutrition Society*, **45**, 114A.

Patrick, J. (1977). Death during recovery from severe malnutrition and its possible relationship to sodium pump activity in the leucocyte. *British Medical Journal*, **i**, 1051–4.

Patrick, J. (1978). The relationship between intracellular and extracellular potassium in normal and malnourished subjects as studied in leukocytes. *Pediatric Research*, **12**, 767–70.

Patrick, J. (1979). Oedema in protein energy malnutrition: the role of the sodium pump. *Proceedings of the Nutrition Society*, **38**, 61–9.

Patrick, J., Golden, B.E. & Golden, M.H.N. (1980). Leucocyte sodium transport and dietary zinc in protein energy malnutrition. *American Journal of Clinical Nutrition*, **33**, 617–20.

Perlmutter, D.H., Dinarello, C.A., Punsai, P.I. & Colten, H.R. (1986). Cachectin/tumour necrosis factor regulates hepatic acute phase gene expression. *Journal of Clinical Investigation*, **78**, 1349–54.

Picou, D. & Phillips, M. (1972). Urea metabolism in malnourished and recovered children receiving a high or low protein diet. *American Journal of Clinical Nutrition*, **25**, 1261–6.

Rand, W.M., Uauy, R. & Scrimshaw, N.S. (ed.) (1984). Protein–energy–requirement studies in developing countries: results of international research. *Food and Nutrition Bulletin*, Suppl. 10, pp. 331–65. United Nations University, Tokyo.

Reed, D.J. & Fariss, M.W. (1984). Glutathione depletion and susceptibility. *Pharmacological Reviews*, **36**, 25S–33S.

Remick, D.G., Kunkel, R.G., Larrick, J.W. & Kunkel, S.L. (1987). Acute *in vivo* effects of human recombinant tumor necrosis factor. *Laboratory Investigation*, **56**, 583–90.

Rerat, A., Simoes-Nunes, C., Mendy, F. & Roger, L. (1988). Amino acid absorption and production of pancreatic hormones in non-anaesthetized pigs after duodenal infusions of a milk enzymic hydrolysis or of free amino acids. *British Journal of Nutrition*, **60**, 121–36.

Rhodes, K. (1957). Two types of liver disease in Jamaican children, Parts 1 and 2. *West Indian Medical Journal*, **6**, 1–93.

Sakuma, K., Ohyama, T., Sogawa, K., Fujii-Kuriyama, Y. & Matsumara, Y. (1987). Low protein–high energy diet induces repressed transcription of albumin mRNA in rat liver. *Journal of Nutrition*, **117**, 1141–8.

Schneider, R.E. & Viteri, F.E. (1974). Luminal events of lipid absorption in protein calorie malnourished children: relationship with nutritional recovery and diarrhoea. I. Capacity of the duodenal content to achieve micellar solubilisation of lipids. *American Journal of Clinical Nutrition*, **27**, 777–87.

Scrimshaw, N.S., Taylor, C.E. & Gordon, J.E. (1968). *Interaction of Nutrition and Infection*. WHO Monograph Series 57, WHO, Geneva.

Scuderi, P., Sterling, K.E., Lam, K.S., Finley, P.R., Ryan, K.J., Ray, C.G., Petersen, E., Slymen, D.J. & Salmon, S.E. (1986). Raised serum levels of tumour necrosis factor in parasitic infections. *Lancet*, ii, 1364–6.

Shizgal, H.M. (1980). The effect of nutritional support on body composition. In: *Practical Nutrition Support*, ed. S.J. Karran & K.G.M.M. Alberti, pp. 190–203. Pitman Medical Publishing, Tunbridge Wells.

Starling, E.H. (1896). On the absorption of fluid from the connective tissue spaces. *Journal of Physiology (London)*, **19**, 312.

Stephen, J.M.L. (1981). Adaptive enzyme changes. In: *Nitrogen Metabolism in Man*, ed. J.C. Waterlow & J.M.L. Stephen, pp. 39–43. Applied Science Publishers, London.

Sutton, G.L.G., Taylor, D. & Jackson, A.A. (1987). *Blood glutathione during intravenous and enteral feeding*. Proceedings of the European Society for Parenteral and Enteral Nutrition.

Tanaka, N., Kubo, K., Siraki, K., Hoishi, H. & Yoshimura, H. (1980). A pilot study on protein metabolism in Papua New Guinea Highlanders. *Journal of Nutritional Science and Vitaminology*, **26**, 247–59.

Trowell, H.C., Davies, J.N.P. & Dean, R.F.A. (1954). *Kwashiorkor*. Edward Arnold, London.

van Soest, P.J. (1988). Fibre in the diet. In: *Comparative Nutrition*, ed. K.L. Blaxter & I. Macdonald, pp. 215–25. John Libbey, London.

Wan, J.M. & Grimble, R.F. (1987). Effect of dietary linoleate content on the metabolic response of rats to *Escherichia coli* endotoxin. *Clinical Science*, **72**, 383–5.

Waterlow, J.C. (1948). *Fatty Liver Disease in Infants in the British West Indies*. Medical Research Council Special Report Series, No. 263. HMSO, London.

Waterlow, J.C. (1961). The rate of recovery of malnourished infants in relation to the protein and calorie levels of the diet. *Journal of Tropical Pediatrics*, **7**, 16–22.

Waterlow, J.C. (1975). Amount and rate of disappearance of liver fat in malnourished infants in Jamaica. *American Journal of Clinical Nutrition*, **28**, 1330–6.

Waterlow, J.C. (1984). Kwashiorkor revisited: the pathogenesis of oedema in kwashiorkor and its significance. *Transactions of the Royal Society of Tropical Medicine and Hygiene*, **78**, 436–41.
Waterlow, J.C. (1986). Metabolic adaptation to low intakes of energy and protein. *Annual Review of Nutrition*, **6**, 495–526.
Waterlow, J.C. & Alleyne, G.A.O. (1971). Protein malnutrition in children: advances in the last ten years. *Advances in Protein Chemistry*, **25**, 117–241.
Waterlow, J.C., Cravioto, J. & Stephen, J.M.L. (1960). Protein malnutrition in man. *Advances in Protein Chemistry*, **15**, 131–238.
Waterlow, J.C. & Payne, P.R. (1975). The protein gap. *Nature*, **258**, 113–17.
Waterlow, J.C. & Stephen, J.M.L. (1981). *Nitrogen Metabolism in Man*. Applied Science Publishers, London.
Wellcome Trust Working Party (1970). Classification of infantile malnutrition. *Lancet*, ii, 302–3.
Williams, C.D. (1933). A nutritional disease of childhood associated with a maize diet. *Archives of Disease in Childhood*, **8**, 423–33.
Winick, M. (1979). *Hunger Disease*, p. 95. Wiley, New York.
Wrong, O.M., Vince, A.J.& Waterlow, J.C. (1985). The contribution of endogenous urea to faecal ammonia in man, determined by ^{15}N labelling of plasma urea. *Clinical Science*, **68**, 193–9.

6 *The iodine deficiency disorders*

BASIL S. HETZEL

Introduction

The term 'iodine deficiency disorders' (IDD) is now used to denote all the effects of iodine deficiency on growth and development that can be prevented by correction of iodine deficiency (Hetzel, 1983). In the past, the term 'goitre' has been used for many years to describe the effects of iodine deficiency. Goitre is indeed the obvious and familiar feature, but our knowledge has greatly expanded in the last 25 years so that it is not surprising that a new term is needed. The term IDD has now been generally adopted in the field of international nutrition and health. This reconceptualisation has been one factor in securing much more attention to the problem of iodine deficiency in the last 5 years.

This paper concerns the new dimension of understanding of the full spectrum of the effects of iodine deficiency on growth and development which include the effects of iodine deficiency on the fetus, the neonate, the child and adolescent, and the adult.

We go on to consider the technology available for the correction of iodine deficiency.

Finally we briefly review the prevention and control of iodine deficiency disorders. A more comprehensive discussion of this aspect, which involves the development of public health programmes, is provided elsewhere (Hetzel, 1987a).

The iodine deficiency disorders

The best known effect of iodine deficiency is endemic goitre. Goitre is a swelling of the thyroid gland which was well known to the Ancient World and has continued to excite interest over the centuries (Langer, 1960). Iodine deficiency is the major primary etiological factor in endemic goitre.

Extensive reviews of the global geographic prevalence of goitre have been published, notably by Kelly & Snedden in the WHO monograph (Clements *et al.*, 1960) and more recently by Stanbury & Hetzel (1980).

Iodine deficiency is demonstrated by determination of urine iodine excretion using either 24 hour samples or more conveniently casual samples with determination of iodine content per gram of creatinine. Normal iodine intake is 100–150 μg/day which corresponds to a urinary iodine excretion in this range (Stanbury & Hetzel, 1980). In general in endemic goitre areas, the intake is well below 100 μg/day and goitre is usually seen when the level is below 50 μg/day (Pretell *et al.*, 1972). The rate increases as the iodine excretion falls so that goitre may be almost universal at levels below 10 μg/day. The iodine content of drinking water is also low in association with endemic goitre (Karmarkar *et al.*, 1974).

Iodine deficiency causes depletion of thyroid iodine stores with reduced daily production of thyroid hormone (T_4). A fall in the blood level of T_4 triggers the secretion of increased amounts of pituitary thyroid stimulating hormone which increases thyroid activity with hyperplasia of the thyroid. An increased efficiency of the thyroid iodide pump occurs with faster turnover of thyroid iodine. This can be demonstrated by an increased thyroidal uptake of radioactive isotopes I^{131} and I^{125}. These features were first demonstrated in the field in the classical observations of Stanbury *et al.* (1954) in the Andes in Argentina.

Goitre also arises from causes other than iodine deficiency – due to a variety of agents known as goitrogens. Goitrogens in general are of secondary importance to iodine deficiency as etiological factors in endemic goitre.

Recent research (Delange, Camus & Ermans, 1972; Bourdoux *et al.*, 1978, 1980a, b; Delange, Iteke & Ermans, 1982) has shown that staple foods from the Third World such as cassava, maize, bamboo shoots, sweet potatoes, lima beans and millets contain cyanogenic glucosides which are capable of liberating large quantities of cyanide by hydrolysis. Not only is the cyanide toxic, but the metabolite in the body is predominantly thiocyanate which is a goitrogen. With the exception of cassava, these glycosides are located in the inedible portions of the plants, or if in the edible portion, in small quantities so that they do not cause a major problem. Cassava on the other hand is cultivated extensively in developing countries and represents an essential source of calories for more than 200 million people living in the tropics (Delange *et al.*, 1982). The role of cassava with iodine deficiency in the etiology of endemic goitre and endemic cretinism has now been demonstrated by Delange *et al.* (1982) from their studies in non-mountainous Zaire. These observations have also been confirmed by Maberly *et al.* (1983) in Sarawak, Malaysia.

Apart from goitre itself, more recent work on the effects of iodine deficiency in man has revealed a great variety of effects on human growth

Table 6.1. *The spectrum of iodine deficiency disorders (IDD) (from Hetzel, 1987a)*

Fetus	Abortions Stillbirths Congenital anomalies Increased perinatal mortality Increased infant mortality Neurological cretinism – mental deficiency deaf mutism spastic diplegia squint Myxedematous cretinism – dwarfism mental deficiency Psychomotor defects
Neonate	Neonatal goitre Neonatal hypothyroidism
Child and adolescent	Goitre Juvenile hypothyroidism Impaired mental function Retarded physical development
Adult	Goitre with its complications Hypothyroidism Impaired mental function Iodine induced hyperthyroidism

and development. These iodine deficiency disorders are best described in relation to four different phases of life (Table 6.1).

Iodine deficiency in the fetus

Iodine deficiency of the fetus is the result of iodine deficiency in the mother. The condition is associated with a greater incidence of stillbirths, abortions and congenital abnormalities, which can be reduced by iodisation. The effects are similar to those observed with maternal hypothyroidism which can be reduced by thyroid hormone replacement therapy (McMichael, Potter & Hetzel, 1980).

Another major effect of fetal iodine deficiency is the condition of endemic cretinism. This condition, which occurs with an iodine intake below 25 μg per day in contrast to a normal intake of 80–150 μg per day, is still widely prevalent, affecting for example up to 10% of the populations living in severely iodine deficient areas of India (Pandav & Kochupillai, 1982), Indonesia (Djokomoeljanto, Tarwotjo & Maspaitella, 1983) and China (Ma *et al.*, 1982). In its most common form, it is characterised by mental deficiency, deaf mutism and spastic

Table 6.2. *Comparative clinical features in neurological and hypothyroid cretinism (Hetzel & Potter, 1983)*

	Neurological cretin	Hypothyroid cretin
Mental retardation	Present, often severe	Present, less severe
Deaf mutism	Usually present	Absent
Cerebral diplegia	Often present	Absent
Stature	Usually normal	Severe growth retardation usual
General features	No physical signs of hypothyroidism	Coarse dry skin, husky voice
Reflexes	Excessively brisk	Delayed relaxation
ECG	Normal	Small voltage 'QRS' complexes and other abnormalities of hypothyroidism
X-ray of limbs	Normal	Epiphyseal dysgenesis
Effect of thyroid hormones	No effect	Improvement

diplegia, which is referred to as the 'nervous' or neurological type in contrast to the less common 'myxedematous' type characterised by hypothyroidism with dwarfism.

These two conditions were first described in modern medical literature by McCarrison (1908) and the differences are summarised in Table 6.2. The conditions still exist in the same areas of the Karakoram Mountains and in the Himalayas (Pandav & Kochupillai, 1982). Neurological, myxedematous and mixed types still occur in the Hetian District of Sinkiang, China, some 300 km east of Gilgit, where McCarrison made his original observations (Ma *et al.*, 1982). In both China and India, the condition occurs more frequently below the mountain slopes in the fertile silt plains that have been leached of iodine by snow waters and glaciation.

Apart from its prevalence in Asia and Oceania (Papua New Guinea), cretinism also occurs in Africa (Zaire) and in South America in the Andean region (Ecuador, Peru, Bolivia and Argentina) (Pharoah *et al.*, 1980). In all these situations, with the exception of Zaire, neurological features are predominant (Buttfield & Hetzel, 1969). In Zaire the myxedematous form is more common, probably due to the high intake of cassava (Delange *et al.*, 1982). However, there is considerable variation in the clinical manifestations of neurological cretinism which include isolated deaf mutism and mental defects of varying degrees. In China the term 'cretinoid' is used to describe these individuals.

The common form of endemic cretinism is not usually associated with severe clinical hypothyroidism as in the case of the so-called sporadic cretinism, although mixed forms with both the neurological and myxedematous features do occur. However, the neurological features are not

Table 6.3. *Pregnancy outcome in the controlled trial of iodised oil in the Western Highlands of Papua New Guinea (Jimi River District) (Pharoah et al.*, 1971)

	Births	Children examined	Normals	Deaths	Cretins[a]
Untreated	534	406	380	97[b]	26[c]
Iodised oil	498	412	405	66[b]	7[c]

[a]Pregnancies were already established when mothers were injected with iodised oil (6 cases) or with saline (5 cases).
[b]$p<0.05$.
[c]$p<0.001$.

reversed by the administration of thyroid hormones unlike hypothyroidism (Fierro-Benitez *et al.*, 1970).

The apparent spontaneous disappearance of endemic cretinism in Southern Europe raised considerable doubts as to the relation of iodine deficiency to the condition. Such a spontaneous disappearance without iodisation was noted by Costa *et al.* (1964) in Switzerland.

It was in these circumstances that it was decided in 1966 to set up a controlled trial in the Western Highlands of Papua New Guinea to see whether endemic cretinism could be prevented by correction of iodine deficiency. This study, carried out in collaboration with the Public Health Department, was based on the use of iodised oil in a single intramuscular injection (4 ml) which provided 2.15 g of iodine. This dose had previously been shown (Buttfield & Hetzel, 1967) to provide satisfactory correction of severe iodine deficiency for a period of 4–5 years. Iodised oil or saline injections were given to alternate families in the Jimi River District at the time of the first census (1966). Each child born subsequently was examined for evidence of motor retardation, as assessed by the usual milestones of sitting, standing, or walking, and for evidence of deafness. Examination was carried out without knowledge as to whether the mother had received an iodised oil injection or saline. Infants presenting with a full syndrome of hearing and speech abnormalities together with abnormalities of motor development with or without squint were classified as suffering from endemic cretinism. Later follow-up confirmed the diagnosis of cretinism in these cases (Pharoah, Buttfield & Hetzel, 1971; Pharoah & Connolly, 1984).

The results of the follow-up are shown in Table 6.3. There were seven cretins born to women who had received iodised oil; in six of these seven cases conception had occurred prior to the iodised oil injections. In the seventh case the mother received the injection on the sixth of October, 1966, or 42–46 weeks later. The lack of precision about this last birth date

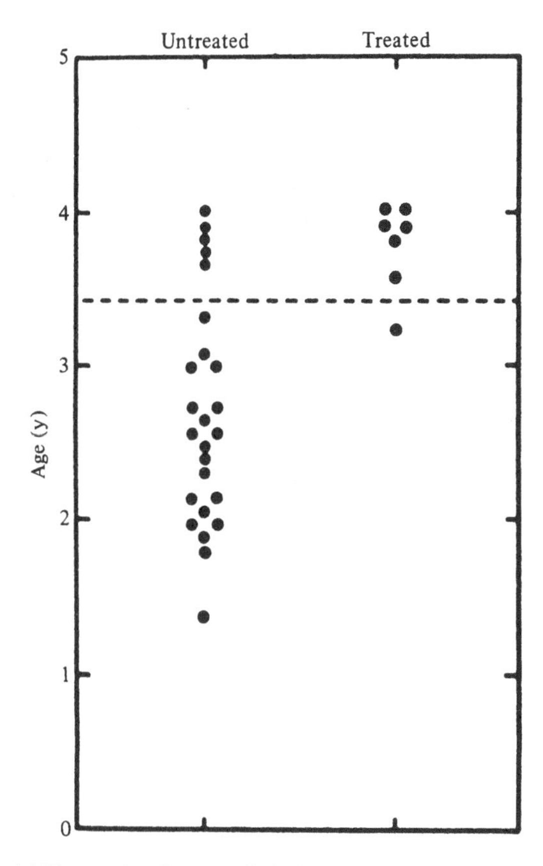

Fig. 6.1 The results of a controlled trial of iodised oil injection in the Jimi River District of the Highlands of Papua New Guinea. Alternate mothers were given an injection of iodised oil and saline in September 1966. All newborn children were followed up for the next five years. Each dot represents a cretin child. The figure shows that mothers given iodised oil injections do not have subsequent cretin children, in comparison with their persistence in the untreated group (from Pharoah *et al.*, 1971).

raised the possibility that conception had occurred prior to treatment in this instance also.

In the untreated group there were 25 endemic cretins born since the trial began. In five of these 25, conception had occurred prior to saline being given. The ages of all the cretins born between 1966 and 1972 in relation to the injection of iodised oil are shown in Fig. 6.1

It was concluded that an injection of iodised oil given prior to pregnancy could prevent the occurrence of the neurological syndrome of endemic cretinism in the infant. The occurrence of the syndrome in those who were pregnant at the time of oil injection indicated that the damage probably occurred during the first half of pregnancy.

The value of iodised oil injection in the prevention of endemic cretinism has been confirmed in Zaire and in South America. Mass injection programs have been carried out in New Guinea in 1971–72 and in Zaire, Indonesia and in China. Recent evaluations of these mass programs in Indonesia and China indicate that endemic cretinism has been prevented where correction of iodine deficiency has been achieved (Ma *et al.*, 1982; Dulberg *et al.*, 1983).

The controlled trial with iodised oil also revealed a significant reduction in recorded fetal and neonatal deaths in the treated group (Table 6.3) which is consistent with other evidence indicating the effect of iodine deficiency on fetal survival.

Further data from Papua New Guinea indicates a relationship between the level of maternal thyroxine and the outcome of current and recent past pregnancies, including mortality and the occurrence of cretinism. The rate of perinatal deaths was twice as high among mothers with very low serum concentrations of total thyroxine (TT_4) (36.0%) as in the other women (16.4%) ($p<0.05$) and the same was true of free thyroxine (FT_4) (Pharoah *et al.*, 1976).

These data indicating the importance of maternal thyroid function to fetal survival and development are complemented by extensive animal data (Hetzel, Chavadej & Potter, 1988).

Iodine deficiency in the neonate

An increased perinatal mortality due to iodine deficiency has been shown in Zaire from the results of a controlled trial, with iodised oil or control injections given in the latter half of pregnancy (Thilly, 1981). There was a substantial fall in perinatal and infant mortality with improved birth weight. Low birth weight (whatever the cause) is generally associated with a higher rate of congenital anomalies and higher risk throughout childhood.

Apart from the question of mortality, the importance of the state of thyroid function in the neonate relates to the fact that at birth the brain of the human infant has only reached about one third of its full size and continues to grow rapidly until the end of the second year (Dobbing, 1974). The thyroid hormone, dependent on an adequate supply of iodine, is essential for normal brain development as has been confirmed by animal studies (Hetzel *et al.*, 1988).

Recently data on iodine nutrition and neonatal thyroid function in Europe have been published. These data confirm the continuing presence of severe iodine deficiency affecting neonatal thyroid function and hence a threat to early brain development (Delange *et al.*, 1986).

There is similar evidence from neonatal observations in Zaire in Africa where incidence of chemical hypothyroidism of 10% has been found (Ermans *et al.*, 1980). In Zaire it has been observed that this hypothyroidism persists into infancy and childhood if the deficiency is not corrected, with resultant retardation of physical and mental development.

These observations suggest a much greater risk of mental defect in severely iodine deficient populations than is indicated by the presence of cretinism.

Iodine deficiency in infancy and childhood

Iodine deficiency in children is characteristically associated with goitre. The classification of goitre which has been standardised by the World Health Organization has been discussed elsewhere (Dunn *et al.*, 1986). The goitre rate increases with age so that it reaches a maximum with adolescence. Girls have a higher prevalence than boys. Observations of goitre rates in school children over the period 8–14 years provides a convenient indication of the presence of iodine deficiency in a community. The availability of children in schools is a great advantage in providing access to a population. Collections of casual samples of urine for detection of urinary iodine can also be conveniently carried out.

Recent studies in school children living in iodine deficient areas in a number of countries indicate impaired school performance and IQ in comparison with matched groups from non-iodine deficient areas. These studies are difficult to set up because of the problem of the control group. There are many possible causes for impaired school performance, and impaired performance on an IQ test, which make the interpretation of any difference that might be observed difficult. The iodine deficient area is likely to be more remote, suffer more social deprivation, with a disadvantage in school facilities, a lower socioeconomic status and poorer general nutrition. All such factors have to be taken into account, as well as the problem of adapting tests developed in Western countries for use in Third World countries.

Initially, studies of psychomotor development, as indicated by tests of motor coordination, revealed differences that could be regarded as, to a large extent, independent of educational status. In Papua New Guinea, children from the controlled trial in the Western Highlands have been tested periodically. Differences in bimanual dexterity were revealed by threading beads and putting pegs into a pegboard, in relation to whether the mother was severely iodine deficient or not. These differences have been significantly related to the level of maternal thyroxine at the time of pregnancy and have persisted up to the age of 10–12 years (Connolly,

Pharoah & Hetzel, 1979; Pharoah *et al.*, 1981; Pharoah & Connolly, 1984).

Differences in motor coordination have also been observed in Indonesia. Also, more recent critical studies by Bleichrodt *et al.* (1987) in Indonesia and in an iodine deficient area in Spain, using a wide range of psychological tests, have shown that the mental development of children from iodine deficient areas lags behind that of children from non-iodine deficient areas. The differences in psychomotor development became apparent only after the age of two and a half years.

In a study from Chile (Muzzo, Leiva & Carrasco, 1986) a comparison was made between children with goitre and those without goitre, from the same area, which demonstrated poorer performance in IQ tests by those children with goitre. Similar data are now becoming available from China, but have not yet been published.

The next question is whether these differences can be affected by correction of the iodine deficiency. In a pioneering study initiated in Ecuador in 1966, Fierro-Benitez *et al.* have reported the long term effects of iodised oil injections by comparison of two Highland villages – one (Tocachi) being treated, the other (La Esperanza) acting as a control (Fierro-Benitez *et al.*, 1986). Particular attention was paid to 128 children aged 8–15 whose mothers had received iodised oil prior to the second trimester of pregnancy and a matched control group of 293 of similar age. All children were periodically examined from birth at key stages in their development. Women in Tocachi were re-injected or injected in 1970, 1974 and 1978. Assessments in 1973, 1978 and 1981 revealed the following: 'Scholastic achievement was better in the children of treated mothers when measured in terms of school year reached, for age, school dropout rate, failure rate, years repeated and school marks. There was no difference between the two groups by certain tests (Terman-Merrill, Wechsler or Goodenough). However, both groups were impaired in school performance – in reading, writing and mathematics, but more notably the children of untreated mothers'. The results indicate the significant role of iodine deficiency, but other factors were also considered to be important in the school performance of these Ecuadorean children, such as social deprivation and other nutritional factors.

In 1982 the results were reported of a controlled trial carried out with oral iodised oil in a small Highland village (Tiquipaya) 2645 metres above sea level in Bolivia (Bautista *et al.*, 1982). The children did not show clinical evidence of protein–calorie malnutrition. The mean serum protein bound iodine had been low (1.7 μg/dl) in 1970. Four hundred and eight children were examined. A hundred boys and 100 girls, all with some degree of thyroid enlargement, were chosen for further study.

They ranged in age from $5\frac{1}{2}$–12 years. Goitre size was estimated using the standard methods and the IQ was measured with the Stanford Binet test in Spanish translation using the short form of the test. The Bender Gestalt test was also used.

Each child in the treatment group received an oral dose of 1.0 ml iodised oil (475 mg iodine) while children in the second group received 1.0 ml of a non-iodine containing mineral oil of a similar brown colour. Subsequent assessment was double blind so that the group of the child was not known to the examiners.

On follow up 22 months later the urinary iodine had increased and goitre size had decreased in both groups. This reflected 'contamination' of the village environment with iodine due, probably, to urine excretion by those who had received the iodised oil injections. There were no differences between the two treatment groups in growth rate or in performance on the tests. However, improvement in IQ could be demonstrated in all those children who showed significant reduction of goitre. This was particularly evident in girls. It was concluded that correction of iodine deficiency may improve the mental performance of school age children, but that a bigger dose should be given.

These studies are now being followed up in a number of countries. More data are required. However, these data do point to significant benefits to school children's mental performance from correction of iodine deficiency.

Iodisation programs have been shown to increase the level of circulating thyroid hormones in children in India (Sooch *et al.*, 1973) and in China (Zhu, 1983). These changes occur whether or not the child is goitrous and indicate a mild degree of hypothyroidism without any apparent symptoms.

The major determinant of brain (and pituitary) triiodothyronine (T_3) is serum thyroxine (T_4) and not T_3 itself (as is true of T_3 levels in the liver, kidney and muscle) (Crantz & Larsen, 1980). Low levels of brain T_3 have been demonstrated in the iodine deficient rat in association with reduced levels of serum T_4 and these have been restored to normal with correction of iodine deficiency (Obregon *et al.*, 1984).

These findings provide a rationale for suboptimal brain function in subjects with endemic goitre and lowered serum T_4 levels and its improvement following correction of iodine deficiency.

Iodine deficiency in the adult

The common effect of iodine deficiency in adults is goitre. Characteristically there is an absence of classical clinical hypothyroidism in adults with endemic goitre. However, laboratory evidence of hypothyroidism with reduced T_4 levels is common. This is often accompanied by normal T_3 levels and raised thyroid stimulating hormone levels (Patel *et al.*, 1973; Goslings *et al.*, 1977; Zhu, 1983).

Iodine administration in the form of iodised salt (Zhu, 1983), iodised bread (Clements *et al.*, 1960) or iodised oil (Buttfield & Hetzel, 1967) have all been demonstrated to be effective in the prevention of goitre in adults. Iodine administration may also reduce existing goitre in adults. This is particularly true of iodised oil injections (Buttfield & Hetzel, 1967). This obvious effect leads to quick acceptance of the measure by people living in iodine deficient communities. A rise in circulating thyroxine can be readily demonstrated in adult subjects following iodisation. As already pointed out, this could mean a rise in brain T_3 levels with improvement in brain function.

In Northern India a high degree of apathy has been noted in populations living in iodine deficient areas. This may even affect domestic animals such as dogs! It is apparent that reduced mental function is widely prevalent in iodine deficient communities, with effects on their capacity for initiative and decision-making.

This means that iodine deficiency is a major block to the human and social development of communities living in an iodine deficient environment. Correction of the iodine deficiency is indicated as a major contribution to development. An instructive and broad example of the possibilities is provided by observations of the effects of an iodised salt program dating only from 1978 in the Northern Chinese village of Jixian in Heilongjiang Province (Li & Wang, 1987).

In 1978 there were 1313 people with a goitre rate of 65%, with 11.4% cretins. The cretins included many severe cases which caused the village to be known locally as 'the village of idiots'. The economic development of the village was retarded – for example, no truck driver or teacher was available. Girls from other villages did not want to marry and live in the village. The intelligence of the student population was known to be low: children aged ten had a mental development equivalent to those aged seven.

Iodised salt was introduced in 1978, after which the goitre rate dropped to 4% by 1982. No cretins had been born since 1978. The attitude of the people had changed greatly – they were much more positive in their approach to life in contrast to their attitude before iodisation. The average income had increased from 43 yuan per head in 1981 to 223 yuan in 1982 and 414 yuan in 1984, which was higher than the average income per capital in the district. In 1983 cereals were exported for the first time. Before iodisation no family had a radio but, by 1986, 55 families had television sets. Forty-four girls had come from other villages to marry boys in Jixian. Seven men had joined the People's Liberation Army, whereas before iodisation they had been rejected for goitre. These effects can be largely, although not entirely, attributed to the correction of community hypothyroidism by iodised salt.

These observations indicate the important social and economic benefits that can result from the correction of severe iodine deficiency.

Iodine induced hyperthyroidism

A mild increase in incidence of hyperthyroidism has now been described following iodised salt programs in Europe and South America and following iodised bread in Holland and Tasmania (Connolly, Vidor & Stewart, 1970; Stewart *et al.*, 1971). A few cases have been noted following iodised oil administration in South America. No cases have yet been described in New Guinea, India or Zaire. This is probably due to the scattered nature of the population and limited opportunities for observation (Larsen *et al.*, 1980). Natural remission also occurs. The condition is largely confined to those over 40 years of age – a smaller proportion of the population in developing countries than in developed countries. Detailed observations are available from the island of Tasmania (Stewart *et al.*, 1971; Vidor *et al.*, 1973).

The condition is readily controlled with antithyroid drugs or radioiodine. Spontaneous remission also occurs. In general, iodisation should be avoided in those over the age of 40 because of the risk of hyperthyroidism (Stanbury *et al.*, 1974). However, the correction of iodine deficiency prevents the formation of an autonomous thyroid and so prevents the condition of iodine induced hyperthyroidism. Hence this condition is included as an 'iodine deficiency disorder'.

Conclusions

This review of the effects of iodine deficiency in man indicates the broad spectrum of consequences on growth and development. Effects on the fetus, the neonate and the infant, are particularly important because of the necessity of normal thyroid gland function for the rapid growth period of early development.

The effects on fetal survival and fetal brain development have been fully documented in experimental animal studies which have been reviewed elsewhere (Hetzel & Potter, 1983; Hetzel *et al.*, 1988).

Methods for correction of iodine deficiency

Iodised salt

Iodised salt has been the major method used since the 1920s when it was first used in Switzerland. Since then, successful programmes have been reported from a number of countries. These include Central and South America (e.g. Guatemala, Colombia), Finland in Europe, China and Taiwan in Asia (Stanbury & Hetzel, 1980).

The difficulties in the production and maintenance of quality to the millions that are iodine deficient, especially in Asia, are vividly demonstrated in India, where there has been a breakdown in supply. The difficulties have been discussed in a detailed report prepared by the Nutrition Foundation of India (The National Goitre Control Program, 1983).

In Asia, the cost of iodised salt production and distribution at present is of the order of three to five cents per person per year (UNICEF, 1984). This must be considered cheap in relation to the social benefits that have been described in the previous section.

However, there is still the problem of the salt actually reaching the iodine deficient subject. There may be a problem with distribution or preservation of the iodine content – it may be left uncovered or exposed to heat. It should be added after cooking to reduce the loss of iodine.

Finally there is the difficulty of actual consumption of the salt. While the addition of iodine makes no difference to the taste of the salt, the introduction of a new variety of salt to an area where salt is already available, familiar and much appreciated as a condiment is likely to be resisted. In the Chinese provinces of Sinjiang and Inner Mongolia, the strong preference of the people for desert salt of very low iodine content led to a mass iodised oil injection program in order to prevent cretinism (Ma *et al.*, 1982).

Iodised oil by injection

The value of iodised oil injection in the prevention of endemic goitre and endemic cretinism was first established in New Guinea with controlled trials involving the use of saline injection as a control. These trials established the value of the oil in the prevention of goitre (McCullagh, 1963) and the prevention of cretinism (Pharoah *et al.*, 1971). Experience in South America (Hetzel *et al.*, 1980) has confirmed the value of the measure. The quantitative correction of severe iodine deficiency by a single intramuscular injection (2–4 ml) has been demonstrated (Buttfield & Hetzel, 1967) for a period of over four years.

The injection of iodised oil can be administered through local health services, where they exist, or by special teams. In New Guinea (Hetzel, 1974) the injection of a population in excess of 100,000 was carried out by public health teams, along with the injection of a triple antigen. In Nepal 2 million injections have now been given by the Expanded Immunisation Programme teams. The obvious disappearance of goitre ensures ready acceptance. Iodised oil is singularly appropriate for the isolated village community so characteristic of mountainous endemic goitre areas.

In a suitable area, the oil should be administered to all females up to the age of 40 years and all males up to the age of 20 years. A repeat of the injection would be required in 3–5 years depending on the dose given and the age. In children the need is greater than in adults and the recommended dose should be repeated after three years if severe iodine deficiency persists (Stanbury *et al.*, 1974).

Iodised walnut oil and iodised soya bean oil are new preparations developed in China since 1980. Preliminary reports on the use of these preparations are available (Liu, 1983; Ouyang *et al.*, 1983).

It is now clear that iodised oil is suitable for use in a mass programme. In Indonesia (Djokomoeljanto *et al.*, 1983) some 1,036,828 injections were given between 1974 and 1978, together with a massive distribution of iodised salt. A further 4.9 million have been given by specially trained paramedical personnel in the period 1979–83. In China in Sinjiang, 707,000 injections were given by barefoot doctors between 1978 and 1981 – and a further 300,000–400,000 were being given in 1982 (Ma *et al.*, 1982). There are considered to be advantages to the use of injections because of the association of injections with the recent successful smallpox eradication campaign.

The disadvantages of the use of injections are the immediate discomfort produced and the infrequent development of abscesses at the site of injection. Sensitivity phenomena have not been reported (Hetzel *et al.*, 1980).

However, the major problem of injections is their cost, although this has been reduced with mass packaging to a similar order of magnitude to the costs of iodised salt. Costs can be kept down if the population to be injected is restricted to women of reproductive age and children, and the primary health care team is available (Hetzel, 1983; SEARO/WHO, 1985; Dunn *et al.*, 1986).

Iodised oil by mouth

There is evidence available of the effectiveness of a single oral administration of iodised oil lasting for one to two years in South America (Watanabe *et al.*, 1974) and in Burma (Kywe-Thein *et al.*, 1978).

More recent studies in India and China reveal that oral iodised oil lasts only half as long as a similar dose given by injection. This has been confirmed by studies in the guinea pig (Li & Wei, 1985).

Oral administration of iodised oil to children could be carried out through the baby health centres and schools – periods of 18 months could be covered at present (2 ml dose). This may well be increased as a result of further research. Cheaper production of iodised oil is readily achievable (Thilly, Delange & Stanbury, 1980) and should be provided in India and other countries with large populations at risk.

The use of other methods of iodisation, such as iodised bread and iodised water, has been reviewed elsewhere (Hetzel, 1983; Dunn *et al.*, 1986). These methods may be indicated in special situations, but the major alternatives for mass use are iodised salt and iodised oil.

The main hazard of iodisation is transient hyperthyroidism seen mainly over the age of 40. It is caused by autonomous thyroid function resulting from long-standing iodine deficiency. It can be minimised by withholding iodisation from those over the age of 40 (Stanbury & Hetzel, 1980; Hetzel, 1983).

Indications for different methods of iodine supplementation

There are three grades of severity of IDD in a population, based on the urinary iodine excretion (Follis, 1964). These are as follows:

(i) *Mild IDD* with goitre prevalence in the range 5–20% (school children) and with median urine iodine levels in excess of 50 μg/g creatinine. Mild IDD can be controlled with iodised salt at a concentration of 10–25 mg/kg. It may disappear with economic development.

(ii) *Moderate IDD* with goitre prevalence up to 30%, and some hypothyroidism, with median urine iodine levels in the range 25–50% μg/g creatinine. Moderate IDD can be controlled with iodised salt (25–40 mg/kg) if this can be effectively produced and distributed. Otherwise, iodised oil, either orally or by injection, should be used through the primary health care system.

(iii) *Severe IDD* indicated by a high prevalence of goitre (30% or more), endemic cretinism (prevalence 1–10%), and median urine iodine below 25 μg/g creatinine. Severe IDD requires iodised oil either orally or by injection for complete prevention of central nervous system defects.

The prevention and control of IDD

Large populations are at risk of IDD because they live in an iodine deficient environment. This iodine deficient environment is char-

Table 6.4. *Prevalence of iodine deficiency disorders in developing countries and numbers of persons at risk (millions). (United Nations Report, 1987)*

	At risk	Goitre	Overt cretinism
South-East Asia	280	100	1.5
Asia (other countries)	400	30	0.9
Africa	60	30	0.5
Latin America	60	30	0.25
Total	800	190	3.15

acterised by soil from which iodine has been leached by glaciation, high rainfall or flood. This occurs most often in mountainous areas as in the Himalayan region, the Andean region and in the vast mountain ranges of China. However, low-lying areas subject to flooding, as in the Ganges Valley in India and Bangladesh, are also severely iodine deficient. This means that all the food grown in such soil is iodine deficient so that iodine deficiency will persist until there is dietary diversification, as occurred in Europe late in the nineteenth century and in the early decades of this century, or alternatively some form of iodine supplement is given.

There is consensus that 800 million are at risk of IDD, of which 190 million are suffering from goitre, more than three million are overt cretins (Table 6.4) and millions more suffer from some intellectual or motor deficit (United Nations, 1987). The global distribution of these populations at risk of IDD is shown in the map (Fig. 6.2.).

The major concentrations of population are in Asia where there has been a major escalation of IDD control programs in the last five years in India, Indonesia, Nepal, Burma and Bhutan.

In Latin America earlier efforts have produced a large measure of control in such countries as Argentina, Brazil, Colombia and Guatemala, but there is evidence of recurrence of the problem in Colombia and Guatemala associated with political and social unrest. Major IDD problems have persisted in Ecuador, Peru and Bolivia, but there has been significant progress in the last three years with the combination of national government initiative and support from international agencies.

In Africa there has been a lag in the development of IDD control programmes in comparison with the other continents. However, new initiatives have begun following a Joint WHO/UNICEF/ICCIDD Regional Seminar held in Yaounde, Cameroon in March 1987. This Seminar set up a joint IDD Task Force which has now initiated comprehensive planning for the prevention and control of IDD in Africa (Hetzel, 1987b).

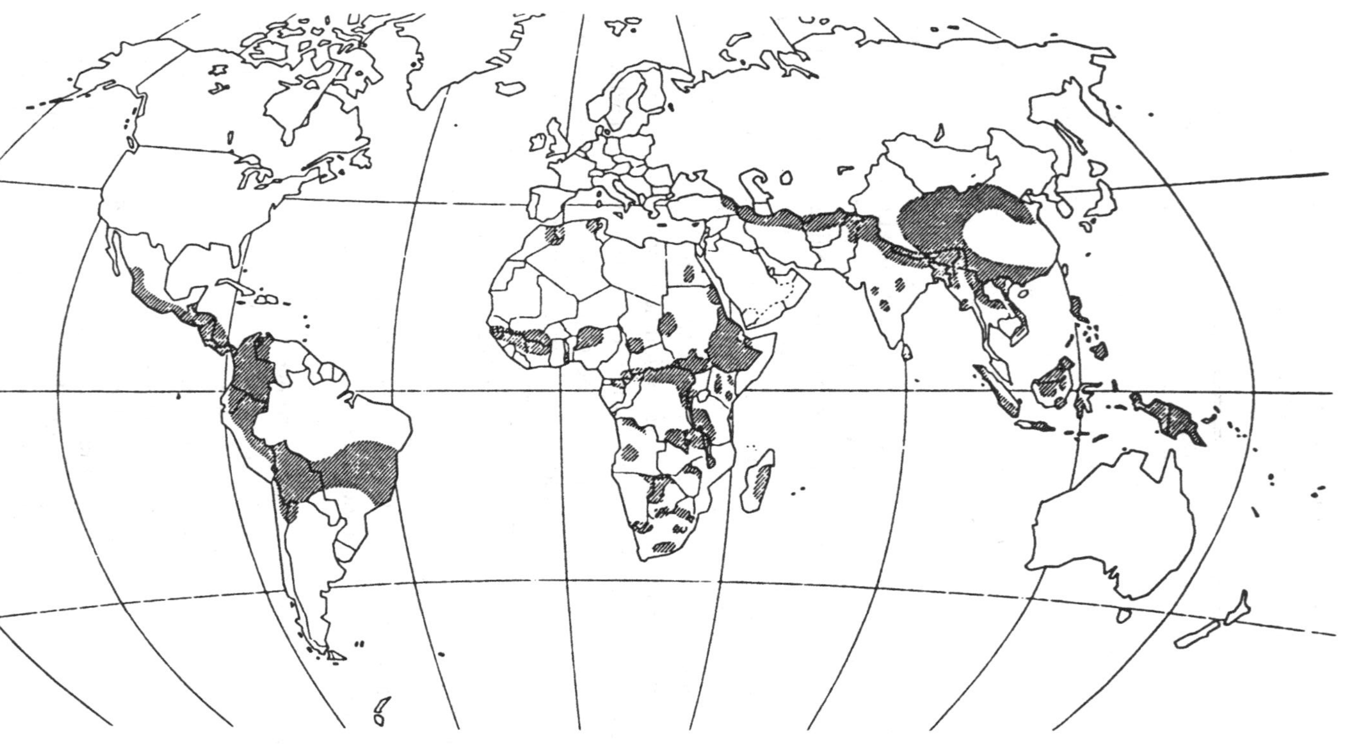

Fig. 6.2 Map showing world wide distribution of iodine deficiency disorders (shaded areas) in developing countries (from World Health Organization, quoted by Hetzel, 1987a).

China has also made rapid progress since the passing of the Cultural Revolution in 1976. One third of the population of China (300 million) is at risk of IDD because of the extensive mountainous areas in that country (Ma *et al.*, 1982).

Both iodised salt and iodised oil are the major mass supplementation measures which have been used on a large scale. In excess of ten million injections of iodised oil have been given in Asia with evidence of successful prevention of IDD.

The great gap between our new knowledge of IDD and the application of this knowledge in national IDD control programmes particularly in developing countries, has led to the formation of the International Council for the Control of Iodine Deficiency Disorders (ICCIDD). The inaugural meeting of this multidisciplinary group of epidemiologists and nutritionists, endocrinologists and chemists, planners and economists was held in Kathmandu, Nepal, in March 1986. A series of papers on all aspects of IDD control programmes presented in Kathmandu was published as a monograph early in 1987 (Hetzel, Dunn & Stanbury). The ICCIDD has now established a global multidisciplinary network of some 300 people with expertise relevant to IDD and IDD control programmes. It works closely with WHO, UNICEF and national governments within the UN system in the development of national programmes (Hetzel, 1987a).

The feasibility of substantial progress in the prevention and control of IDD in the next five to ten years was endorsed in a World Health Assembly Resolution in 1986 (World Health Assembly, 1986).

A Global Strategy for a 'Ten Year Program' has now been adopted by the UN agencies (United Nations, 1987; Hetzel, 1987b).

These various developments encourage the hope that significant progress can be made in the prevention and control of IDD within the next decade, with great benefits to the quality of life of the many millions affected.

References

Bautista, A., Barker, P.A., Dunn, J.T., Sanchez, M. & Kaiser, D.L. (1982). The effects of oral iodized oil on intelligence, thyroid status, and somatic growth in school-age children from an area of endemic goiter. *American Journal of Clinical Nutrition*, **35**, 127–34.

Bleichrodt, N., Garcia, I., Rubio, C., Morreale de Escobar, G. & Escobar Del Rey, F. (1987). Developmental disorders associated with severe iodine deficiency. In: *The Prevention and Control of Iodine Deficiency Disorders*, ed. B.S. Hetzel, J.T. Dunn & J.B. Stanbury, pp. 65–84. Elsevier Biomedical, Amsterdam.

Bourdoux, P., Delange, F., Gerard, M., Mafuta, A., Hanson, A. & Ermans,

A.M. (1978). Evidence that cassava ingestion increases thiocyanate formation: A possible etiologic factor in endemic goiter. *Journal of Clinical Endocrinological Metabolism*, **46**, 613–21.
Bourdoux, P., Delange, F., Gerard, M., Mafuta, A., Hanson, A. & Ermans, A.M. (1980a). Antithyroid action of cassava in humans. In: *Role of Cassava in the Etiology of Endemic Goiter and Cretinism*, ed. A.M. Ermans, pp. 61–8. International Development Research Center, Ottawa.
Bourdoux,P., Mafuta, A., Hanson, A. & Ermans, A.M. (1980b). Cassava toxicity: the role of linamarin. In: *Role of Cassava in the Etiology of Endemic Goiter and Cretinism*, ed. A.M. Ermans, pp. 15–27. International Development Research Center, Ottawa.
Buttfield, I.H. & Hetzel, B.S. (1967). Endemic goitre in Eastern New Guinea with special reference to the use of iodised oil in prophylaxis and treatment. *WHO Bulletin*, **36**, 243–62.
Buttfield, I.H. & Hetzel, B.S. (1969). Endemic cretinism in Eastern New Guinea. *Australasian Annals of Medicine*, **18**, 217–21.
Clements, F.W. *et al.* (1960). (Contributors). *Endemic Goitre*. WHO, Geneva.
Connolly, K.J., Pharoah, P.O.D. & Hetzel, B.S. (1979). Fetal iodine deficiency and motor performance during childhood. *Lancet*, **ii**, 1149–51.
Connolly, R.J., Vidor, G.I. & Stewart, J.C. (1970). Increase in thyrotoxicosis in endemic goiter area after iodation of bread. *Lancet*, **i**, 500–2.
Costa, A., Cottino, F., Mortara, M. & Vogliazzo, U. (1964). Endemic cretinism in Piedmont. *Panminerva Medical*, **6**, 250–9.
Crantz, F.R. & Larsen, P.R. (1980). Rapid thyroxine to 3,5,3′-triiodothyronine conversion binding in rat cerebral cortex and cerebellum. *Journal of Clinical Investigation*, **65**, 935–8.
Delange, F., Camus, M. & Ermans, A.M. (1972). Circulating thyroid hormones in endemic goitre. *Journal of Clinical Endocrinology and Metabolism*, **34**, 891–5.
Delange, F., Heidemann, P., Bourdoux, P., Larsson, A., Vigneri, R., Klett, M., Beckers, C. & Stubbe, P. (1986). Regional variations of iodine nutrition and thyroid function during the neonatal period in Europe. *Biology of the Neonate*, **49**, 322–30.
Delange, F., Iteke, F.B. & Ermans, A.M. (1982). Nutritional factors involved in the goitrogenic action of Cassava. International Development Research Center, Ottawa.
Djokomoeljanto, R., Tarwotjo, I. & Maspaitella, F. (1983). Goitre control program in Indonesia. In: *Current Problems in Thyroid Research*, vol. 605, ed. N. Ui, K. Torizuka, S. Nagataki & K. Miyai, pp. 403–5. Excerpta Medica, Amsterdam.
Dobbing, J. (1974). The later development of the brain and its vulnerability. In: *Scientific Foundations of Paediatrics*, ed. J. Davis & J. Dobbing, pp. 565–77. Heinemann Medical, London.
Dulberg, E.M., Widjaja, K., Djokomoeljanto, R., Hetzel, B.S. & Belmont, L. (1983). Evaluation of the iodisation program in Central Java with reference to the prevention of endemic cretinism and motor coordination defects. In: *Current Problems in Thyroid Research*, vol. 605, ed. N. Ui, K. Torizuka, S. Nagataki & K. Miyai, pp. 19–22. Excerpta Medica, Amsterdam.
Dunn, J.T., Pretell, E.A., Daza, C.H. & Viteri, F.E. (ed.) (1986). *Towards the Eradication of Endemic Goitre, Cretinism, and Iodine Deficiency*. Scientific Publication No. 502, Pan American Health Organization, Washington.

Ermans, A.M., Bourdoux, P., Lagasse, R., Delange, F. & Thilly, C. (1980). Congenital hypothyroidism in developing countries. In: *Neonatal Thyroid Screening*, ed. N. Borrow, pp. 61–73. Raven Press, New York.

Fierro-Benitez, R., Cazar, R., Stanbury, J.B., Rodriguez, P., Garces, F., Fierro-Renoy, F. & Estrella, E. (1986). Long-term effect of correction of iodine deficiency on psychomotor and intellectual development. In: *Towards the Eradication of Endemic Goiter, Cretinism and Iodine Deficiency*, ed. J.T. Dunn, E.A. Pretell, C.H. Daza & F.E. Viteri, pp. 182–200. Pan American Health Organization, Washington.

Fierro-Benitez, R., Stanbury, J.B., Querido, A., De Groot, L., Alban, R. & Endova, J. (1970). Endemic cretinism in the Andean Region of Ecuador. *Journal of Clinical Endocrinology and Metabolism*, **30**, 228–36.

Follis, R.H. (1964). Recent studies in iodine malnutrition and endemic goitre. *Medical Clinics of North America*, **48**, 1919–24.

Goslings, B.M., Djokomoeljanto, R., Doctor, R., Van Hardeveld, C., Hennemann, G., Smeenk, D. & Querido, A. (1977). Hypothyroidism in an area of endemic goitre and cretinism in Central Java, Indonesia. *Journal of Clinical Endocrinology and Metabolism*, **44**, 481–90.

Hetzel, B.S. (1974). The epidemiology, pathogenesis and control of endemic goitre and cretinism in New Guinea. *New Zealand Medical Journal*, **80**, 482–4.

Hetzel, B.S. (1983). Iodine deficiency disorders (IDD) and their eradication. *Lancet*, **ii**, 1126–9.

Hetzel, B.S. (1987a). An overview of the prevention and control of iodine deficiency disorders. In: *The Prevention and Control of Iodine Deficiency Disorders*, ed. B.S. Hetzel, J.T. Dunn & J.B. Stanbury, pp. 7–31. Elsevier Biomedical, Amsterdam.

Hetzel, B.S. (1987b). Progress in the prevention and control of iodine deficiency disorders. *Lancet*, **ii**, 266.

Hetzel, B.S., Chavadej, J. & Potter, B.J. (1988). The brain in iodine deficiency. *Neuropathology and Applied Neurobiology*, **14**, 93–104.

Hetzel, B.S., Dunn, J.T. & Stanbury, J.B. (ed.) (1987). *The Prevention and Control of Iodine Deficiency Disorders*. Elsevier, Amsterdam.

Hetzel, B.S. & Potter, B.J. (1983). Iodine deficiency and the role of thyroid hormones in brain development. In: *Neurobiology of the Trace Elements*, ed. I. Dreosti & R.M. Smith, pp. 83–133. Humana Press, New Jersey.

Hetzel, B.S., Thilly, C.H., Fierro-Benitez, R., Pretell, E.A., Buttfield, I.H. & Stanbury, J.B. (1980). Iodized oil in the prevention of endemic goiter and cretinism. In: *Endemic Goiter and Endemic Cretinism*, ed. J.B. Stanbury & B.S. Hetzel, pp. 513–32. Wiley, New York.

Karmarkar, M.G., Deo, M.G., Kochupillai, N. & Ramalingaswami, V. (1974). Pathophysiology of Himalayan endemic goitre. *American Journal of Clinical Nutrition*, **27**, 96–103.

Kelly, F.C. & Snedden, W.A. (1960). Prevalence and geographical distribution of endemic goitre. In: *Endemic Goitre*, pp. 27–333. WHO, Geneva.

Konig, M.P. & Veraguth, P. (1961). Studies of thyroid function in endemic cretins. In: *Advances in Thyroid Research*, ed. R. Pitt-Rivers, pp. 294–8. Pergamon Press, London.

Kywe-Thein, Tin-Tin-Oo, Khin-Maung-Niang, Wrench, J. & Buttfield, I.H. (1978). A study of the effect of intramuscular and oral iodised poppy seed oil in the treatment of iodine deficiency. In: *Current Thyroid Problems in South-East Asia and Oceania*, ed. B.S. Hetzel, M.L. Wellby & R. Hoschl, pp. 78–82. Proceedings of Asia and Oceania Thyroid Association Workshops on Endemic Goitre and Thyroid Testing, Singapore.

Langer, P. (1960). History of Goitre. In: *Endemic Goitre*, pp. 9–25. WHO, Geneva.

Larsen, P.R., Silva, J.E., Hetzel, B.S. & McMichael, A.J. (1980). Monitoring prophylactic programs: general consideration. In: *Endemic Goiter and Endemic Cretinism*, ed. J.B. Stanbury & B.S. Hetzel, pp. 551–66. Wiley, New York.

Li Jianqun & Wang Xin (1987). Jixian: A success story in IDD control. *IDD Newsletter*, **3**, 3–4.

Li Jianqun & Wei Jun (1985). Studies of the effect of iodised oil in guinea pigs. *Nutritional Reports International*, **31**, 1085–7.

Liu, Z. (1983). Study of prophylaxis and treatment of endemic goitre by oral iodised soybean oil. In: *Current Problems in Thyroid Research*, ed. N. Ui, K. Torizuka, S. Nagataki & K. Miyai, pp. 410–17. Excerpta Medica, Amsterdam.

Ma Tai, Lu Tizhang, Tan Uybin, Chen Bingshong & Zhu, H.I. (1982). The present status of endemic goiter and endemic cretinism in China. *Food and Nutrition Bulletin*, **4**, 13–19.

Maberly, G.F., Eastman, G., Waite, K.V., Corcoran, J. & Rashford, V. (1983). The role of Cassava. In: *Current Problems in Thyroid Research*, ed. N. Ui, K. Torizuka, S. Nagataki & K. Miyai, pp. 341–4. Excerpta Medica, Amsterdam.

McCarrison, R. (1908). Observations on endemic cretinism in the Chitral and Gilgit Valleys. *Lancet*, **ii**, 1275–80.

McCullagh, S.F. (1963). The Huon Peninsula endemic. I. The effectiveness of an intramuscular depot of iodised oil in the control of endemic goitre. *Medical Journal of Australia*, **1**, 769–77.

McMichael, A.J., Potter, J.D. & Hetzel, B.S. (1980). Iodine deficiency, thyroid function, and reproductive failure. In: *Endemic Goiter and Endemic Cretinism*, ed. J.B. Stanbury & B.S. Hetzel, pp. 445–60. Wiley, New York.

Muzzo, S. Leiva, L. & Carrasco, D. (1986). Influence of a moderate iodine deficiency upon intellectual coefficient of schoolage children. In: *Iodine Deficiency Disorders and Congenital Hypothyroidism*, ed. G. Medeiros-Neto, R. Maciel & A. Halpern, pp. 40–5. Ache, Sao Paulo.

Obregon, M.J., Santisteban, P., Rodriguez-pena, A., Pascual, A., Cartagena, P., Ruiz-Marcos, A., Lamas, L., Escobar del Rey, F. & Morreale de Escobar, G. (1984). Cerebral hypothyroidism in rats with adult-onset iodine deficiency. *Endocrinology*, **115**, 614–24.

Ouyang, A., Wang, P.O., Liu, Z.T., Lin, F.F. & Wang, H.M. (1983). Progress in the prevention and treatment of endemic goitre with iodised oil in China. In: *Current Problems in Thyroid Research*, ed. N. Ui, K. Torizuka, S. Nagataki & K. Miyai, pp. 418–25. Excerpta Medica, Amsterdam.

Pandav, C.S. & Kochupillai, N. (1982). Endemic goitre in India: prevalence, etiology, attendant disability and control measures. *Indian Journal of Pediatrics*, **50**, 259–71.

Patel, Y., Pharoah, P.O.D., Hornabrook, R. & Hetzel, B.S. (1973). Serum triiodothyronine, thyroxine and thyroid stimulating hormone in endemic goitre. A comparison of goitrous and non-goitrous subjects in New Guinea. *Journal of Clinical Endocrinology*, **37**, 783–9.

Pharoah, P.O.D., Buttfield, I.H. & Hetzel, B.S. (1971). Neurological damage to the fetus resulting from severe iodine deficiency during pregnancy. *Lancet*, **i**, 308–10.

Pharoah, P.O.D. & Connolly, K.C. (1984). A controlled trial of iodinated oil for the prevention of endemic cretinism: a long term follow up. *International Journal of Epidemiology*, **16**, 68–73.

Pharoah, P.O.D., Connolly, K.J., Hetzel, B.S. & Ekins, R.P. (1981). Maternal thyroid function and motor competence in the child. *Developmental Medicine and Child Neurology*, **23**, 76–82.

Pharoah, P.O.D., Delange, F., Fierro-Benitez, R. & Stanbury, J.B. (1980). Endemic cretinism. In: *Endemic Goiter and Endemic Cretinism*, ed. J.B. Stanbury & B.S. Hetzel, pp. 395–421. Wiley, New York.

Pharoah, P.O.D., Ellis, S.M., Ekins, R.P. & Williams, E.S. (1976). Maternal thyroid function, iodine deficiency and fetal development. *Clinical Endocrinology*, **5**, 159–63.

Pretell, E.A., Torres, T., Zenten, V. & Comejo, M. (1972). Prophylaxis of endemic goitre with iodised oil in rural Peru. *Advances in Experimental Medicine and Biology*, **30**, 249–65.

SEARO/WHO (1985). *Iodine Deficiency Disorders in South East Asia*. SEARO Regional Papers No. 10, WHO, Delhi (Regional Office for South East Asia).

Sooch, S.S., Deo, M.G., Karmarkar, M.G., Kochupillai, N., Ramachandran, K. & Ramalingaswami, V. (1973). Prevention of endemic goitre with iodised salt. *WHO Bulletin*, **49**, 307–12.

Stanbury, J.B., Brownell, G.L., Riggs, D.S., Perinetti, H., Itoiz, J. & Del Castillo, E.B. (1954). *The Adaptation of Man to Iodine Deficiency*, pp. 11–209. Harvard University Press, Cambridge, MA.

Stanbury, J.B., Ermans, A.M., Hetzel, B.S., Pretell, E.A. & Querido, A. (1974). Endemic goitre and cretinism: public health significance and prevention. *WHO Chronicle*, **28**, 220–8.

Stanbury, J.B. & Hetzel, B.S. (ed.) (1980). *Endemic Goitre and Endemic Cretinism: Iodine Nutrition in Health and Disease*. Wiley, New York.

Stewart, J.C., Vidor, G.I., Buttfield, I.H. & Hetzel, B.S. (1971). Epidemic thyrotoxicosis in Northern Tasmania: studies of clinical features and iodine nutrition. *Australia and New Zealand Journal of Medicine*, **1**, 203–11.

The National Goitre Control Program (1983). A blueprint for its intensification. *Nutrition Foundation of India Scientific Report 1*.

Thilly, C.H. (1981). Goitre et crétinisme endémiques: rôle étiologique de la consommation de manioc et stratégie d'éradication. *Bulletin et Memoires de l'Academie Royale de Medezine de Belgique*, **136**, 389–412.

Thilly, C.H., Delange, F. & Stanbury, J.B. (1980). Epidemiologic surveys in endemic goiter and cretinism. In: *Endemic Goiter and Endemic Cretinism*, ed. J.B. Stanbury & B.S. Hetzel, pp. 157–84. Wiley, New York.

United Nations Administrative Coordinating Committee Sub-Committee on Nutrition (1987). A global strategy for the prevention and control of iodine deficiency disorders – proposal for a ten-year program of support to countries. UN, Rome.

UNICEF (1984). *Report*. Regional Office for South-Central Asia, Delhi.

Vidor, G.I., Stewart, J.C., Wall, J.R., Wangel, A. & Hetzel, B.S. (1973). Pathogenesis of iodine-induced thyrotoxicosis. Studies in Northern Tasmania. *Journal of Clinical Endocrinology and Metabolism*, **37**, 901–9.

Watanabe, T., Moran, D., El Tamer, E., Staneloni, L., Salvaneschi, J., Altschuler, N., DeGrossi, O. & Niepominiszcse, H. (1974). Iodised oil in the prophylaxis of endemic goiter in Argentina. In: *Endemic Goiter and Cretinism: Continuing Threats to World Health*, ed. J.T. Dunn & G. Medeiros-Neto, pp. 192–231. Pan-American Health Organization, Washington.

World Health Assembly (1986). *The 39th World Health Assembly, Resolution 29*. WHO, Geneva.

Zhu, X.Y. (1983). Endemic goiter and cretinism in China with special reference to changes of iodine metabolism and pituitary-thyroid function two years after iodine prophylaxis in Gui-Zhou. In: *Current Problems in Thyroid Research*, ed. N. Ui, K. Torizuka, S. Nagataki & K. Miyai, pp. 13–18. Excerpta Medica, Amsterdam.

7 *Nutritional status and susceptibility to infectious disease*

S.J. ULIJASZEK

Introduction

The study of the interaction of nutritional status with disease is complex, and several comprehensive reviews have emphasised the clinical, epidemiological and experimental studies of risks associated with malnutrition (Scrimshaw, Taylor & Gordon, 1968; Chandra, 1983; Tomkins & Watson, 1989). In addition there have been reviews of research carried out into the effects of malnutrition on the individual components of the immune system (Suskind, 1977; Beisel, 1982; Gershwin, Beach & Hurley, 1985; Chandra, 1988a, b).

One problem in this area of study is the definition of malnutrition. There are major scientific and conceptual problems to address before any group of nutritionists agree on a definition of malnutrition (Payne, 1988). In this paper, the word 'malnutrition' will refer to the effects of any nutrient deficiency, including energy, protein and micronutrients. The term 'protein–energy malnutrition' (PEM) will be taken to mean 'a condition in which deficiencies of major body nutrients, resulting from a diet which is generally inadequate in energy and protein, frequently accompanied by deficiencies of micronutrients' (Tomkins, 1986). A child suffering from PEM may experience an impairment of growth which is often primarily due to a deficiency of energy. Conversely, a child may suffer a specific micronutrient deficiency without experiencing PEM and growth faltering. Marasmus and kwashiorkor are clinically overt syndromes, probably caused by a range of severe nutrient deficiencies, mainly of protein and energy.

Malnutrition, of whatever sort, has some very specific effects upon the immune system. Amongst them are hypoplasia of the lymphoid system and a reduction in circulating lymphocytes and phagocytic activity (Weir, 1986). In addition, nutritional status can affect the integrity of skin and mucous membranes, which normally afford a high degree of protection against pathogens. Gastric acidity is reduced in children with marasmus or kwashiorkor (Gracey, Cullity & Suharjono, 1977).

Fig. 7.1 gives a simplified model of the cycle of infection (Tomkins &

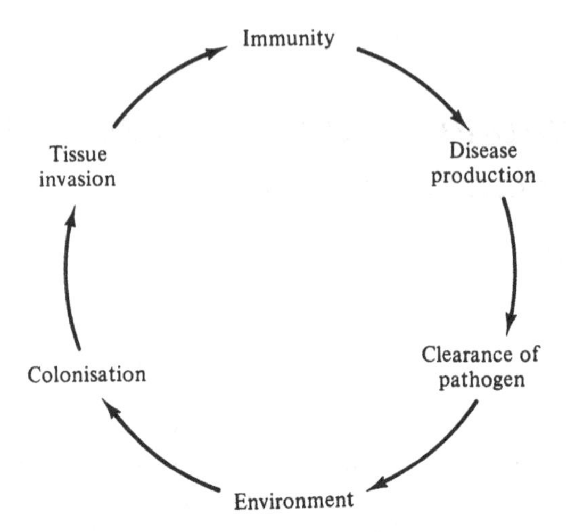

Fig. 7.1 The infection cycle (Tomkins & Watson, 1989).

Watson, 1989). Pathogens have to overcome a variety of host defence mechanisms if they are to colonise, invade a tissue, and multiply sufficiently to cause disease. In the well-nourished host, the immune defence systems are remarkably effective against colonisation and tissue invasion by pathogenic agents. If the host's innate immunity is overwhelmed by pathogens and disease ensues, an adaptive immune response specific for the inducing antigen is mounted. This response involves humoral and cell-mediated factors, both of which depend upon cells of the lymphoid system. The outcome of this response usually results in the clearance of the pathogen, and a restoration of well-being. The outcome may, however, be the death of the host. After novel exposure to a pathogen, the host may actively acquire immunity, the duration of which may vary depending upon the nature of the infection. Some infections may depress the host's immune responsiveness, leaving them susceptible to opportunistic infections. Further colonisation of the host will depend on environmental factors which affect the transmission of potentially pathogenic organisms. Malnutrition has a bearing on the host's response at several points in this cycle. There may be changes in the physiological response to pathogens in the intestinal mucosa, which may have profound effects on whether the pathogen is eliminated from the tissue, or stays (Tomkins & Watson, 1989). Subjects with marasmus or kwashiorkor have reduced secretory immunoglobulin A (IgA) (Tomkins & Watson, 1989) and vitamin A deficiency affects glycoprotein synthesis and mucus production (Rojanapo, Lamb & Olson, 1980).

Once in the bloodstream, there may be ecological struggles between

Table 7.1. *Infectious diseases influenced by nutritional status (Chandra, 1983)*

	Influence		
	Definite	Variable	Light
Bacterial	Tuberculosis Bacterial diarrhoea Cholera Leprosy Pertussis Respiratory infections	Diphtheria Staphylococcus Streptococcus	Typhoid Plague Tetanus Bacterial toxins
Viral	Measles Rotavirus diarrhoea Respiratory infections Herpes	Influenza	Smallpox Yellow fever ARBO[a] encephalitis Poliomyelitis
Parasitic	Pneumocystis carini Intestinal parasites Trypanosomiasis Leishmaniasis Schistosomiasis	Giardia Filariasis	Malaria
Fungal	Candida Aspergillus	Mould toxins	
Other		Syphilis Typhus	

[a]ARBO = arthropod-borne virus.

host and pathogen nutrition, and in some circumstances malnutrition may confer an advantage on the host (Murray & Murray, 1977; Tomkins & Watson, 1989). However, this advantage is balanced against the effect of malnutrition on the host's immune response. The immune response during established infection may determine the duration of infection, and a range of nutritional states can influence the rate of clearance of a pathogen. Marginal malnutrition due to deficiencies of iron and zinc may be protective of the host during the early stages of an infection. Under such conditions, low levels of plasma iron and zinc inhibit bacterial growth. However, the presence of adequate iron stores may enhance the bacteriocidal effects of free radicals produced by stimulated leucocytes (Tomkins & Watson, 1989). However, there is no evidence that malnutrition is protective during recovery (Tomkins & Watson, 1989).

Table 7.1 gives a list of infectious diseases believed to be influenced by nutritional status (Chandra, 1983). In malnutrition, mucosal response to pathogens is often impaired, and the diseases which predominate are those that infect the gastro-intestinal and respiratory tracts. The diseases

Table 7.2. *Immune response in PEM (Chandra, 1983; 1988a, b; Hoffman-Goetz, 1986; 1988)*

Most consistent changes in immunocompetence in PEM:
Cell-mediated immunity
Bacterial function of neutrophils
Secretory IgA
Complement system (C3 and factor B)
Interleukin 1

that have received most attention by researchers are tuberculosis (Jayalakshmi & Goapan, 1958), measles (Laditan & Reeds, 1976; Tomkins *et al.*, 1983), malaria (Edington, 1967), various intestinal parasites (Stehenson, 1987), diarrhoea (Tomkins, 1981; Black, Brown & Becker, 1984; Koster *et al.*, 1987) and non-specific respiratory infections (Rowland, Cole & Whitehead, 1977; Mata, 1978).

Nutrition and the immune response

Table 7.2 lists the most consistent changes in immunocompetence occurring as a direct result of PEM. These include reduction in cell-mediated immunity (CMI), bacteriocidal function of polymorphonuclear neutrophils, secretory IgA antibody response, activation of the C3 and factor B components of the complement system (Chandra, 1983, 1988a, b) and decreased responsiveness of lymphocytes to Interleukin 1 (Hoffman-Goetz, 1988).

Human malnutrition is usually a composite syndrome of multiple nutrient deficiencies. However, observations of laboratory animals deprived of one dietary element, and findings in rare patients with a single nutrient deficiency, have confirmed the crucial role played by several vitamins and trace elements in immunocompetence. These are summarised in Tables 7.3 and 7.4 (Beisel, 1982; Bendich, 1988; Bendich & Cohen, 1988; Bhaskaram, 1988; Cunningham-Rundles & Cunningham-Rundles, 1988; Fletcher *et al.*, 1988; Watson & Rybski, 1988). Deficiencies of vitamin A and pyridoxine lead to impaired CMI and reduced antibody and phagocytic responses. Deficiencies of vitamin B12, folic acid and vitamin C result in impaired phagocytic activity, whilst deficiencies of riboflavin and pantothenic acid cause impaired humoral immunity. The evidence for vitamin B12 and vitamin C deficiency in the reduction of humoral immunity is equivocal, as is the evidence for vitamin B12 deficiency in the reduction of CMI. Vitamin E deficiency appears to cause a reduction in cell-mediated and humoral immunity,

Table 7.3. *Effects of vitamin deficiencies on immune response (from Beisel, 1982; Bendich, 1988; Bendich & Cohen, 1988)*

	CMI	Humoral	Phagocytic	Comments/other
Vitamin A	↓	↓	↓	Serum haemolytic complement may increase
Thiamin	ND	ND	ND	
Riboflavin	ND	ND	↓	
Vitamin B12	E	E	↓	
Folic acid	ND	ND	↓	
Pyridoxine	↓	↓	↓	
Panthothenic acid	Normal	↓	ND	
Vitamin C	ND	E	↓	Interaction with vitamin E
Vitamin E	ND	ND	ND	Increased humoral response on supplementation

ND = No data
E = Equivocal

Table 7.4. *Effects of trace elements on immune response (from Beisel, 1982; Bhaskaram, 1988; Cunningham-Rundles, 1988; Fletcher et al.*, 1988)

	CMI	Humoral	Phagocytic	Comments/other
Iron	E	Normal	↓	Increased susceptibility to intestinal infections
Zinc	↓	↓	↓	
Magnesium	ND	↓	ND	
Calcium	ND	ND	ND	May have regulatory effects of lymphocytic and phagocytic cell functions
Copper	ND	↓	↓	
Cobalt (ionic)	ND	ND	ND	Excess of Co stimulates macrophage activity
Manganese	ND	ND	ND	Excess of Mn stimulates macrophage activity

ND = No data
E = Equivocal

although evidence for a depression in phagocytic responsiveness is equivocal. Vitamins E and C may interact in the facilitation of immune response.

Immune functions have been studied in both patients and experimental animals with varying degrees of iron-deficiency anaemia. Although most authors agree that iron deficiency produces a wide range of adverse immunological consequences, there is some disagreement about which specific immune functions are impaired (Beisel, 1982; Kuvibidila, 1987). The disagreement probably stems from a lack of control of confounding variables such as PEM and/or deficiencies of micronutrients other than iron in many of the studies. There is general agreement that iron deficiency causes increased susceptibility to intestinal infections (Jenkins *et al*, 1977); Baggs & Miller, 1973) and decreased phagocytic activity (Arbeter *et al.*, 1971), but has little or no effect on humoral activity (Chandra & Saraya, 1975). The evidence for depressed CMI due to iron deficiency is equivocal (Beisel, 1982; Bhaskaram, 1988).

Zinc deficiency has widespread effects on the immune system, including depression of CMI (Cunningham-Rundles *et al.*, 1979; Pekarek *et al.*, 1979) and of humoral (Cunningham-Rundles *et al.*, 1979; Castillo-Duran *et al.*, 1987) and phagocytic (Patrick, Golden & Golden, 1980; Briggs *et al.*, 1981) activity. Studies of humoral immunity in magnesium-deficient laboratory animals have shown reductions in humoral response (Beisel, 1982), whereas copper deficiency has been shown to cause a depression in both humoral immunity (Lukasewycz & Prohaska, 1981) and phagocytic activity (Gross & Newberne, 1980). Calcium may have regulatory effects on the lymphocytic and phagocytic cell functions, but effects of deficiency have not been shown in either man or animals (Beisel, 1982). Although there is no evidence of immune suppression as a result of either cobalt or manganese deficiencies, a moderate excess of either of these trace elements can stimulate macrophage activity (Rabinovitch & Destefano, 1973).

An excess of some nutrients may also increase susceptibility to infectious disease. The most notable of these is iron. Bacteria need adequate quantities of iron in order to achieve their full potential for growth, replication and for the production and release of exotoxins. During infection, the availability of iron for acquisition by microorganisms is limited. Iron is bound to transport proteins which withhold it from bacterial siderophores (chemicals secreted by bacteria which chelate iron in the environment and increase its availability for uptake by the bacteria). In conditions of PEM in association with iron excess, invading microorganisms are able to grow and initiate a more virulent septic process in the host than they would otherwise be able to do (Tomkins & Watson, 1989).

Malnutrition and infection as a public health problem

Table 7.5 gives the probable prevalence and extent of various types of malnutrition in the developing world (Miller, 1979). Anaemias due to iron or folic acid deficiencies and PEM are by far the most common

Table 7.5. *Probable prevalence and extent of malnutrition in the developing world (from Miller, 1979)*

	Prevalence (%)	People affected (millions)
Anaemia (iron and folate)	30	525
Undernourished (PEM)	25	434
Goitre	10	175
Dental caries	10	175
Obesity	3	52
Heart disease	2	35
Xerophthalmia	1	18

Table 7.6. *Diseases in Africa, Asia and Latin America (from Walsh & Warren, 1979).*

	Infections (1000 cases/year)	Deaths (1000 cases/year)
Diarrhoeas	3–5 000 000	5–10 000
Ascariasis	0.8–1 000 000	20
Tuberculosis	1 000 000	400
Hookworm	7–9 000	50–60
Malaria	800 000	1200
Trichuriasis	500 000	low
Amoebiasis	400 000	30
Filariasis	250 000	low
Giardiasis	200 000	very low
Schistosomiasis	200 000	500–1000
Measles	85 000	900
Polio	80 000	10–20
Whooping cough	70 000	250–450
Onchocerciasis	30 000	20–50

nutritional disorders in the developing world. They are also disorders which impinge on the normal functioning of the immune system. Xerophthalmia, due to vitamin A deficiency, has a comparatively low prevalence. However, subclinical vitamin A deficiency may cause depression of the immune system, as may subclinical deficiencies of other vitamins and trace elements, which are found in association with PEM or with anaemia.

Table 7.6 gives the probable morbidity and mortality due to infectious diseases in Africa, Asia and Latin America (Walsh & Warren, 1979). Diarrhoeas due to bacterial or viral infection have the highest prevalence, followed by infections caused by intestinal parasites, tuberculosis,

Table 7.7. *Maternal malnutrition and immune defects in intrauterine growth retarded (IUGR) animals (Neumann, 1986)*

Nutrient deficiency	Effects on progeny
Protein	Thymolymphatic athrophy
Energy	Imparied AB response to T-cell-dependent antigen
Pyridoxine	Thymic size reduction
Zinc	Cellular immunity decreased
Lipotropes	Impaired ontogenesis of IgA (intergenerational)
Choline	Thymolymphatic atrophy
Methionine	Increased deaths from infection

and malaria. Intestinal parasitic infection is a far less common cause of death than are diarrhoeas, tuberculosis and malaria. These are all diseases which interact with nutritional status, and have high public health priority in developing countries. Measles is also considered to be a high priority because of the extremely high risk of death upon infection.

Groups at risk

In developing countries, the groups at greatest risk of infection as a result of malnutrition are infants and young children. The immune system develops *in utero* and during the first few months after birth. If the infant is born prematurely or exhibits growth retardation as a result of a number of environmental factors, including maternal malnutrition or infection, then immunocompetence is reduced. Table 7.7 shows the effects of maternal malnutrition on the immune system of intrauterine growth retarded animals (Neumann, 1986). Cell mediated immunity is the system most affected and T-lymphocyte numbers are reduced (Neumann, 1986). The preterm low-birth-weight infant generally recovers its ability to mount an immune response by the age of three months. However, a small-for-gestational-age (SGA) infant may continue to show reduced cell mediated immunity for several months, or even years.

Figure 7.2 compares the morbidity of full-term infants with that of SGA low-birth-weight infants (Chandra, 1988a). The first peak of morbidity represents 70% of SGA infants, and their distribution overlaps with that of healthy full-term infants. The second peak represents the remainder of all SGA infants, who experience a three-fold higher morbidity. This latter group experience a much higher frequency of immunological abnormalities in comparison with both the other groups of SGA infants, and the control group.

Healthy babies are born with low serum levels of immunoglobulin M

Table 7.8. *Spectrum of antimicrobial activity in human colostrum and milk (Ogra, Fishaut & Theodore, 1979)*

Bacteria	Viruses	Fungi
Enterobacteriacae	Enteroviruses	*Candida*
E. coli enterotoxin	Polio 1, 2, 3	
Clostridium tetani	Coxsackie A, B	
Diphtheria	Echo 6, 9	
Streptococcus pneumoniae	Rotavirus	
Salmonella	Herpes simplex	
Staphylococcus aureus	Influenza	
Vibrio cholerae	Arboviruses	
	Simliki Forest	
	Ross River	
	Japanese B	
	Dengue	
	Rubella	
	Respiratory syncytial	

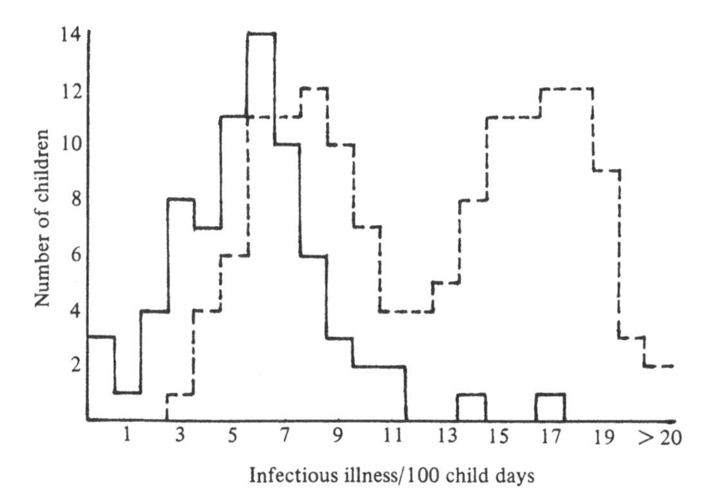

Fig. 7.2 Morbidity of healthy full-term infants (continuous line) and that of small for gestational age low birth weight infants (interrupted line) (Chandra, 1988a).

(IgM) and IgA, but they are capable of active immune responses *in utero* (Hayward, 1986). IgA in the mucous secretions of the lungs and gastro-intestinal tract may inhibit the binding of potentially pathogenic microorganisms to the mucous membrane, and thus prevent infections. Maternal IgA in breast milk is protective with a broad spectrum of antimicrobial activity (Table 7.8) (Ogra, Fishaut & Theodore, 1979). There is very little evidence that diarrhoea has any effect on the growth of exclusively breastfed infants in the Gambia (Rowland, Rowland &

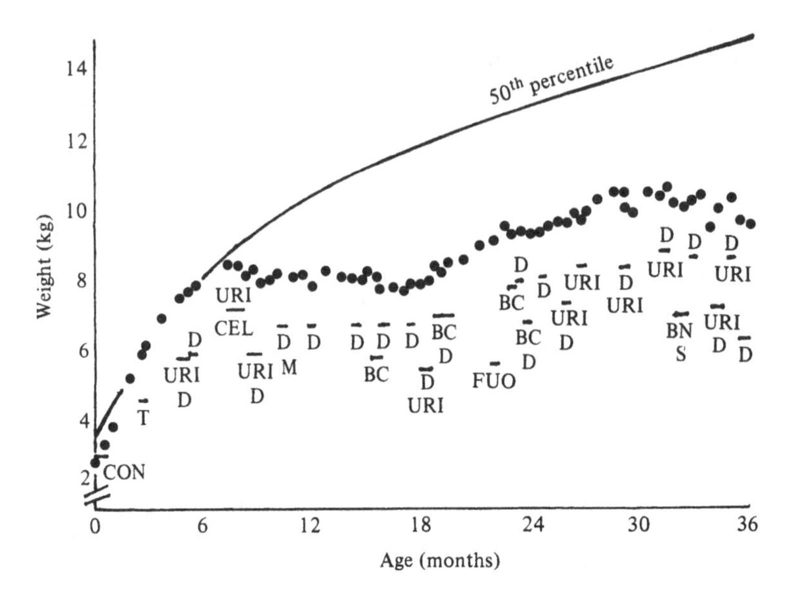

Fig. 7.3 Weight and infectious disease in one male child. The length of each horizontal line indicates the duration of infectious diseases: BC, bronchitis; BN, bronchopneumonia; CEL, cellulitis; CON, conjunctivitis; D, diarrhoea; FUO, fever of unknown origin; M, measles; S, stomatitis; T, oral thrush; URI, upper respiratory infections (from Mata, Urrutia & Lechtig, 1971).

Cole, 1988) and in Sudan (Zumrawi, Diamond & Waterlow, 1987). In developed countries some studies (Cunningham, 1977; Fergusson *et al.*, 1978; Eaton-Evans & Dugdale, 1987), but not all (Adebonojo, 1972; Gurwith, Wenman & Hinde, 1981), have shown that the incidence of diarrhoea and/or vomiting is lower in breast fed infants than in those not receiving breast milk. Lower respiratory tract infections are also reduced (Cunningham, 1977; Fergusson *et al.*, 1978), but upper respiratory tract infections are not (Holmes, Hassanein & Miller, 1983).

In developing countries, most mothers are unable to produce sufficient milk to sustain adequate child growth after six months postpartum, and supplementation will often begin around or before that time (Nabarro, 1984). Delayed supplementation or extreme lack of breast milk production will lead to growth faltering and put the child in danger of undernutrition, rendering it more susceptible to infectious diseases. Early dietary supplementation may provide adequate nutrient intake but put the child in danger of diarrhoea (Rowland & Rowland, 1986; Rowland *et al.*, 1988).

The infectious disease experience of the child after the introduction of solid food is a product of its past infectious disease experience and its past

Table 7.9. *Effects of protein–energy malnutrition on immunisation outcome*

	Vaccine	Reduced response	Country	Reference
Mild or moderate malnutrition	tetanus	no	India	1
	tetanus	no	India	2
	tetanus	no	Lebanon	3
	diphtheria	no	India	2
	diphtheria	no	Lebanon	3
	BCG	yes	Columbia	4
	BCG	yes	India	5
Severe malnutrition	tetanus	no	India	6
	tetanus	no	Lebanon	3
	diphtheria	no	Lebanon	3
	measles	yes, temporarily	Tanzania	7
	measles	yes, permanently	India	8
	typhoid	yes	India	9
	polio	yes	India	8
	BCG	yes	India	5
	BCG	yes	India	10

References:
1. Kielmann *et al.*, 1976
2. Reddy *et al.*, 1976
3. Awdeh, Kanawati & Alami, 1977
4. McMurray *et al.*, 1981
5. Ziegler & Ziegler, 1975
6. Chandra, Chandra & Gupta, 1984
7. Wesley, Coovadia & Watson, 1979
8. Chandra, 1975
9. Chandra, 1972
10. Sinha & Bang, 1976.

and present nutritional status. Fig. 7.3 shows the body weight and episodes of infectious disease in a male Guatemalan child (Mata, Urrutia & Lechtig, 1971). From birth to six months of age, the child experienced normal weight gain, and few bouts of seriously debilitating infectious disease. Between six and twelve months of age he experienced repeated bouts of diarrhoea, upper respiratory tract infections, and one episode of measles, which served to halt his weight gain, and resulted in malnutrition. From 12 to 36 months the child suffered repeated bouts of diarrhoea, upper respiratory tract infections and bronchitis.

Infection in infancy and early childhood has both immunological and nutritional consequences. Malnutrition as a result of infection causes further depression of the immune system rendering a child susceptible to further bouts of infection (Coovadia 1986; Tomkins & Watson, 1989).

Immunisation

Table 7.9 shows the effects of PEM on immunisation outcome. Mild or moderate malnutrition has no effect on specific antibody responses to tetanus or diphtheria toxoids (Kielmann *et al.*, 1976; Reddy *et al.*, 1976; Awdeh, Kanawati & Alami, 1977). However, moderately malnourished children have an impaired cellular response to BCG vaccination (Ziegler & Ziegler, 1975; McMurray, Watson & Reyes, 1981). In severe malnutrition there is depressed cell-mediated response to BCG vaccine (Sinha & Bang, 1976; Chandra, 1981) and some reduced secretory antibody response to measles (Chandra, 1975; Wesley, Coovadia & Watson, 1979), diphtheria (Awdeh *et al.*, 1977) and tetanus (Awdeh *et al.*, 1977; Chandra, Chandra & Gupta, 1984) vaccines. The antibody response to typhoid toxoid, however, still provides adequate protection despite some depression. Wesley *et al.* (1979) reported delay rather than inhibition in humoral response in severely malnourished Tanzanian children administered measles vaccine, whereas Chandra (1975) reported inhibition in response to the same vaccine in India. In general, immunisations that produce antibodies will in most cases be satisfactory, albeit sometimes delayed, whatever the nutritional status of the child, whereas immunisation responses of cellular systems are likely to be affected, even in moderate malnutrition (Tomkins, 1986).

Direct public health measures

Despite the lack of detailed understanding of the malnutrition–immunity–infectious disease complex, the present state of knowledge is adequate for some effective public health action. Tests of immunocompetence can be used as sensitive functional indices of nutritional status (Chandra & Scrimshaw, 1980). Immunisation policies have certainly been affected by our understanding of immune responsiveness in different nutritional states. On the one hand, response to prophylactic immunisation to tuberculosis and measles can be improved if nutritional support is provided before, and possibly even after, the administration of the vaccine (Chandra, 1988b). On the other hand, seroconversion response to tetanus and diphtheria toxoids and poliovirus is likely to be adequate in all but the most serious cases of malnutrition. Breast feeding confers some protection against infection and limits the young infants' exposure to pathogenic organisms.

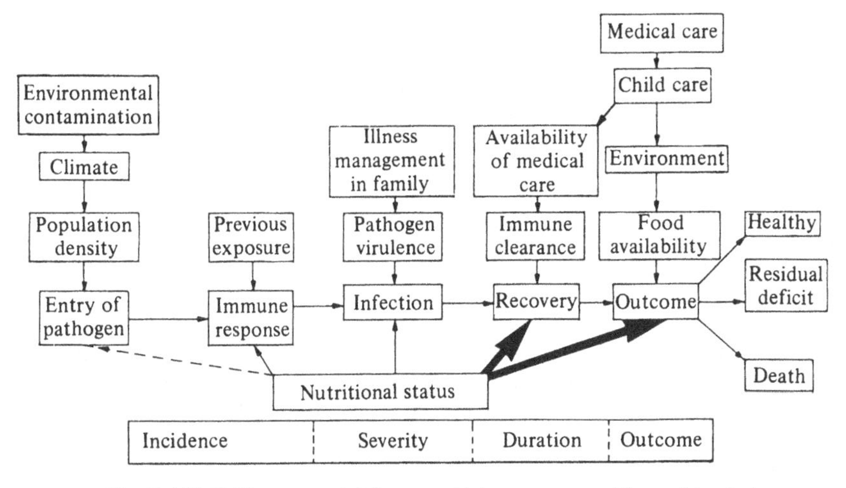

Fig. 7.4 Model incorporating factors which co-operate with nutrition in increasing the complications and mortality from infectious disease (Tomkins, 1986).

Research needed

Chandra (1988a) has outlined gaps in existing knowledge of nutritional status in relation to immunity. Priorities for research include:

(i) Greater understanding of the molecular basis of immunodeficiency in malnutrition;
(ii) Study of the influence of nutrition on antigen-non-specific mechanisms of host resistance, such as fever, production of endogenous pyrogen, interferon and mucus;
(iii) The functional implications of impaired immunity secondary to nutritional deficiency in fetal life;
(iv) Definition of clinically sufficient thresholds of malnutrition and immunodeficiency.

The malnutrition–immunity–infection complex does not exist in isolation. Cultural, behavioural, educational, economic, environmental and ecological factors all have a bearing on the prevalence and outcome of infection. Tomkins (1986) has proposed a model for the study of the interaction of malnutrition and infection which incorporates such factors (Fig. 7.4). Researchers in this area should attempt to explain why the relationship between malnutrition and infection varies between continents, communities and even households.

These aspects of research are important, and should be conducted in parallel. Chandra's recommendations, if pursued, will lead to better understanding of innate and acquired immunity in the malnourished infant and young child. This should lead to improved control and management of infectious diseases in this risk group. The use of Tomkins' model would of necessity incorporate information on innate and acquired immunity, as well as a wide range of behavioural and ecological information. An understanding of the interaction of factors leading to disease outcome should prove invaluable in the public health management of infectious diseases within communities.

References

Adebonojo, F.O. (1972). Artificial vs breast feeding. *Clinical Pediatrics*, **11**, 25–9.

Arbeter, A., Echeverri, L., Franco, D., Munson, D., Velez, H. & Vitale, J.J. (1971). Nutrition and infection. *Federation Proceedings*, **30**, 1421–8.

Awdeh, Z.L., Kanawati, A.K. & Alami, S.Y. (1977). Antibody response in marasmic children during recovery. *Acta Pediatrica Scandinavica*, **66**, 689–92.

Baggs, R.B. & Miller, S.A. (1973). Nutritional iron deficiency as a determinant of host resistance in the rat. *Journal of Nutrition*, **103**, 1554–60.

Beisel, W.R. (1982). Single nutrients and immunity. *American Journal of Clinical Nutrition*, **35**, Suppl., 417–68.

Bendich, A. (1988). Antioxidant vitamins and immune responses. In: *Nutrition and Immunology*, ed. R.K. Chandra, pp. 125–47. A.R. Liss, New York.

Bendich, A. & Cohen, M. (1988). B vitamins: effects on specific and non-specific immune responses. In: *Nutrition and Immunology*, ed. R.K. Chandra, pp. 101–23. A.R. Liss, New York.

Bhaskaram, P. (1988). Immunology of iron-deficient subjects. In: *Nutrition and Immunity*, ed. R.K. Chandra, pp. 149–68. A.R. Liss, New York.

Black, R.E., Brown, K.H. & Becker, S. (1984). Malnutrition is a determining factor in diarrheal duration, but not incidence, among young children in a longitudinal study in rural Bangladesh. *American Journal of Clinical Nutrition*, **39**, 87–94.

Briggs, W.A., Pederson, M., Mahajan, S., Sillix, D., Rabbani, P., McDonald, F. & Prasad, A. (1981). Nonuclear and polymorphonuclear cell function in zinc (ZN) deficient hemodialysis patients (HD PTS). *American Journal of Clinical Nutrition*, **34**, 628.

Castillo-Duran, C., Heresi, E., Fisberg, M. & Uauy, R. (1987). Controlled trial of zinc supplementation during recovery from malnutrition: effects on growth and immune function. *American Journal of Clinical Nutrition*, **45**, 602–8.

Chandra, R.K. (1972). Immunocompetence in undernutrition. *Journal of Pediatrics*, **81**, 1194–200.

Chandra, R.K. (1975). Reduced secretory antibody response to live attenuated measles and poliovirus vaccine in malnourished children. *British Medical Journal*, **2**, 583–5.

Chandra, R.K. (1981). Serum thymic hormone activity and cell mediated immunity in healthy neonates, preterm infants and small for gestational age infants. *Pediatrics*, **67**, 407–11.

Chandra, R.K. (1983). Nutrition immunity and infection: present knowledge and future directions. *Lancet*, **i**, 688–91.

Chandra, R.K. (1988a). Nutrition, immunity and outcome; past, present and future. Eleventh Goplan gold medal oration. *Nutrition Research*, **8**, 225–37.
Chandra, R.K. (ed.) (1988b). *Nutrition and Immunology*. A.R. Liss, New York.
Chandra, R.K., Chandra, S. & Gupta, S. (1984). Antibody affinity and immune complexes after immunisation with tetanus toxoid in protein–energy malnutrition. *American Journal of Clinical Nutrition*, **40**, 131–4.
Chandra, R.K. & Saraya, A.K. (1975). Impaired immunocompetence associated with iron deficiency. *Journal of Paediatrics*, **86**, 899–902.
Chandra, R.K. & Scrimshaw, N.S. (1980). Immunocompetence in nutritional assessment. *American Journal of Clinical Nutrition*, **33**, 2694–7.
Coovadia, H.M. (1986). Nutritional and immunological consequences of infection. In: *Proceedings of the XIII International Congress of Nutrition*, ed. T.G. Taylor & N.K. Jenkins, pp. 736–40. John Libbey, London.
Cunningham, A.S. (1977). Morbidity in breast-fed and artificially fed infants. *Journal of Pediatrics*, **90**, 726–9.
Cunningham-Rundles, S. & Cunningham-Rundles, W.F. (1988). Zinc modulation of immune response. In: *Nutrition and Immunology*, ed. R.K. Chandra, pp. 197–214. A.R. Liss, New York.
Cunningham-Rundles, W., Cunningham-Rundles, S., Garafolo, J., Iwata, T., Incefy, G., Twomy, J. & Good, R. (1979). Increased T-lymphocyte function and thymopoietin following zinc repletion in man. *Federation Proceedings*, **38**, 1222.
Eaton-Evans, J. & Dugdale, A.E. (1987). Effects of feeding and social factors on diarrhoea and vomiting in infants. *Archives of Disease in Childhood*, **62**, 445–8.
Edington, G.M. (1967). Pathology of malaria in West Africa. *British Medical Journal*, **1**, 715–18.
Fergusson, D.M., Horwood, L.J., Shannon, F.T. & Taylor, B. (1978). Infant health and breast-feeding during the first 16 weeks of life. *Australian Paediatric Journal*, **14**, 254–8.
Fletcher, M.P., Gershwin, M.E., Keen, C.L. & Hurley, L. (1988). Trace element deficiencies and immune responsiveness in humans and animal models. In: *Nutrition and Immunology*, ed. R.K. Chandra, pp. 215–39. A.R. Liss, New York.
Gershwin, M.E., Beach, R.S. & Hurley, L.S. (1985). *Nutrition and Immunity*. Academic Press, London.
Gracey, M., Cullity, G.J. & Suharjono, S. (1977). The stomach in malnutrition. *Archives of Disease in Childhood*, **52**, 325–7.
Gross, R.L. & Newberne, P.M. (1980). Role of nutrition in immunologic function. *Physiological Reviews*, **60**, 188–302.
Gurwith, M., Wenman, W. & Hinde, D. (1981). A prospective study of rotavirus infection in infants and young children. *Journal of Infectious Disease*, **144**, 218–24.
Hayward, A.R. (1986). Immunity development. In: *Human Growth*, ed. F. Falkner & J.M. Tanner, pp. 377–90. Plenum Press, London.
Hoffman-Goetz, L. (1986). Malnutrition and immunological function with special reference to cell-mediated immunity. *Yearbook of Physical Anthropology*, **29**, 139–59.
Hoffman-Goetz, L. (1988). Lymphokines and monokines in protein–energy

malnutrition. In: *Nutrition and Immunology*, ed. R.K. Chandra, pp. 9–24. A.R. Liss, New York.

Holmes, G.E., Hassanein, K.M. & Miller, H.C. (1983). Factors associated with infections among breast-fed babies and babies fed proprietary milks. *Pediatrics*, **72**, 300–6.

Jayalakshmi, V.T. & Goapan, C. (1958). Nutrition and tuberculosis. I. An epidemiological study. *Indian Journal of Medical Research*, **46**, 87–92.

Jenkins, W.M.M., McFarlane, T.W., Ferguson, M.M. & Mason, D.K. (1977). Nutritional deficiency in oral candidosis. *International Journal of Oral Surgery*, **6**, 204–10.

Kielmann, A.A., Uberoi, I.S., Chandra, R.K. & Mehra, V.L. (1976). The effect of nutrition status on immune capacity and immune responses in pre-school children in a rural community in India. *Bulletin of the World Health Organization*, **54**, 477–83.

Koster, F.T., Palmer, D.L., Chakraborty, J., Jackson, T. & Curlin, G.C. (1987). Cellular immune competence and diarrheal morbidity in malnourished Bangladeshi children: a prospective field study. *American Journal of Clinical Nutrition*, **46**, 115–20.

Kuvibidila, S. (1987). Iron deficiency, cell-mediated immunity and resistance against infections: present knowledge and controversies. *Nutrition Research*, **7**, 989–1003.

Laditan, A.A.O. & Reeds, P.J. (1976). A study of the age of onset, diet and the importance of infection in the pattern of severe protein–energy malnutrition in Ibadan, Nigeria. *British Journal of Nutrition*, **36**, 411–19.

Lukasewycz, O.A. & Prohaska, J.R. (1981). Dietary copper deficiency suppresses the immune response of C58 mice. *Federation Proceedings*, **40**, 918.

Mata, L.J. (1978). *The Children of Santa Maria Cauque: A Prospective Field Study of Health and Growth*. MIT Press, Cambridge, Mass.

Mata, L.J., Urrutia, J.J. & Lechtig, A. (1971). Infection and nutrition of children of a low socioeconomic rural community. *American Journal of Clinical Nutrition*, **24**, 249–59.

McMurray, D.N., Watson, R.R. & Reyes, M.A. (1981). Effect of renutrition of humoral and cell-mediated immunity in severely malnourished children. *American Journal of Clinical Nutrition*, **34**, 2117–26.

Miller, D.S. (1979). The prevalence of nutritional problems in the world. *Proceedings of the Nutrition Society*, **38**, 197–206.

Murray, M.J. & Murray, A.B. (1977). Starvation suppression and refeeding activation of infection. An ecological necessity? *Lancet*, **i**, 123–5.

Nabarro, D. (1984). Social, economic health and environmental determinants of nutritional status. *Food and Nutrition Bulletin*, **6**, 18–32.

Neumann, C.G. (1986). Intrauterine growth retardation and immune responses. In: *Proceedings of the XIII International Congress of Nutrition*, ed. T.G. Taylor & N.K. Jenkins, pp. 721–5. John Libbey, London.

Ogra, P.L., Fishaut, M. & Theodore, C. (1979). Immunology of breast milk: maternal neonatal interactions. In: *Human Milk: Its Biological and Social Value*, ed. S. Freier & A.I. Eidelman, pp. 115–24. Excerpta Medica, Amsterdam.

Patrick, J., Golden, B.E. & Golden, M.H.N. (1980). Leucocyte sodium transport and dietary zinc in protein–energy malnutrition. *American Journal of Clinical Nutrition*, **33**, 617–20.

Payne, P.R. (1988). Undernutrition: measurement and implications. *WIDER Seminar on Malnutrition and Poverty*, Helsinki.
Pekarek, R.S., Sandstead, H.H., Jacob, R.A. & Barcome, D.F. (1979). Abnormal cellular immune responses during acquired zinc deficiency. *American Journal of Clinical Nutrition*, **32**, 1466–71.
Rabinovitch, M. & Destefano, M.J. (1973). Macrophage spreading in vitro. II. Manganese and other metals as inducers or as co-factors for induced spreading. *Experimental Cell Research*, **79**, 423–30.
Reddy, V., Jagadeesan, V., Ragharamulu, N., Bhaskaram, C. & Srikantia, S.G. (1976). Functional significance of growth retardation in malnutrition. *American Journal of Clinical Nutrition*, **29**, 3–7.
Rojanapo, W., Lamb, A.J. & Olson, J.A. (1980). The prevalence, metabolism and migration of goblet cells in rat intestine following the induction of rapid synchronous vitamin A deficiency. *Journal of Nutrition*, **110**, 178–88.
Rowland, M.G.M., Cole, T.J. & Whitehead, R.G. (1977). A quantitative study into the role of infection in determining nutritional status in Gambian village children. *British Journal of Nutrition*, **37**, 441–50.
Rowland, M.G.M. & Rowland, S.G.J.G. (1986). Growth faltering in diarrhoea. In: *Proceedings of the XIII International Congress of Nutrition*, ed. T.G. Taylor & N.K. Jenkins, pp. 115–19. John Libbey, London.
Rowland, M.G.M., Rowland, S.G.J.G. & Cole, T.J. (1988). Impact of infection on the growth of children from 0 to 2 years in an urban West African community. *American Journal of Clinical Nutrition*, **47**, 134–8.
Scrimshaw, N.S., Taylor, C.I. & Gordon, J.E. (1968). *Interactions of Nutrition and Infection*. World Health Organization, Geneva.
Sinha, D.P. & Bang, F.B. (1976). Protein and calorie malnutrition, cell-mediated immunity, and BCG vaccination in children from rural West Bengal. *Lancet*, **ii**, 31–4.
Stephenson, L.S. (1987). *The Impact of Helminth Infections on Human Nutrition*. Taylor & Francis, London.
Suskind, R.M. (1977). *Malnutrition and the Immune Response*. Raven Press, New York.
Tomkins, A.M. (1981). Nutritional status and severity of diarrhoea among pre-school children in rural Nigeria. *Lancet*, **i**, 860–2.
Tomkins, A.M. (1986). Protein–energy malnutrition and risk of infection. *Proceedings of the Nutrition Society*, **45**, 289–304.
Tomkins, A.M., Garlick, P.J., Schofield, W.N. & Waterlow, J.C. (1983). The combined effects of infection and malnutrition on protein metabolism in children. *Clinical Science*, **65**, 313–24.
Tomkins, A.M. & Watson, F. (1989).*Interaction of Nutrition and Infection*. World Health Organization, Rome.
Walsh, J.A. & Warren, K.S. (1979). Selective primary health care. An interim strategy for disease control in developing countries. *New England Journal of Medicine*, **301**, 967–74.
Watson, R.R. & Rybski, J.A. (1988). Immunological response modification by vitamin A and other retinoids. In: *Nutrition and Immunology*, ed. R.K. Chandra, pp. 87–100. A.R. Liss, New York.
Weir, D.M. (1986). *Immunology*, fifth edition. Churchill Livingstone, Edinburgh.

Wesley, A., Coovadia, H.M. & Watson, A.R. (1979). *Transactions of the Royal Society of Tropical Medicine and Hygiene*, **73**, 719–25.

Ziegler, H.D. & Ziegler, P.B. (1975). Depression of tuberculin reaction in mild and moderate protein calorie malnourished children following BCG vaccination. *John Hopkins Medical Journal*, **137**, 59–64.

Zumrawi, F.Y., Diamond, H. & Waterlow, J.C. (1987). Effects of infection on growth in Sudanese children. *Human Nutrition: Clinical Nutrition*, **41C**, 453–61.

8 *Nutrition and illness in the aged*

H.M. HODKINSON

Major surveys of the diet of the elderly in Britain were mainly directed at excluding malnutrition as a major problem in this age group (DHSS, 1972, 1979). They found very little malnutrition, showing rather that the elderly had a diet which was very similar in composition, though somewhat smaller in total amount, to that of younger adults.

However, such surveys did uncover a more striking relationship between the general health of the subjects and dietary intake than there was between age and dietary intake (although more impaired health was found in older age groups). This can well be illustrated by Table 8.1, which derives from the findings of the first DHSS nutrition survey (1972). This studied 879 people over 65 from six separate areas of Britain who were living alone or with spouses or friends. They were not in any kind of institutional accommodation and were identified by stratified random sampling of the populations of Health Service Executive Council lists so as to achieve approximately equal numbers of men and women and of those under and over 75 years of age. The findings can thus be taken to be reasonably representative of the 'free living' elderly of Britain at that time.

As can be seen from Table 8.1, there was a strong association between a clinician's rating of the subject's general health and total calorie intake (assessed by a seven-day diary of food intake). This association was found in each of the age groups in both sexes. Despite the generally healthy nature of the survey sample, the illness effect on calorie intake is of similar magnitude to the age effect. When we turn to patients, as opposed to free living elderly subjects, the effects are correspondingly larger. Thus, for example, Morgan and his colleagues (1986) found major differences in energy intakes and in anthropometric measurements and serum albumin values between elderly women who were either active, needing to attend a day centre, geriatric day hospital patients or acute or long-stay geriatric inpatients (Table 8.2). The differences across health categories are far larger here, as would be expected, where many far more ill patients are included in the analysis.

The findings of Evans & Stock (1971) are consistent with those of

Table 8.1. *Daily energy intakes (kcal) by sex, age and health status from DHSS (1972), numbers of subjects in parentheses*

	Men		Women	
Health status	65–74	75+	65–74	75+
better than average	2411 (141)	2210 (106)	1817 (129)	1670 (92)
worse than average	2197 (67)	1918 (68)	1739 (93)	1582 (94)

Table 8.2. *Nutritional status of elderly women (from Morgan* et al., *1986)*

	N	Average age	Weight (kg)	Body fat (kg)[a]	Albumin (g/l)	Energy intake (kcal/d)
Active elderly	57	73	64	22	40	1864
Elderly day centre attenders	75	81	60	19	41	—
Day hospital patients	46	81	55	16	40	—
Long-stay patients	71	84	48	13	37	1541
Acutely ill patients	20	82	48	13	35	1269

[a]By skinfold thickness using Durnin & Womersley's (1974) equations.

Morgan *et al.* They studied patients in a geriatric department and found energy intakes to be lowest in the admission wards, somewhat better in rehabilitation wards and somewhat better again in the long-stay wards, corresponding to the degree of illness of the patients. Kemm & Allcock (1984) studied admissions to a geriatric department and confirmed that both anthropometric and biochemical nutritional parameters (albumin, retinol and retinol binding prealbumin) were often impaired and were predictors of death occurring during the hospital admission. These findings which can be taken to indicate that the sicker patients had poorer anthropometric and biochemical indices. They also found poor indices to be associated with poor appetite.

Hodkinson (1981) also found low serum albumin and thyroxine binding protein to be powerful predictors of mortality in hospital for geriatric patients. Burr, Lennings & Milbank (1982) looked at weight change longitudinally in a community sample of the elderly and found that surviving individuals declined in weight, but that in the over 70s the survivors were heavier when originally seen than those who died during the eight year period of follow-up. This is consistent with the view that lower weight characterises those subjects who were already in poorer health when originally surveyed.

So, to summarise, we can say that energy intake tends to fall with age in the absence of illness, probably in step with falling body weight and more particularly falling lean body mass. However, more striking falls in body weight are associated with illness and low body weight is an adverse prognostic feature for survival both in the long-term (Burr *et al.*, 1982) and short-term (Kemm & Allcock, 1984). Low serum proteins (albumin, prealbumin and thyroxine binding globulin) are also associated with sickness and poor short-term prognosis (Hodkinson, 1981; Kemm & Allcock, 1984), although albumin does not appear to have long term prognostic significance in community samples (Hodkinson & Exton-Smith, 1976). Thus hypoproteinaemias are probably merely markers of severity of illness, perhaps more associated with the degree of acute phase response than acting as clear cut nutritional markers (Kenny *et al.*, 1984).

The association of impaired nutritional parameters with illness is thus a complex one. Clearly anorexia, particularly that occurring in association with acute illness, is relevant. Restriction of activity with consequent fall in lean body mass may also result from chronic illness and disability. Poor nutrition may indeed favour the development of illness by its effects on body defences, although one's impression is rather that poor nutrition is more often the result rather than the cause of illness.

A further important possibility, however, is that illness may not merely lead to poorer intake of energy, but may have important energy costs.

I believe this aspect has been relatively ignored by those working with the elderly and warrants far greater attention than it has received. We need to consider many different ways in which such a relationship could come about. I will illustrate the way such relationships may have importance by results from the studies in my own department (carried out in association with Dr Andrew Tomkins and Mr Michael Cox of the Human Nutrition Department of the London School of Hygiene and Tropical Medicine). We have been able to show that acute chest infection and Parkinson's disease, both common clinical problems in old age, are associated with increased resting energy expenditure as assessed by a ventilated hood oxygen consumption technique.

Table 8.3 shows our findings for chest infection where a 23% elevation of resting energy expenditure was found – far greater than that explicable from the modest one degree elevation of body temperature. In part this energy cost may relate to the metabolic work associated with the acute phase response and other body defence mechanisms for, with the chest infection patient group, we found a significant correlation between

Table 8.3. *Effect of acute chest infection on resting energy expenditure in elderly patients*

	Infected patients ($n = 19$)	Control patients ($n = 10$)
percentage of women	79	80
average age (y)	84.9	81.3
average weight (kg)	48.8	54.9
oral temperature (°C)	37.2 ± 0.6	36.2 ± 0.5[a]
resting energy expenditure (kcal/kg)	23.3 ± 4.3	18.9 ± 3.1[b]

[a] $p < 0.001$
[b] $p < 0.01$

resting metabolism corrected for body weight and serum C-reactive protein ($r_s = 0.86$, $p < 0.05$), one of the acute phase proteins. The extra energy expenditure during infection is particularly significant given that food intake may be substantially reduced at this time. There is an important possibility that a nutritional deficit during the acute period of illness may be relevant to subsequent recuperation and recovery.

Similarly, we have been able to demonstrate that treated patients with Parkinson's disease nonetheless have elevated resting metabolic rates (+17%) in comparison with elderly controls and that elevation can be related both to muscle rigidity and to involuntary movements, both of which must clearly require energy expenditure. Our Parkinsonian patients were 6 kg lighter than control patients, indicating that extra energy costs have a practical significance.

Conclusion

There are complex interactions between illness and nutrition in old age. In these interactions, the possible importance of energy costs of both acute and chronic illness have yet to be properly explored and may emerge as having considerable practical importance.

References

Burr, M.L., Lennings, C.I. & Milbank, J.E. (1982). The prognostic significance of weight and of Vitamin C status in the elderly. *Age and Ageing*, **11**, 249–55.

DHSS (1979). *A Nutrition Survey of the Elderly*. Reports on Health & Social Subjects No. 3. HMSO, London.

DHSS (1979). *Nutrition and Health in Old Age*. Reports on Health & Social Subjects No. 16. HMSO, London.

Durnin, J.V.G.A. & Womersley, J. (1974). Body fat assessed from total body density and its estimation from skinfold thickness: measurements on 481 men and women aged from 16 to 72 years. *British Journal of Nutrition*, **32**, 77–97.

Evans, E. & Stock, A.L. (1971). Dietary intakes of geriatric patients in hospital. *Nutrition and Metabolism*, **13**, 21–35.

Hodkinson, H.M. (1981). The value of admission profile tests for prognosis in elderly patients. *Journal of the American Geriatrics Society*, **29**, 206–10.

Hodkinson, H.M. & Exton-Smith, A.N. (1976). Factors predicting mortality in the elderly in the community. *Age and Ageing*, **5**, 110–15.

Kemm, J.R. & Allcock, J. (1984). The distribution of supposed indicators of nutritional status in elderly patients. *Age and Ageing*, **13**, 21–8.

Kenny, R.A., Hodkinson, H.M., Cox, M.L., Caspi, D. & Pepys, M.B. (1984). Acute phase protein response in elderly patients. *Age and Ageing*, **13**, 89–94.

Morgan, D.B., Newton, H.M.V., Schorah, C.J., Jewitt, M.A., Hancock, M.P. & Hullin, R.P. (1986). Abnormal indices of nutrition in the elderly: a study of different clinical groups. *Age and Ageing*, **15**, 65–76.

9 *Growth and growth charts in the assessment of pre-school nutritional status*

JERE D. HAAS AND JEAN-PIERRE HABICHT

Introduction

For many years measures of child growth have been used to assess health and nutritional status under very different circumstances and for very different purposes. The premise for the use of growth as an indicator of health and nutritional status has remained unchanged: 'poor' growth performance reflects, for the majority of children who are so designated, deviations from optimal environmental conditions that support growth. The major environmental conditions which are of interest to biomedical researchers, medical and health care practitioners, and health planners in less developed countries are infectious disease and inadequate food and nutrients. To be more effective in dealing with these public health problems, we need to clarify the use of growth indicators with regard to: (i) screening individual children for treatment or participation in intervention programs, (ii) identifying subpopulations for targeting interventions and resources, and (iii) conducting research in the clinical setting, the community setting or the larger population. All these uses of growth performance have different requirements for the type of measures used, the quality of the measurements and the analytical techniques applied to the data collected. In this paper we will deal with two of the principle applications of growth for the assessment of nutritional status. These are (i) the screening of individual children to identify the poorly nourished, and (ii) the use of growth to quantify malnutrition in a population.

This review will focus on several issues related to the use of child growth in the assessment of nutritional status. Specific attention will be given to the use of growth reference charts in the assessment process, since assessment of 'poor' growth performance implies comparison to 'normal' growth inferred by the growth charts.

Screening individual children

The most common use of growth charts is to identify specific children whose achieved growth or growth pattern is inadequate, indicating that they are likely to benefit from a nutritional or health intervention. The notion of benefit in this context implies that small children are at risk of diminished functional capacity (e.g. immunocompetence, mental development) which would improve following appropriate treatment. However, even if there were no deficit in function, a child who is stunted or wasted would grow better if he had more food and better health care, and access to such levels of food and health care are a matter of equity.

Through screening it is assumed that interventions are more effectively targeted, or that more expensive or invasive diagnostic procedures and treatments are reserved for those who truly need the additional attention. If this assumption is wrong, screening is not a useful exercise. Examples of this are abundant. For example, past malnutrition will lead to stunting or short stature which, if diagnosed later in life, may not be corrected fully by any interventions. Another example is that children who are small for genetic reasons will exist in all populations and no nutritional intervention will improve their growth. Also children diagnosed as malnourished because they are light and wasted will not respond to nutritional intervention if their reduced growth is due to tuberculosis.

Another situation is the identification through anthropometric screening of children who are starving, and giving their mothers nutrition education which is irrelevant because the family has no resources to apply the new nutrition knowledge. Inappropriate and unavailable interventions are one of the major barriers to successful growth monitoring, a topic which we will discuss later.

General principles

When using growth as a screening tool one must consider the potential problems associated with misclassification of children and its cost effectiveness in the program in which it is used. Misclassification concerns the failure of a growth indicator to distinguish a well-nourished from a malnourished child, using a specific level or cutoff of achieved growth. No diagnostic test using growth as an indicator of malnutrition exactly reflects the true reality of concern. Thus, as shown in Table 9.1, some misclassification will occur whereby malnourished children are diagnosed as well-nourished (false negatives) and well-nourished children are diagnosed as malnourished (false positives). The correctly diagnosed are true positives (well nourished) and true negatives (malnourished).

Table 9.1. *Relationships between the malnourished and their diagnoses (adapted from Habicht, Meyers & Brownie, 1982)*

		Malnourished	
		Yes	No
Diagnosis of malnourishment	Yes	TP	FP
	No	FN	TN

$$\text{Sensitivity (Se)} = \text{TP}/(\text{TP}+\text{FN})$$
$$\text{Specificity (Sp)} = \text{TN}/(\text{FP}+\text{TN})$$
$$\text{Positive predictive value } (V+) = \text{TP}/(\text{TP}+\text{FP})$$
$$= 1/[1+(1-\text{Sp})(1-P)/(P\text{Sc})]$$
$$\text{Measured prevalence } (p) = (\text{TP}+\text{FP})/\text{TP}+\text{FP}+\text{FN}+\text{TN})$$
$$= P\text{Se}+(1-P)(1-\text{Sp})$$
$$\text{True prevalence } (P) = (\text{TP}+\text{FN})/(\text{TP}+\text{FP}+\text{FN}+\text{TN})$$
$$= (p+\text{Sp}-1)/(\text{Se}+\text{Sp}-1)$$

where TP = true positive diagnosis; FP = false positive; FN = false negative; TN = true negative.

Choosing the best cutoff value for an indicator

In efforts to select the best diagnostic tests it might seem logical that the cutoff point with the least misclassification (least false negatives and false positives) would be the most appropriate. Unfortunately, as one changes the cutoff point over the useful range for screening to decrease the number of false positives who do not need treatment, one inevitably increases the number of false negatives who will not get the treatment (Galen & Gambino, 1975). Therefore, choosing a cutoff point results in a tradeoff between treating children who don't need it and missing children who do. The ratio of children who receive the treatment over all those who need the treatment is called the sensitivity (see Table 9.1) and determines coverage. The ratio of children who need the treatment to all those diagnosed for treatment is called the positive predictive value and determines the cost-effectiveness of treatment.

It would appear logical that once the positive predictive value of a cutoff point has been determined for a population, this cutoff point will deliver the same positive predictive value in another population. Unfortunately, this is not true. The positive predictive value varies with variation in the true prevalence of malnutrition. More stable characteristics of the classification scheme are the sensitivity, or the ratio of those classified as malnourished who are actually malnourished, and the specificity, or the ratio of those classified as well-nourished who are actually well-nourished (see Table 9.1).

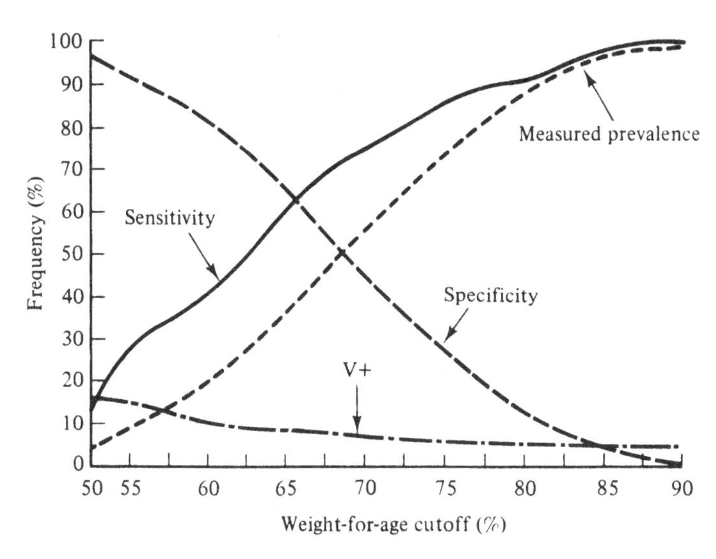

Fig. 9.1 Percentage frequency distribution of sensitivity, specificity and positive predictive value (V+) at different cutoff values of weight-for-age to predict death during a 24-month demographic surveillance following measurement of weight. See Table 9.1 for computation of values. The death rate was 5%. The weight reference standard is from NCHS (Hamill *et al.*, 1977). (Data from Cogill, 1982.)

Fig. 9.1 demonstrates that changing cutoff points results in similar tradeoffs for sensitivity and specificity because false negatives and false positives are the respective numerators for these ratios. An increase in the positive predictive value and therefore an increase in the cost effectiveness of treatment requires an increased specificity which results in a decreased sensitivity and thus poor coverage. Therefore, for a stable prevalence, there is an unavoidable trade-off between coverage and cost-effectiveness. As indicated above, an increase in the true prevalence of malnutrition will result in an increase in the positive predictive value. Thus, for the positive predictive value to be equal at high and low prevalences (as would occur with seasonal changes in food supply or in response to intervention), the specificity at low prevalence would have to be greater than the specificity at high prevalence. This could come about if the cutoff point is lowered, which would also lead to a reduction in sensitivity. As prevalence falls the appropriate cutoff point must be set lower for equal positive predictive values. Often the cutoff points fall to a level on the growth charts where percentile estimates are imprecise. Improving the precision of these estimates of extreme percentiles to accommodate screening in these low prevalence populations would require enormous sample sizes in constructing the reference standards.

As the cutoff point is lowered, the delivery of the intervention becomes more cost-effective. Fortunately, countries which have a low true prevalence of malnutrition can afford the higher cost of the intervention which results from raising the cutoff and lowering the positive predictive value. Poor countries should use a cutoff which selects the number of children which most closely approximates the number for which treatment resources are available (Habicht, 1980). That selection of a cutoff point can be made as easily on the basis of an international reference standard (like the 1983 WHO standard), as any other standard. However, one needs to avoid prescribing various cutoff points for various purposes, such as one for the rainy season and one for the dry season. Too many options would just cause confusion in primary health care programs.

Because sensitivity and specificity are more stable than is the positive predictive value under conditions of changing prevalence, they are the criteria one should use to choose the best indicator of malnutrition.

Choosing the best indicator

The validity of an anthropometric indicator in identifying the malnourished can be evaluated in several ways. Of primary interest in this evaluation is the nature of the nutritional reality one wishes the indicator to reflect. Habicht *et al.* (1982) described three such realities, (i) prediction of risk, (ii) prediction of benefit, and (iii) normative.

The prediction of risk indicator is based on the consequence or functional outcome (impaired performance) associated with the indicator. The inverse relation of weight-for-age to mortality risk as shown in Fig. 9.2 from a study in Bangladesh (Chen *et al.*, 1980) is an example of this reality. The second reality is the prediction of the results of a direct nutritional intervention. This reality is seen in nutrient intervention trials whereby an indicator of nutritional status such as hemoglobin predicts response to iron supplementation, as shown in Fig. 9.3 for a study in Ecuador (Freire, 1982).

A third reality is one of normality whereby the 'abnormal' individuals fall outside of the normal healthy population distribution of the indicator. It is this last reality that underlies the use of growth reference charts. For example the WHO (1983) growth curves are derived from a presumably healthy, well-nourished population. If one uses this reference curve, it is only possible to classify a child relative to some percentile of a healthy population distribution or percentage deviation from its mean; therefore, one can only assess specificity as shown in Fig. 9.4. The use of this standard does not allow an estimate of the child's likelihood of being unhealthy or malnourished. To be of use in public health work one

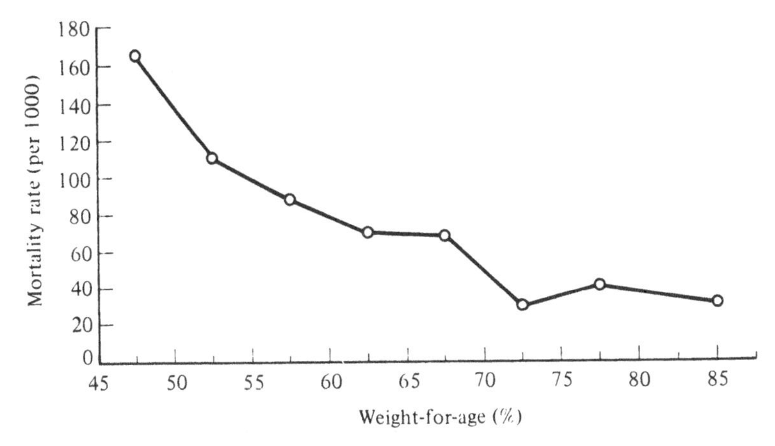

Fig. 9.2 The relation of early childhood mortality (over 24 months) and weight-for-age in Bangladeshi children. (Redrawn from Chen, Chowdhury & Huffman, 1980.)

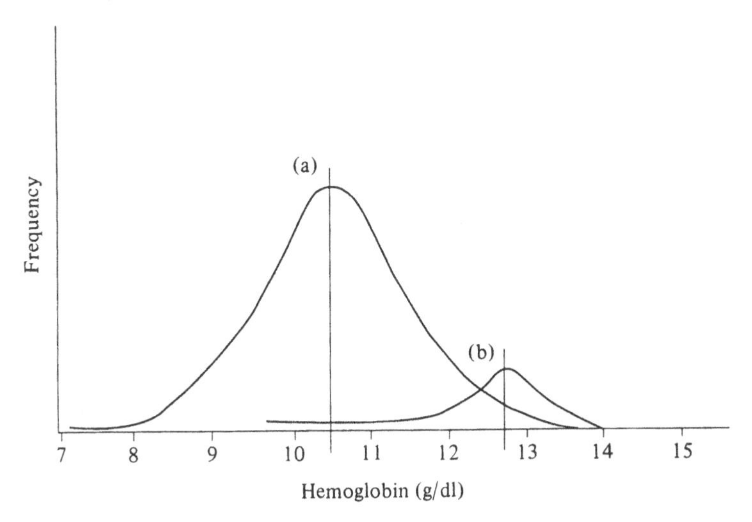

Fig. 9.3 Distribution of (a) initial hemoglobin values in iron deficient subjects and (b) initial hemoglobin values in non-iron deficient subjects, as judged by response to an iron supplementation trial in Ecuadorian school girls. (From Freire, 1982.)

needs to know the sensitivity distribution as well. The normative standards are therefore incomplete and of little theoretical meaning for public health. The predictive realities of risk (functional consequence) and of benefit (nutritional determinant) permit the sensitivity distribution to be estimated and serve as the basis for selecting among several potential indicators of nutritional status. They also permit a normal or

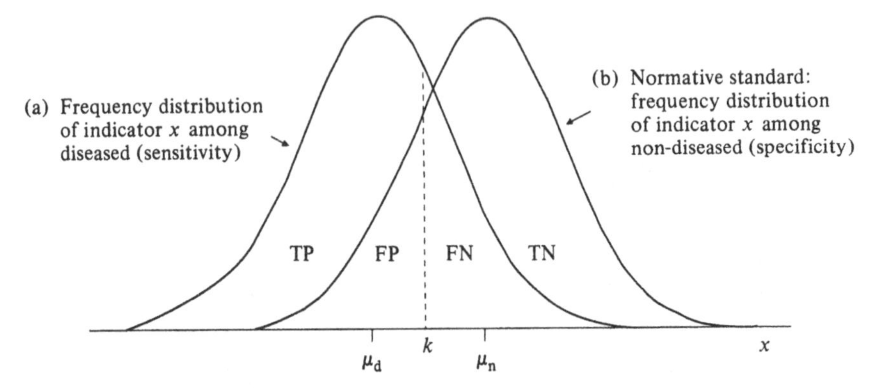

Fig. 9.4 Distribution of a measurement (x) among subpopulations of those diagnosed as diseased (a) and of those considered normal or non-diseased (b). An arbitrary cutoff value of x is indicated by k. Without knowledge of the sensitivity distribution (a) representing the diseased subpopulation, one cannot determine the number of false negatives or false positives as indicated in Table 9.1. All normative growth curves currently in common use are based on knowledge of the specificity distribution (b) of the non-diseased subpopulation.

specificity population to be identified as the basis for defining a reference population standard and appropriate screening cutoff values based on functional criteria.

The interpretation of cross-sectional data

Most screening of children is based on a single observation where some aspect of attained size is interpreted. There are several issues in the interpretation of single observation or cross-sectional data that deserve some discussion. They deal with the choice of cutoff points, the choice of indicators, the choice of reference standards, the statistical expression of data and several operational issues.

Choice of cutoff points and indicators

The choice of a cutoff point for screening in terms of the criteria one might use has been addressed briefly above. A major concern regarding consistency in cutoff points across populations or among various indicators is the need to maintain similar positive predictive values. However, this is impossible if there is much variation in prevalence, because prevalence influences predictive values. Some users may wish to achieve a high specificity, perhaps 95% of the true negatives identified, and then compare indicators according to their sensitivities. This approach also tends to identify cutoff points with higher positive predictive values. An easy approach, although flawed, is to use the maximum sum of sensitivity

Table 9.2. *Cutoff values (% of NCHS median) for various indicators to best predict death (from three studies). The 'best' cutoff is based on maximum sum of sensitivity and specificity*

Population (reference)	Age (mo.)	Wt–Age (%)	Ht–Age (%)	Wt–Ht (%)	AC–Age (%)	AC–Ht (%)
Bangladesh (Cogill, 1982)	12–24	65.0	85.0	72.5	75.0	77.5
Indonesia (Jahari, 1982)	12–60	67.5	87.5	87.5	82.5	85.5
Tanzania (Yambi, 1988)	6–30	80.2	91.7	90.1	—	—

Wt = weight, Ht = height or recumbent length, AC = arm circumference.

Table 9.3. *Misclassification criteria for three common indicators of nutritional status at fixed cutoff values of NCHS median (three studies, sexes combined)*

Population	P	p	Se	Sp	Se + Sp	$V+$
	Weight-for-age, cutoff = 80%					
Bangladesh	0.05	0.88	0.90	0.12	1.02	0.05
Indonesia	0.06	0.79	0.96	0.22	1.19	0.07
Tanzania	0.04	0.37	0.59	0.64	1.23	0.06
	Height-for-age, cutoff = 90%					
Bangladesh	0.05	0.61	0.79	0.40	1.19	0.06
Indonesia	0.06	0.61	0.79	0.40	1.19	0.06
Tanzania	0.04	0.27	0.50	0.74	1.24	0.07
	Weight-for-height, cutoff = 80%					
Bangladesh	0.05	0.24	0.30	0.76	1.05	0.06
Indonesia	0.06	0.23	0.38	0.78	1.16	0.09
Tanzania	0.04	0.08	0.10	0.92	1.02	0.05

P = True prevalence of sensitivity population, those who died within 1–2 years of measurement.
p = Measured prevalance or the percentage who fall below a specified anthropometric cutoff value.
Se = sensitivity, Sp = specificity, $V+$ = positive predictive value.

plus specificity as a criterion for general screening. More correct, albeit complicated, procedures can be found in Brownie & Habicht (1984).

Table 9.2 presents data on cutoff points for various anthropometric indicators of nutritional status from three different studies. Each study employed a similar approach whereby the sensitivity distribution of the indicator was based on the subpopulation of children who died during a specified period of time after the measurements were taken, and the specificity distribution was based on survivors. It is clear that within each column representing a different indicator there is considerable variation

Table 9.4. *Rank ordering (1 = best) of various anthropometric indicators based on maximum sum of Se + Sp in three populations (sexes combined). Numbers in brackets are ranking excluding arm circumference*

Indicator	Population Bangladesh ranking of indicator	Indonesia	Tanzania
Weight-for-age	2 (2)	3 (1)	(1)
Height-for-age	1 (1)	4 (2)	(2)
Weight-for-height	5 (3)	5 (3)	(3)
Arm circ.-for-age	3	1	—
Arm circ.-for-height	4	2	—

in the value of the best cutoff point, if the maximum sum of sensitivity plus specificity is the criterion for selecting the cutoff point. The values for Bangladesh and Indonesia are more similar to each other and lower than the values for Tanzania for weight-for-age and height-for-age. However, Bangladesh differs considerably from Indonesia in the cutoff points for weight-for-height and for the two indicators using arm circumference.

One can examine indicators in these same three populations slightly differently by comparing the three most common indicators at the cutoff points which are often recommended in the literature. This comparison is shown in Table 9.3 where true and measured prevalence, sensitivity, specificity, their sum, and positive predictive value are presented. In all three populations true prevalence is low, with 4–6% mortality observed. All of the anthropometric indicators in each population greatly overestimated the measured prevalence of those who would die. Weight-for-age at 80% of the WHO reference and height-for-age at 90% of reference are close to the 'best' determined cutoffs in Tanzania (see Table 2). None of these common cutoff values perform well if one examines their positive predictive values, which are uniformly low.

One can also examine the rankings of indicators for each population, where the best indicator is the one with the highest sum of sensitivity plus specificity as shown in Table 9.4. Here the three studies yield more consistent results, at least among the three more common indicators: weight-for-age, height-for-age and weight-for-height. There is relatively little difference among the three indicators in each study with regard to the maximum sum of Se+Sp. However, as an indicator of prospective risk of death, weight-for-age and height-for-age are better than weight-for-height. Arm circumference indicators were superior to all weight and height indicators in Indonesia but not in Bangladesh.

The explanation for these variable findings among three different populations requires further investigation. Ages varied somewhat in the three studies. The Bangladesh study had a narrower age range with a two year period of demographic surveillance. The Indonesian and Tanzanian studies monitored deaths for only one year. Mortality rates over the observation period are comparable. Causes of death are similar in Tanzania and Bangladesh, with diarrhoea the leading cause at 45 and 57% respectively. Diarrhoea at 30% was second to convulsions and fever at 62% in Indonesia as the major cause of death. Measles and fever also accounted for a large number of deaths in Tanzania and Bangladesh. Primary health care is likely to have differed across the groups. It is likely that these small but measurable differences in characteristics of the three study populations contributed to the differences in performance of the various indicators of nutritional status. It would be useful to examine these indicators related to mortality in other populations, where environmental and perhaps genetic factors differ from the three cases presented. Also, it would be important to study the indicators as they relate to other less severe outcomes, such as morbidity, immunocompetence or mental development. However, none of these outcomes is as easily measured as mortality and thus create problems in defining the indicator's distribution in the sensitivity population.

Choice of growth reference standard

Another issue which has received considerable attention is the choice of a single growth reference standard to assess all children. We do not propose to present a comprehensive statement on the issue here. However, a few comments regarding the application of sensitivity and specificity analysis to the reference standards is appropriate. There is considerable information to support the contention that healthy, well-off populations of different ethnic and genetic backgrounds have similar achieved growth during the pre-school period (Habicht *et al.*, 1974; Martorell, 1984). This can be seen in Fig. 9.5 where the mean values for healthy, well-nourished children representing the specificity distributions are fairly similar across very different population groups. The main exceptions are the six Asian population groups which have mean heights around the twenty-fifth percentile of the WHO reference standard. The consistency of this observation across all the Asian groups warrants further study to determine whether the smaller stature is related to environmental or genetic characteristics of the population. This is generally all the information we have to make an evaluation of the use of

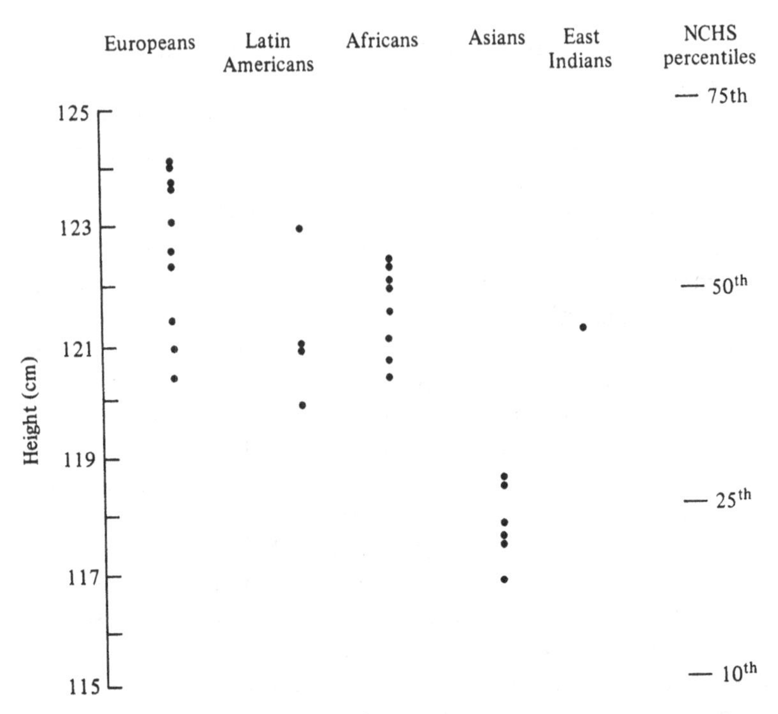

Fig. 9.5 Mean heights of selected samples of well-off seven-year-old boys of various ethnic origins. The NCHS (WHO) percentiles represent predominantly white children from the US where the fiftieth percentile is the same as the mean. (From Martorell, 1984.)

local versus international references. Imprecision due to differences among these groups pales in comparison to our lack of information on the sensitivity distribution. We have few studies employing similar sampling and analytical techniques to allow sensitivity distributions, based on a common outcome like death, to be constructed for different populations. Moreover, there is practically no information about relative prevalences of a deleterious outcome at different ages. Without this information choice of cutoff points, which is more important than choice of the reference standard, is arbitrary.

Expression of data

All of the examples presented thus far have used the percentage of reference median to express the data. Other expressions that take account of the age effect on growth, or the height effect on weight, use percentiles and Z-scores (deviations from the mean in the standard deviation units). Inconsistencies exist in the comparisons of percentage of reference median with percentiles and Z-scores. This is apparent for

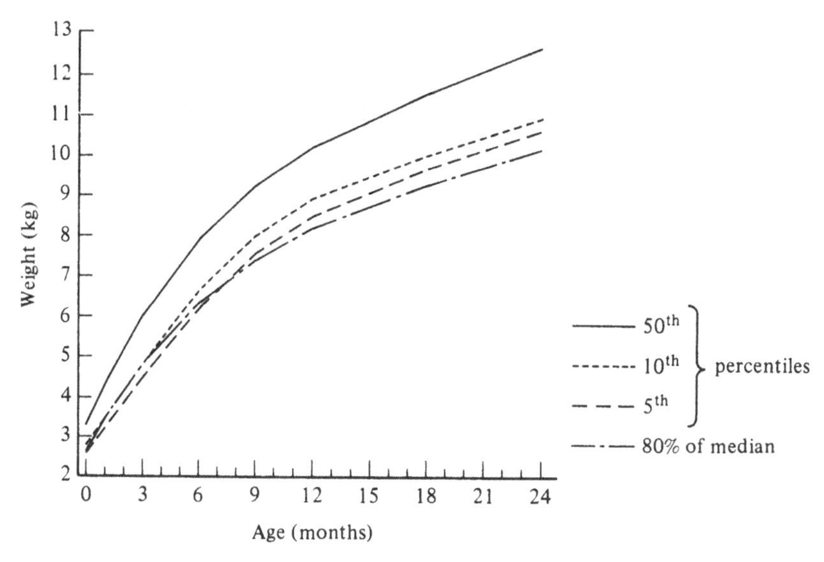

Fig. 9.6 Comparison of a percentage of median with WHO percentiles for weight of boys. The 80% of median weight is not at the same percentile for all ages.

children under six months compared with older children, as seen in Fig. 9.6. Whereas 80% of median weight-for-age is about the tenth percentile for children under four months, it is close to the fifth percentile at six months and below the third percentile after one year. Thus, compared to a statistical distribution, younger children at 80% of median are better off than older children at 80% of median. This is because the coefficient of variation of weight is 30–40% greater in younger than older children.

Other problems make the choice of method for expressing data less important than one might expect. Because the precision of the cutoff values to predict need for treatment is so poor, the small differences between percentage of reference median, percentiles and *Z*-scores are immaterial for screening purposes. Also, *Z*-scores are flawed conceptually unless a perfect Gaussian distribution is assumed, and percentiles are generally useless because in both the scarce and generous resource situations cited above, the cutoffs are usually so low (for different reasons) that percentiles cannot be estimated with any precision even if they were the appropriate cutoffs, which they are not. Thus, there is little justification to choose one method over another for screening purposes; however, the choice may be important for standardization across studies, age groups and indicators in order to permit valid comparisons.

From a practical point of view, one should just use whatever one has been using in the past. This is because the decision as to what cutoff value to use in screening individuals is complex and depends on the resources available to treat the malnourished, the prevalence of malnutrition in the population and the sensitivity, specificity and positive predictive value of the indicators of nutritional status (Habicht *et al.*, 1982). With scarce resources it is necessary to choose a cutoff that selects exactly as many children as there are resources to treat them. The most cost-effective cutoff is the one with the highest positive predictive value, or the highest probability of benefiting those who are diagnosed as malnourished. If resources are plentiful, then choosing a cutoff with a high sensitivity will allow the intervention to reach the most number of people who will benefit, but may be at high cost due to treating those who will not benefit (false positive). If costs are fixed per recipient of an intervention, then the overall cost per child benefited will depend only on the positive predictive value, and that can be obtained as easily by using international standards as by using local standards. The choice as to whether to express these cutoff points as percentiles, *Z*-scores or percentages of reference median again should be made for practical as opposed to theoretical reasons.

Operational issues

Finally, there are several operational issues regarding the use of growth charts that deserve brief comment. Of importance are issues related to consistency and measurement error. Consistency problems arise in the choice of charts that may be very different even among users within the same health agency and, more commonly, across agencies. This creates obvious confusion which should be avoided. Also, inconsistencies within the charts create problems. This is probably best seen in the use of both recumbent length and standing height in the two to three year old references from the NCHS (Fig. 9.7), and the use of data from various sources to construct the references. Dibley *et al.* (1987a, b) recently reviewed the development of these charts and the implications of these inconsistencies.

Measurement error is more complicated to understand and to resolve. There are errors in the data that were used to construct the reference charts; these are probably inconsequential. There are greater problems with both random and systematic measurement errors among the anthropometrists in the field who do the actual screening. Better training of health workers would reduce these errors somewhat. Errors in age estimation can be very serious when weight-for-age and height-for-age are used as indicators. Verification of birth dates and more diligent questioning of parents or care-givers could reduce these errors somewhat. Knowledge of the degree of measurement error and age bias

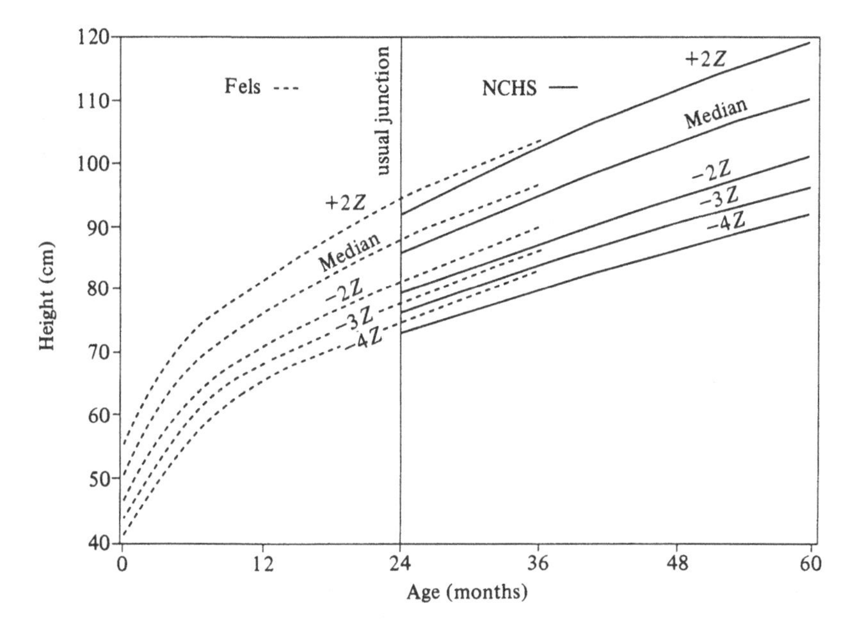

Fig. 9.7 Height-for-age *Z*-scores for the two samples (Fels and NCHS) of pre-school age boys used to construct the WHO reference curves; the Fels sample is a longitudinal series of rural Ohio children, the NCHS sample is cross-sectional from many settings in the US. (From Dibley *et al.*, 1987b).

permits correction of estimates of prevalence in population level analysis. This will be discussed later.

The interpretation of longitudinal growth data

Growth monitoring

Screening with longitudinal growth information is generally carried out with programs devoted to 'growth monitoring.' In these programs children are weighed and measured periodically, usually monthly or quarterly, and their growth is plotted on a chart. The growth trend is interpreted by a health worker who identifies children whose growth may 'falter' or deviate from a prescribed pattern, and treatment or another intervention is initiated just as in a cross-sectional, single-time screening. Everything mentioned above related to cross-sectional screening and the lack of useful information to define normal and abnormal achieved growth applies to the assessment of longitudinal growth patterns. In the case of longitudinal growth we have far less information on the specificity distribution than for achieved growth. Moreover, the sensitivity distributions are less understood and the relevant prevalences are completely unknown. There is a lot of room for research here.

One thing, however, is clear. Growth monitoring does not fail because of the lack of information on sensitivity, specificity, prevalence and biological interpretation of the growth data. Imperfect standards, imprecise measurements and inadequate conversion of the information to screening decisions pale in comparison to the inadequacies of other prerequisites of useful growth monitoring. These inadequacies include: (i) inappropriate or lack of intervention resources, (ii) inadequate training to apply the resources, and (iii) inadequate follow-up of targeted children. Until these problems can be addressed, the research on indicators for screening that use longitudinal growth data is mostly academic.

Identifying growth faltering and resumption of adequate growth in a population

There has been considerable concern recently over the validity of existing growth standards when used to identify individuals and groups of children whose growth is 'faltering' during the first year of life. The major issue is whether western growth standards derived from primarily bottle-fed children are appropriate for examining the growth trends associated with variation in feeding practices of primarily breast-fed infants from developing countries (Whitehead & Paul, 1984). As these authors and others (Fomon *et al.*, 1971; Persson, 1985; Chandra, 1982) have shown for developed countries, bottle-fed infants have lower weight gains than infants who are exclusively breast-fed for the first three months of life, but the trend reverses for the next three months in favor of faster growth for the bottle-fed infants, presumably due to greater energy intakes. Whitehead and Paul argued that if one wishes to observe when infant feeding practices become abnormal in a population that usually breast-feeds, such as found in most of the less developed countries, use of western growth reference standards to identify growth faltering will result in premature action to change feeding patterns. They called for reference standards based on infants whose feeding patterns reflect contemporary and locally appropriate breast-feeding practices.

Fundamental to the diagnosis of growth faltering is the method used to define it. For individuals it may be possible to apply the techniques used in growth monitoring. Implicit in any definition of faltering is a measurable deterioration of growth status through serial measurements. Thus growth velocity is compared to some reference. Tanner (1986) discussed the use of velocity growth charts for individual diagnosis, but concentrated on their application during adolescence, after the early ages when

Table 9.5. *Effect of changing age on apparent catch-up growth*

Child at 12 months weighs 7.35 kg

thus, $\frac{7.35\ \text{kg} \times 100}{10.15\ \text{kg}} = 72.4\%$ of reference median, this is a 2.8 kg difference.

Same child at 24 months weighs 9.79 kg or a gain of 2.44 kg

thus, $\frac{9.79\ \text{kg} \times 100}{12.59\ \text{kg}} = 77.8\%$ of reference median, also a 2.8 kg difference.

Note that the reference in weight also changes by 2.44 kg in same period.

growth faltering is commonly defined. Other approaches rely upon the change in a child's age adjusted weight or length expressed as a percentage of a standard, *Z*-score or percentile ranking. In all of these approaches there are fundamental questions regarding such issues as: (i) choosing the best indicator of growth faltering, considering length of monitoring and sensitivity of various velocity measurements; (ii) allowing for the effect of variable time intervals between measurements, especially since normal growth rates change with age during the period when most early faltering is observed; (iii) determining an acceptable deviation from normal growth velocity; which may include (iv) taking into account initial size differences between children who otherwise have similar rates of growth; and (v) determining the most valid way of standardizing longitudinal growth data.

These and many other issues in growth faltering have been unstudied. The last item, for example, can be resolved in part by recognizing the spurious apparent catch-up growth that occurs when one uses percent of an age-specific anthropometric measure as is common in nutritional assessment. Because the denominator in these calculations (Table 9.5) is a reference measurement such as median weight, it will increase with age. If a child's absolute deficiency in size remains unchanged over time, the numerator (achieved size) and the denominator (reference size) will increase at an equal rate and the ratio or the percentage of standard will increase, indicating spurious catch-up growth.

Catch-up growth following treatment or intervention suffers from similar problems of definition to those of growth faltering. Since it is impossible to know what an individual's growth potential really is, it is difficult to determine if complete catch-up has occurred. One could use such criteria as achievement of a percentile level similar to a child's presumed pre-stress level, or deceleration of growth to a stable 'normal' growth rate. However, these conditions assume that pre-stress and post-catch-up environmental conditions are fully supportive of optimal growth – a situation that is unlikely to occur in less developed countries and among the poor of some more developed countries. Some of these problems, can be resolved in population studies where an appropriate reference standard or control group can be identified and a pattern of optimal growth inferred.

Quantifying malnutrition in a population

Information on the growth and nutritional status of individual children is often analysed for population characteristics. While many of the previously discussed issues on individual screening for nutritional status are relevant for quantifying malnutrition in populations, this level of analysis has some unique problems which relate to the use of growth charts to estimate population prevalence of malnutrition, and in the comparison of nutritional status within and across populations.

Estimating prevalence of malnutrition

Malnutrition based on growth is quantified in a population using either the prevalence of individuals falling outside a specified 'normal' range, or the population mean for a specific anthropometric measure of growth. Prevalence estimates are often arbitrary given the unknown proportion of false positives and negatives resulting from the use of the anthropometric cutoff values to distinguish the malnourished from the well-nourished. The obvious solution to improve the estimate of prevalence is to subtract false positives and add false negatives; this is possible if one knows the sensitivity and specificity of the cutoff points.

Other approaches which examine the characteristics of the distribution of growth measures in a population are promising, but to date have been underutilized for anthropometry and growth. The method of distributional analysis has been applied to hemoglobin values to estimate the prevalence of anemia in various populations (Meyers *et al.*, 1983; Tufts *et al.*, 1985). This approach assumes that the distribution of values for the normal healthy segment of a population can be normalized to meet Gaussian criteria. Untransformed values for hemoglobin and probably for stature fit such a Gaussian distribution, and when plotted on probability paper the cumulative frequency represents a straight line as shown in Fig. 9.8. If the population is represented by two subpopulations, a large normal and healthy one and another small population of malnourished individuals, the cumulative probability plot will be expressed as a straight line, which represents the normal children, and a curved 'tail' which represents the malnourished plus some normal

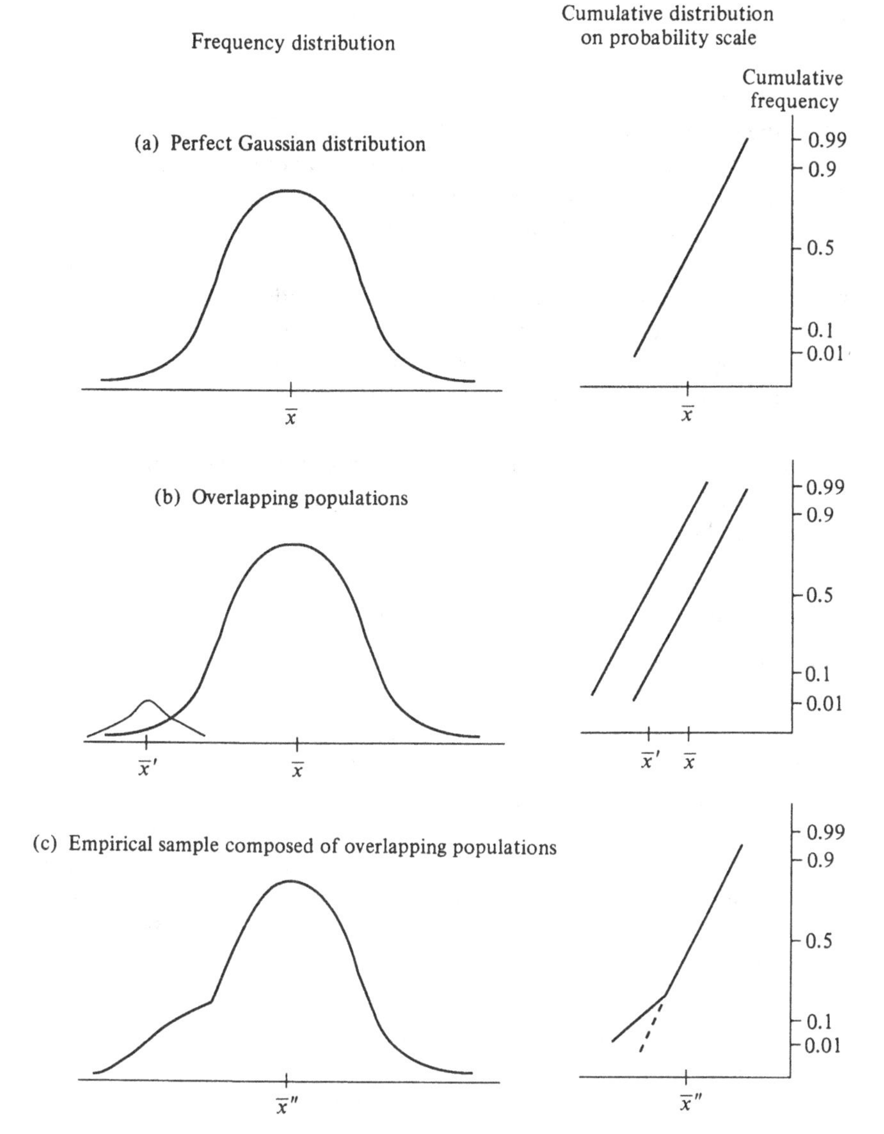

Fig. 9.8 Distribution frequencies and probability plots for hemoglobin values in hypothetical subpopulations. (a) The frequency distribution (on left) and cumulative probability plot (on right) are of a normal (Gaussian) population with no anemia. (b) The same non-anemic population as in (a) is shown as well as another smaller Gaussian population of anemic individuals. Each population is represented by a separate linear probability plot of similar shape but with different means ($\bar{x}'$ for anemic, $\bar{x}$ for non-anemic). (c) The comingling of the two distributions is shown as one might find in a population survey which has not been previously screened for anemia. Note that the population mean $\bar{x}''$ in (c) is between $\bar{x}$ and $\bar{x}'$, the distribution is skewed to the left and the probability plot is not linear at the lower values of hemoglobin. The dashed line represents an extrapolation of the 'normal' or linear portion of the probability line to simulate the distribution as if it were normal.

children. Estimates of prevalence of malnutrition are made by extrapolating the linear portion of the curve to the x-axis and estimating the area between the extrapolated line and the empirically observed curved line. This method makes no assumption about cutoff values, but is influenced by the prevalence of malnutrition in the population. High prevalence (between 10 and 90%) may cause distortion in the slope of the presumed normal population probability plot causing an underestimation of true prevalence. The degree of difference in anthropometric values between the two subpopulations also influences the plot. Computer simulation of the proposed distributions has been applied using maximum likelihood estimation procedures which compute the prevalence, mean and standard deviation of the subpopulations (Meyers *et al.*, 1983; Tufts *et al.*, 1985). There is a need for research on the application of this promising method to child growth data.

Population means of attained growth

The examination of population means for anthropometry relative to published values for other populations is a common practice. These comparative studies which analyze interpopulation variation have been particularly useful to our understanding the contribution of environmental and ethnic/genetic factors to growth (Habicht *et al.*, 1974; Eveleth & Tanner, 1976). For example, they serve as the basis for much of the debate over the choice of the appropriate growth reference standard which should be used to assess nutritional status in populations around the world (Habicht *et al.*, 1974; Goldstein & Tanner, 1980; Martorell & Habicht, 1986).

Decisions regarding local versus universal reference standards

By placing population means for various indicators of growth between the extremes of published values, one gains considerable insight as to the extent of environmental deprivation a group of children might suffer. The consistency of the large difference in mean achieved growth between the well-off and the poor populations living in developing countries is a striking contrast to the relatively small difference that exists in enhanced growth among the well-off across many populations. Data summarized by Habicht *et al.* (1974), Eveleth & Tanner (1976) and Martorell (1984) are evidence of these trends and argue for the use of a single growth reference for assessing pre-school child nutritional status in all populations. As Martorell & Habicht (1986) pointed out, 'the variation (in growth) that can be attributed to environmental factors, at least in developing countries, far overshadows that which can be attributed to genetics.' Even the 3–4 cm difference between heights of healthy seven year old US and Japanese children is insignificant in comparison to the 11–14 cm deficit in mean height between impoverished Mexican Zapotec Indian children and the same Japanese and US children. They went on to state that 'what is sometimes forgotten is that nature does not produce populations as short as the Zapotec Indians outside an environment of poor nutrition, high infection rates, and high rates of mortality' (p. 259).

Comparing growth across populations

At this population level a major concern is the use of child growth data to document a change in nutritional status. This change may be seen over short periods, such as from season to season, or over long periods, such as during the course of a five-year health or nutrition intervention program. A related application of growth data is to compare growth indicators of nutritional status across locations, such as adjacent communities, ecological zones or countries. There are generally two approaches to these comparative analyses – one examines differences in mean values for growth measures and the other examines differences in prevalence of some degree of growth retardation as a classification of malnutrition.

It appears that testing for population differences in means or prevalences gives somewhat similar results if one wishes to rank populations, but tests of means are usually more precise (Brownie, Habicht & Cogill, 1986). For example, this is the case when one tests for the relative discriminatory power of weight-for-age and weight-for-height (Wt/Ht^2) in separating Bangladeshi children who die from those who survive during a two-year surveillance. There is a need to test these observations in different sampling designs such as pre-intervention and post-intervention measures for intrapopulation responsiveness and across different socio-economic groups, or ecological zones. Moreover, much of the comparability of change in means versus change in prevalence depends on whether the distributions of the measure in both groups are Gaussian (Brownie *et al.*, 1986). One additional advantage to comparing age and sex specific means rather than prevalence is that a reference growth standard is not required. However, much of the research in nutritional epidemiology that utilizes growth data, expresses that data in the form of prevalences of malnutrition.

Whether in clinical practice, public health programs or research, measurements of growth always include a certain amount of measurement error. The effects of this error on individual screening has been previously discussed. The effects on population estimates of malnutrition is equally important.

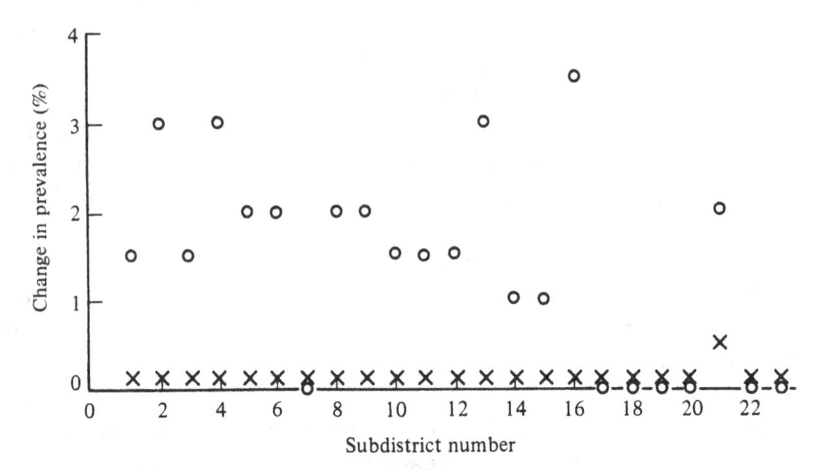

Fig. 9.9 The effects of random measurement error in weight (crosses) and age (open circles) on the overestimation of prevalence of low weight-for-age (below 80% of WHO reference) for 23 Indonesian subdistricts. The change in prevalence is calculated after Bairagi (1986). Random errors in weight measurement have little if any effect compared to random errors in age reporting. There is considerable variation across subdistricts in the degree to which prevalence of low weight-for-age was over-estimated. (From Haas, Marks & Jus'at, 1986.)

Correcting for measurement error

Measurement error not only influences screening precision but also affects the distribution of values and hence the prevalences of malnutrition based on the number of individuals who fall below a certain cutoff.

Systematic error, or bias, will increase or decrease prevalence depending on the direction of the bias. The corrections in prevalence for systematic measurement error have been estimated by Bairagi (1986) for various anthropometric measurements and age. For example, a 0.2 kg systematic error in weight *or* a 1 month error in age results in a 5% bias in prevalence of low weight-for-age. Random measurement error adds to the total variance and thus affects the shape of the distribution of a measured trait in a population. Bairagi (1986) also estimated corrections for random measurement error. Increased random error results in increased over-estimation of prevalence in the population. A test–retest measurement variance of 0.6 kg^2 for weight or 40 $months^2$ for age results in an overestimation of 3% for low weight-for-age. Obviously, one needs to have done the test–retest reliability study to estimate the magnitude of the systematic and random errors before these corrections can be made to one's data. This has been done for nutritional surveillance data collected in Indonesia, as shown in Fig. 9.9. Each point represents the correction in prevalence for each of 23 subdistricts for which data were collected. Measurement error was estimated for each anthropometry team working in separate subdistricts. Adjustments in prevalence are necessary to avoid errors in ranking subdistricts for the purpose of delivering health and nutrition services to the neediest.

Conclusion

This highly selective review of issues related to the use of growth to screen individual children and quantify malnutrition in a population should demonstrate that confusion still exists in the interpretation of a long established method for clinical and public health practice. There is still considerable research that needs to be done to clarify such issues as appropriate indicators for specific types of undernutrition and specific cutoff values for specific types of intervention. There is a comparable research agenda to be proposed for the use of these and other indicators to quantify the type and magnitude of malnutrition in populations.

While emphasis in some quarters may be towards the use of more sophisticated techniques employing modern technology to assess nutritional status, we believe that more can be achieved with judicious application of existing methods and techniques after considering some of the problems presented in this review.

References

Bairagi, R. (1986). Effects of bias and random error in anthropology and in age on estimation of malnutrition. *American Journal of Epidemiology*, **123**, 185–91.

Brownie, C., Habicht, J.-P. & Cogill, B. (1986). Comparing indicators of health or nutritional status. *American Journal of Epidemiology*, **124**, 1031–44.

Brownie, C. & Habicht, J.-P. (1984). Selecting a screening cut-off point or diagnostic criterion for comparing prevalences of disease. *Biometrics*, **40**, 675–84.

Chandra, R.K. (1982). Physical growth of exclusively breast-fed infants. *Nutrition Research*, **2**, 275–6.

Chen, L.S., Chowdhury, A.K.M.A. & Huffman, S.L. (1980). Anthropometric assessment of energy–protein malnutrition and subsequent risk of mortality among preschool aged children. *American Journal of Clinical Nutrition*, **33**, 1836–45.

Cogill, B. (1982). *Ranking Anthropometric Indicators using Mortality in Rural Bangladeshi Children*. MSc thesis. Cornell University, Ithaca, NY.

Dibley, M.J., Goldsly, J.B., Staehling, N. & Trowbridge, F.L. (1987a). Development of normalized curves for the international growth reference: historical and technical considerations. *American Journal of Clinical Nutrition*, **46**, 736–48.

Dibley, M.J., Staehling, N., Nieburg, P. & Trowbridge, F.L. (1987b). Interpretation of *z*-score anthropometric indicators derived from the international growth reference. *American Journal of Clinical Nutrition*, **46**, 749–62.

Eveleth, P.G. & Tanner, J.M. (1976). *Worldwide Variation in Human Growth*. Cambridge University Press, Cambridge.

Fomon, S.J., Thomas, L.N., Filer, L.J., Ziegler, E.E. & Leonard, M.T. (1971). Food consumption and growth of normal infants fed milk-based formulas. *Acta Paediatrica Scandinavica*, Suppl. **223**, 1–326.

Friere, W. (1982). *Use of Hemoglobin Levels to Determine Iron Deficiency in High Prevalence Areas of Iron Deficiency Anemia*. PhD thesis. Cornell University, Ithaca, NY.

Galen, R.S. & Gambino, S.R. (1975). *Beyond Normality: The Predictive Value and Efficiency of Medical Diagnoses*. Wiley, New York.

Goldstein, H. & Tanner, J.M. (1980). Ecological considerations in the creation and use of child growth standards. *Lancet*, i, 582–5.

Haas, J.D., Marks, G.C. & Jus'at, I. (1986). *Recommendations for the Implementation of the Nutritional Status Monitoring System: Pilot Project Final Report – Part B*. Cornell Nutritional Surveillance Program, Ithaca, NY.

Habicht, J.-P. (1980). Some characteristics of indicators of nutritional status for use in screening and surveillance. *American Journal of Clinical Nutrition*, **33**, 531–5.

Habicht, J.-P., Yarbrough, C., Martorell, R., Malina, R.M. & Klein, R.E. (1974). Height and weight standards for preschool children: how relevant are ethnic differences in growth potential? *Lancet*, i, 611–15.

Habicht, J.-P., Meyers, L.D. & Brownie, C. (1982). Indicators for identifying and counting the improperly nourished. *American Journal of Clinical Nutrition*, **35**, 1241–54.

Hamill, P.V.V., Drizd, T.A., Johnson, C.L., Reed, R.B., & Roche, A.F. (1977). *NCHS Growth Curves for Children from Birth to 18 Years: United States*, Publ. No. PHS 78–1650: Vital and Health Statistics Series 11, No. 165. US Department of Health, Education and Welfare, Hyattsville, MD.

Jahari, A.B. (1982). *Anthropometric Measurements for Monitoring Risk of Death in Children Under Five Years of Age*. MSc thesis. Cornell University, Ithaca, NY.

Martorell, R. (1984). Genetics, environment and growth: issues in the assessment of nutritional status. In: *Genetic Factors in Nutrition*, ed. S. Velásquez & H. Bourges, pp. 373–92. Academic Press, New York.

Martorell, R. & Habicht, J.-P. (1986). Growth in early childhood in developing countries. In: *Human Growth: A Comprehensive Treatise*, 2nd edn., vol. 3, ed. F. Falkner & J.M. Tanner, pp. 241–62. Plenum Press, New York.

Meyers, L.D., Habicht, J.-P., Johnson, C.L. & Brownie, C. (1983). Prevalences of anemia and iron deficiency anemia in black and white women in the United States estimated by two methods. *American Journal of Public Health*, **73**, 1042–9.

Persson, L.Å. (1985). Infant feeding and growth – a longitudinal study in three Swedish communities. *Annals of Human Biology*, **12**, 41–52.

Tanner, J.M. (1986). Use and abuse of growth standards. In: *Human Growth: A Comprehensive Treatise*, 2nd edn., vol. 3, ed. F. Falkner & J.M. Tanner, pp. 95–109. Plenum Press, New York.

Tufts, D.A., Haas, J.D., Beard, J.L. & Spielvogel, H. (1985). The distribution of hemoglobin and functional consequences of anemia in adult males at high altitudes. *American Journal of Clinical Nutrition*, **42**, 1–11.

Whitehead, R.G. & Paul, A.A. (1984). Growth charts and the assessment of infant feeding practices in the western world and in developing countries. *Early Human Development*, **9**, 187–207.

WHO (1983). *Measuring Change in Nutritional Status. Guidelines for Assessing the Nutritional Impact of Supplementary Feeding Programmes for Vulnerable Groups*. WHO, Geneva.

Yambi, O. (1988). *Nutritional Status and the Risk of Death: A Prospective Study of Children Six to Thirty-six Months Old in Iringa Region, Tanzania*. PhD thesis. Cornell University, Ithaca, NY.

10 *Nutritional vulnerability of the child*

P.S. NESTEL

Introduction

This paper looks at the nutritional vulnerability of children from the perspective of *food security*. This has been defined as 'adequate access to food for all sections of the population at all times and is concerned with production, distribution and consumption of food from the household to the national level' (Maxwell, 1988). The World Bank uses a similar definition and goes on to differentiate between two kinds of *food insecurity*, namely chronic and transitory (World Bank, 1986). Chronic food insecurity 'is a continuously inadequate diet caused by the inability to acquire food. It affects households that persistently lack the ability either to buy enough food or to produce their own'. Transitory food insecurity, on the other hand, 'is a temporary decline in a household's access to enough food. It results in instability in food production and household incomes, and in its worst form produces famine'.

The extent of chronic and/or transitory food insecurity in a community can vary and be dependent on many factors including food production, marketing and consumption patterns. The outcome is, in part, reflected in the nutritional status of children.

Using data from Sudan this paper examines the link between food insecurity and nutritional status. Sudan is interesting as a case study since it has recently emerged from what was probably its worst nutritional crisis this century, namely the 1983–85 drought. Following this drought two major studies were carried out. The first was a national nutrition survey, covering some 80,000 children under the age of five years, which was undertaken by the Nutrition Department of the Ministry of Health. The second was a national food security study carried out under the auspices of the Food Security Unit of the Ministry of Agriculture.

This paper utilises the findings of both these studies and covers the period May–July 1986 to May–July 1987. It is focused on north Sudan since the political situation in the south precluded any data collection there.

Table 10.1. *Economic indicators for Sudan*

Indicator	Level	Year	Source
Government deficit (US$ m)	−754.8	86/87	World Bank, 1988
Debt as % GNP	77.2	84	
Real GDP growth (%)	3.2	86/87	
Consumer price index ('80 = 100)	504.2	86	
Average annual growth rate GNP/capita (%)	−4.2	80–85	
GNP per capita ($)	300	85	
GDP in agriculture (%)	33	84	
Labour force in agriculture (%)	80	80	
Population (million)	22.4	86	CBS, 1987[a]
Population growth rate (%)	2.8	73–84	
Life expectancy (years)	50	86	UNICEF, 1988
Infant mortality (per 1000)	108	86	

[a]Extrapolated from 1983 census data.

Economic characteristics of Sudan

Agriculture absorbs 80% of the labour force in Sudan and accounts for 33% of the national GDP (Table 10.1). Per capita GNP in 1986 was US$ 300 and had declined at 4% per annum during the period 1980–85. The 1986 national current account had a negative balance of US$ 755 million and the external public debt was equivalent to 77% of the GNP. Inflation has been high for a number of years, the consumer price index in 1986 was over five times its level of 1980.

The Sudanese population totals 22.4 million of whom approximately 70% live in rural areas, 20% are urban and 10% nomads; Sudan also hosts a large number of refugees (approximately 1.2 million people), 70% of whom originate from Ethiopia (United Nations High Commission for Refugees, 1988). The population growth rate is 2.8%, life expectancy is 50 years and the infant mortality rate is estimated to be 108/1000.

In addition to the refugees, there are a large number of internally displaced Sudanese. Some 8–10 million people were affected by the 1983–85 drought, particularly those in Darfur, Kordofan, and Red Sea provinces (Fig. 10.1) where some 1.8 million were displaced, although the majority of these people have now returned home. By mid 1988 it was estimated that there were over 2 million displaced people in the North, most of whom were from the South.

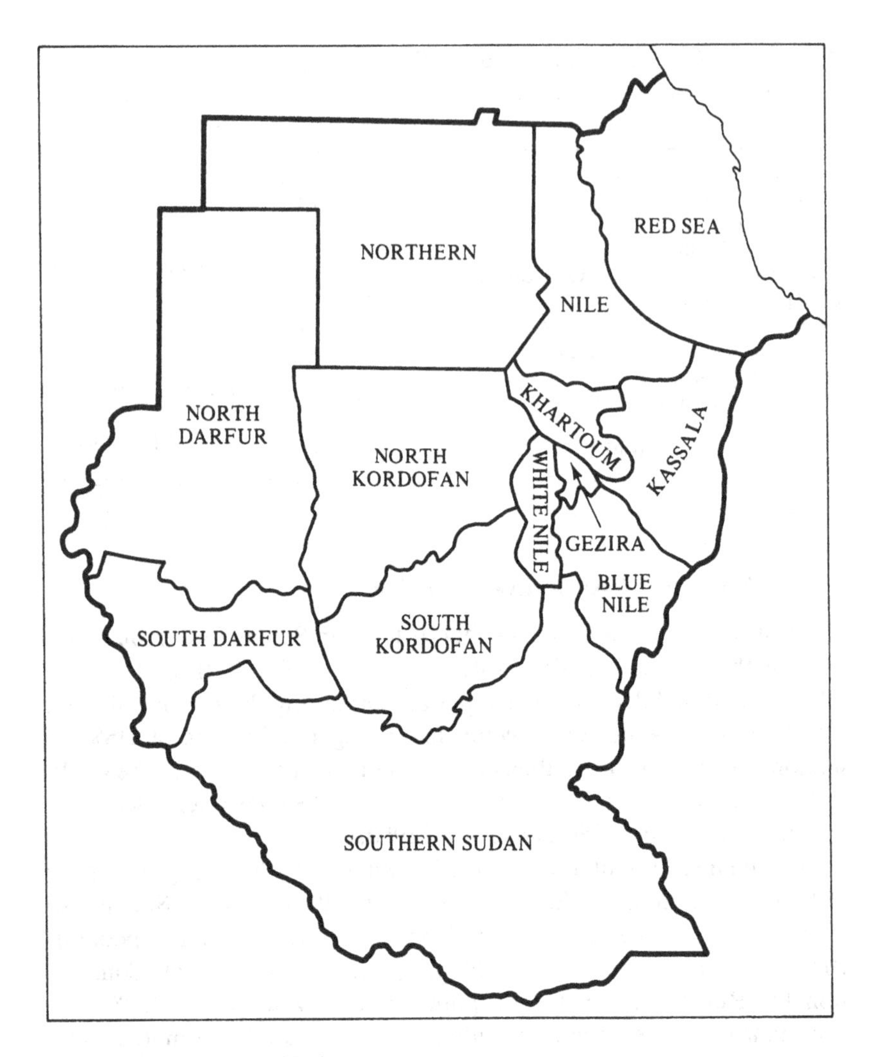

Fig. 10.1 Map of Sudan.

Cereal production

The total land mass of Sudan is 2.5 million square kilometres of which only 32% is potentially cultivatable (D'Silva, 1985). The cropped area is divided into three main sectors, namely irrigated, mechanised rainfed and traditional rainfed. In 1986/87 some 20 million feddans (8.4 million hectares) were cultivated; of this 10% were in the irrigated, 46% in the mechanised rainfed and 44% in the traditional rainfed areas. Of the cultivatable area, 78% was under cereals of which 59% was sorghum, 18% millet and less than 1% wheat (Ministry of Agriculture, 1988).

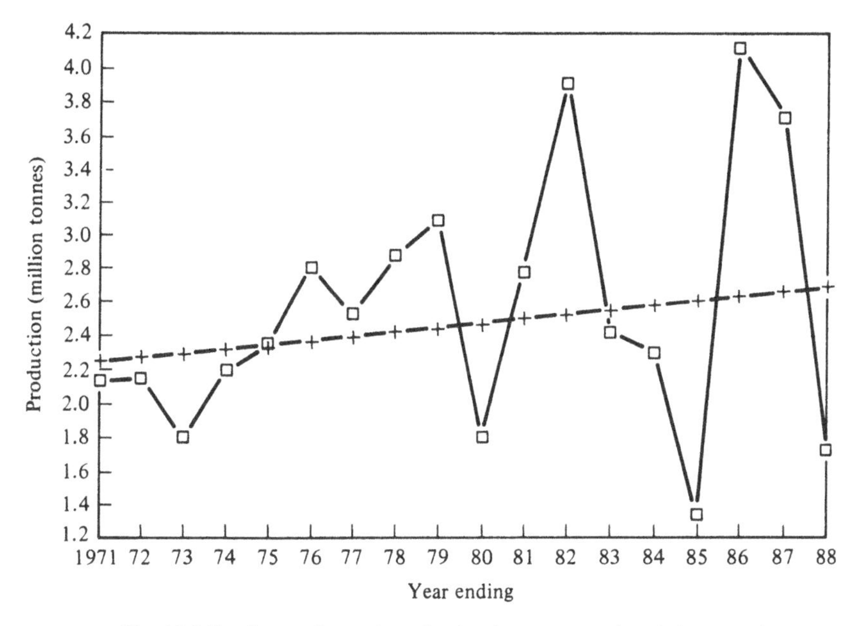

Fig. 10.2 Total annual cereal production (open squares) and the trend (crosses) over the period 1970/71–1987/88 (from Maxwell, 1988).

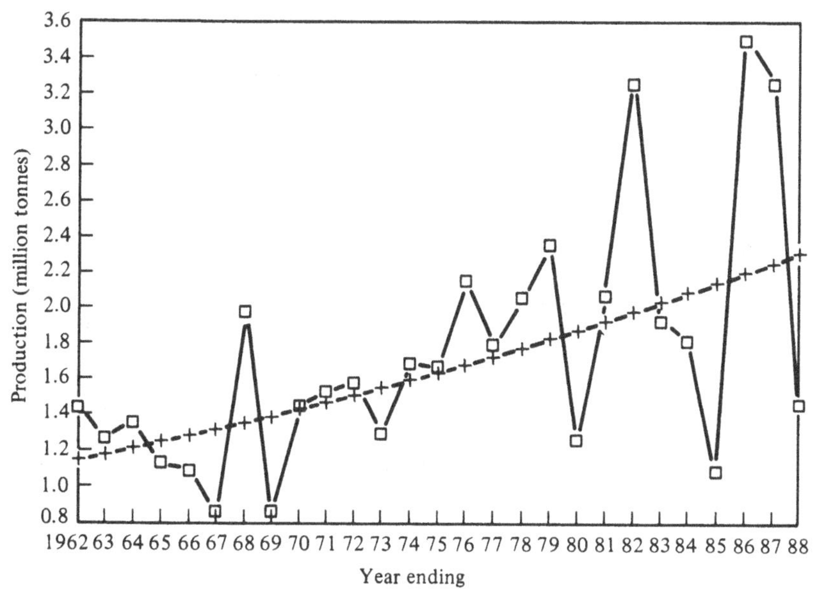

Fig. 10.3 Sorghum output in Sudan: actual (open squares) and trend (crosses) 1961/62–1987/88 (from Maxwell, 1988).

Table 10.2. *Crop production (thousand tonnes) in North Sudan 1986/87 (from Ministry of Agriculture, 1988)*

	Sorghum		Millet		Wheat		Total	
Province	Wt	%	Wt	%	Wt	%	Wt	%
Khartoum	0	0.0	0	0.0	0	0.0	0	0.0
Northern	15	0.5	0	0.0	32	20.4	47	1.3
Nile	18	0.6	0	0.0	5	3.2	23	0.6
Red Sea	12	0.4	5	1.8	0	0.0	17	0.5
Kassala	1078	34.6	2	0.7	15	9.6	1095	30.8
Blue Nile	1135	36.4	9	3.2	0	0.0	1144	32.2
Gezira	271	8.7	1	0.4	90	57.3	362	10.2
White Nile	311	10.0	6	2.2	15	9.6	332	9.3
North Kordofan	150	4.8	98	35.4	0	0.0	248	7.0
South Kordofan	75	2.4	11	4.0	0	0.0	86	2.4
North Darfur	7	0.2	68	24.5	0	0.0	75	2.1
South Darfur	44	1.4	77	27.8	0	0.0	121	3.4
North Sudan total	3116	100.0	277	100.0	157	100.0	3550	100.0

Table 10.3. *Cereal balance for North Sudan 1986/87 (Sources: population – CBS, 1987, 1983 – extrapolated, production – Ministry of Agriculture, 1988)*

Province	Population (thousands)	Production (thousand t)	Demand[a] (thousand t)	Balance (thousand t)
Khartoum	1958	0	254.5	−254.5
Northern	461	47	60.0	−13.0
Nile	706	23	92.0	−69.0
Red Sea	756	17	98.3	−81.3
Kassala	1643	1095	213.6	881.4
Blue Nile	1148	1144	149.2	994.8
Gezira	2200	362	286.0	76.0
White Nile	1026	332	133.4	198.6
North Kordofan	1962	86	255.1	−169.1
South Kordofan	1399	248	181.9	66.1
North Darfur	1443	75	187.6	−112.6
South Darfur	1918	121	249.3	−128.3
North Sudan total	16620	3550	2160.9	1389.1

[a]130 kg/capital/year

Cereal production has increased, by an average of 1% per annum, since 1970/71, although there has been considerable variation between years (Fig. 10.2) (Maxwell, 1988). Within the cereal sector sorghum production has increased, on average, 3.2% per annum, while millet production has declined considerably and wheat production remained stable. Fig. 10.3 shows that sorghum production has fluctuated greatly and, as Maxwell (1988) pointed out, the amplitude of the fluctuation is increasing. Much of the past growth in sorghum production has been in the mechanised sector, largely due to increases in the area under production rather than through increased yields.

Table 10.2 presents 1986/87 cereal production on a provincial basis. Nearly two thirds of production is represented by sorghum produced in Blue Nile and Kassala provinces. Millet production is more important than sorghum in both North and South Darfur, although North Kordofan is also an important millet producing area. Wheat production is primarily in Gezira and Northern provinces and to a lesser extent in Kassala and White Nile provinces.

Cereal marketing

A cereal balance sheet for 1986/87 is shown in Table 10.3. The final column shows that the volume of grain available for moving from the surplus to the deficit areas was of the order of 2.2 million tonnes. The major food flows are from the surplus areas of Kassala, Blue Nile and White Nile provinces into the deficit areas of Khartoum, the north and the west (Fig. 10.1). Little of the millet produced in the west is traded since millet is the preferred staple in this area. Sudan also receives over 700,000 tonnes of wheat imports annually. Most of this goes to the urban areas.

Transport of commodities in Sudan is severely hampered by the poor quality of the roads, and many routes are impassable during the rainy season when food needs are highest. In addition, the vast distances to be covered mean that transport costs are high. Indeed the price of grain in Darfur, during the rainy season, can be up to three times higher than prices in the east, because of transportation costs. The alternative to road transport is rail. Sudan Railway, however, has serious maintenance problems and the railways only operate at about 10% of their capacity.

There are no national data on the volume of cereals stored by the private sector. The government's storage capacity, however, is currently some 300,000 tonnes (Agricultural Bank of Sudan, 1987). Storage losses are high (20%) because of the poor physical state of the warehouses and/or inadequate management, i.e. poor pest control etc.

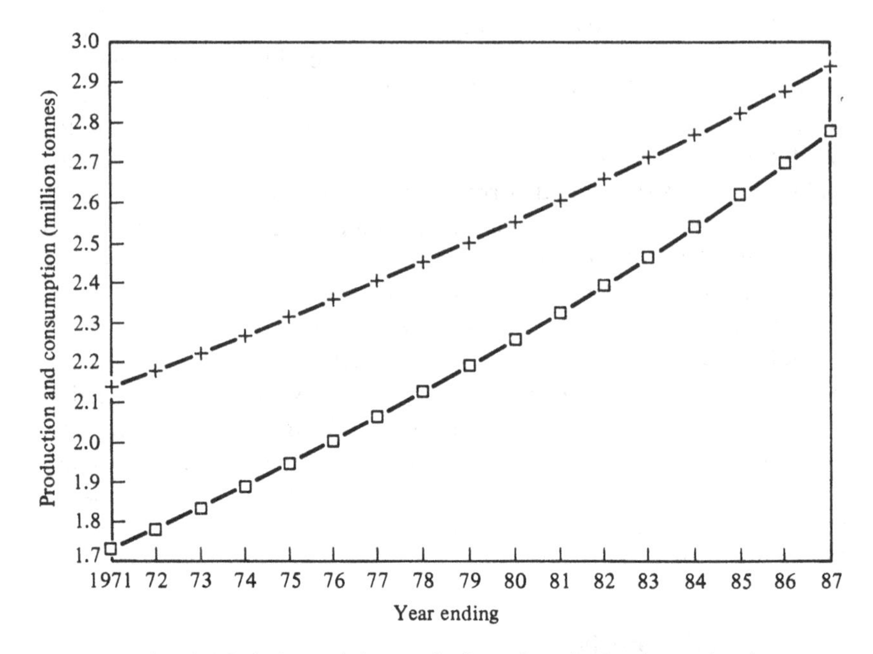

Fig. 10.4 Sudan's trends in growth of cereal production (crosses) and consumption (open squares) (from Maxwell, 1988).

Cereal consumption

Total consumption of cereals has risen at a rate of 3% per annum since 1970/71 (Maxwell, 1988). This is marginally higher than the 2.8% increase per annum in population. Average current consumption of cereals is 130 kg per capita per year, although this average disguises a wide range of intake levels.

Overall the country has surplus grain in most years, although there is always a deficit in wheat for which Sudan has to depend on imports. As Maxwell (1988) pointed out, it is unlikely that the country will always have a surplus of grain since production is only growing at 1% per annum while consumption is increasing by 3% per annum. Extrapolation of Fig. 10.4 shows that, unless production increases at a faster rate than the past trend, Sudan is likely to become an importer of its staple food by the end of this century.

The most detailed information on access to food is that collected by the Ministry of Health (Ministry of Health, 1988a, b) over four periods during 1986/87 from some 50,000 households and includes information on 80,000 children under the age of five years. These data provide the basis of the rest of this report.

Table 10.4. *Consumption of bread and grain in North Sudan (from Ministry of Health, 1988b)*

	Bread		Grain		
Province	No.	%	No.	%	Total
Khartoum	1886	28.9	4651	71.1	6537
Northern	170	12.6	1182	87.4	1352
Nile	374	18.1	1699	81.9	2073
Red Sea	270	12.7	1847	87.3	2117
Kassala	351	7.0	4664	93.0	5015
Blue Nile	51	1.5	3244	98.5	3295
Gezira	148	2.6	5511	97.4	5659
White Nile	72	2.6	2718	97.4	2790
North Kordofan	87	1.6	5270	98.4	5357
South Kordofan	11	0.3	4166	99.7	4177
North Darfur	10	0.2	5026	99.8	5036
South Darfur	17	0.3	6118	99.7	6135
North Sudan					
Urban	2693	23.4	8838	76.6	11531
Rural	745	2.1	35551	97.9	36296
Nomadic	9	0.5	1708	99.5	1717
Total	3447	7.0	46096	93.0	49543

No. refers to number of households surveyed.
Significance of differences $p<0.001$ among provinces and among locations.

Of the households surveyed in north Sudan, 93% used grain as their staple food, the residual 7% used bread (Table 10.4). Bread consumption is highest in Khartoum, Nile, Northern and Red Sea, indicating that in these provinces there is better access to wheat from either local production or imports. Bread consumption was highest in urban areas where just under one quarter of the households surveyed consumed it as their staple.

Type of cereal consumed

Among households dependent on grain, 72% ate sorghum, 23% millet and less than 5% wheat and other cereals (Table 10.5). However, there were considerable provincial differences. For instance, the majority of households in both North and South Darfur were dependent on millet. Millet was also an important staple in Kordofan, more so in North than in South Kordofan. Wheat was the staple for over 90% of households in Northern Province. In all other provinces over three quarters of the households depended on sorghum. These provincial differences are, in

Table 10.5. *Type of grain consumed in North Sudan (from Ministry of Health, 1988b)*

	Sorghum		Millet		Other		
Province	No.	%	No.	%	No.	%	Total
Khartoum	3213	96.2	61	1.8	65	1.9	3339
Northern	72	7.7	7	0.7	860	91.6	939
Nile	1152	92.3	25	2.0	71	5.7	1248
Red Sea	1018	77.5	106	8.1	189	14.4	1313
Kassala	3078	94.9	37	1.1	127	3.9	3242
Blue Nile	2237	98.2	15	0.7	27	1.2	2279
Gezira	3797	98.7	13	0.3	36	0.9	3846
White Nile	1815	91.4	128	6.4	44	2.2	1987
North Kordofan	2380	65.5	1229	33.9	22	0.6	3631
South Kordofan	2364	80.7	561	19.1	11	0.2	2936
North Darfur	1521	41.3	2132	57.9	32	0.9	3685
South Darfur	964	22.7	3259	76.8	21	0.5	4244
North Sudan							
Urban	4510	73.0	1181	19.1	490	7.9	6181
Rural	18298	72.2	6055	23.9	987	3.9	25340
Nomadic	803	68.8	337	28.9	28	2.4	1168
Total	23611	72.2	7573	23.2	1505	4.6	32689

No. refers to number of households surveyed.
Significance of differences $p < 0.001$ among provinces and among locations.

part, a reflection of the respective provincial cereal production.

Besides the inter-provincial differences there was also intra-provincial variation depending on whether the household was located in an urban or rural area or was nomadic. Millet consumption was highest among the nomads followed by rural households, while wheat consumption was highest among urban households. These findings reflect both the access to and the prices of cereals in the different locations.

Per caput consumption of cereals

In terms of grain availability per caput (Table 10.6) during 1986/87, one half of the households had less than 300 g per caput per day and just over a third had less than 200 g per caput per day. The availability of grain (per caput per day) was, on average, lowest in Red Sea province where agricultural production is very limited, followed by Northern and Khartoum provinces where bread consumption was highest. The difference in the average amount of grain available per caput per day between the

Table 10.6. *Availability of grain in 1986/87 (gms/person/day) in North Sudan (from Ministry of Health, 1988b)*

Grain (g/person/d)	<200		200–299		300–399		400–499		>500		
Province	No.	%	No.	%	No.	%	No.	%	No.	%	Total
Khartoum	2435	52.6	1019	22.0	618	13.3	273	5.9	287	6.2	4632
Northern	432	36.6	307	26.0	201	17.0	100	8.5	141	11.9	1181
Nile	997	58.8	409	24.1	198	11.7	52	3.1	39	2.3	1695
Red Sea	1060	59.8	552	31.1	115	6.5	43	2.4	4	0.2	1774
Kassala	1839	39.6	997	21.4	642	13.8	619	13.3	552	11.9	4649
Blue Nile	1170	36.1	367	11.3	713	22.0	283	8.7	706	21.8	3239
Gezira	2087	37.9	867	15.8	1349	24.5	513	9.3	685	12.5	5501
White Nile	926	34.1	342	12.6	597	22.0	222	8.2	629	23.2	2716
North Kordofan	1764	33.5	444	8.4	898	17.1	896	17.0	1266	24.0	5268
South Kordofan	1259	30.2	355	8.5	597	14.3	793	19.1	1159	27.8	4163
North Darfur	1372	27.4	273	5.5	782	15.6	706	14.1	1867	37.3	5000
South Darfur	1882	30.8	308	5.0	1271	20.8	935	15.3	1710	28.0	6106
North Sudan											
Urban	4355	49.8	1617	18.5	1371	15.7	626	7.2	777	8.9	8746
Rural	12237	34.5	4484	12.6	6457	18.2	4559	12.8	7744	21.8	35481
Nomadic	631	37.1	139	8.2	154	9.1	250	14.7	524	30.9	1698
Total	17223	37.5	6240	13.6	7982	17.4	5435	11.8	9045	19.7	45925

No. refers to number of households surveyed.
Significance of differences $p<0.001$ among provinces and among locations.

other provinces was due to the combined effects of the lack of purchasing power and the limited availability of subsistence cereals.

The amount of grain available per person per day was also dependent on location. On average, nomadic households had more grain available per caput, followed by rural households and then urban households. This trend was true in all provinces except Red Sea and North Darfur, where the availability of grain was greater among rural households, and Northern province where there were no locational differences. These findings highlight the relationship between access to different foods and location.

Seasonality inevitably influenced the amount of grain available per caput. Grain availability per caput was low from May through July 1986 prior to harvest, but higher once harvest began in September. From January 1987 onwards the overall availability of grain per caput gradually decreased. This was probably related to decreased subsistence food stocks, rather than to price fluctuations, since the association between the availability of grain per caput and price was not very strong.

Season also affected the locational differences in grain availability; urban households had more grain during the summer months and rural ones had less. This suggests that during the summer months cereals became more important in the diet of urban people because other foodstuffs are no longer available in the market. The decline in the availability of grain in rural areas during the summer months reflects the fact that access to food is limited, because subsistence food stocks are low and grain prices high.

Source of cereals

Dependency on subsistence, or home produced cereals, was high. Indeed one half of the households surveyed by the Ministry of Health purchased cereals while the other half used only home produced cereals (Table 10.7). Dependency on subsistence cereals was highest in Northern, Kassala, South Kordofan, Blue Nile and South Darfur provinces.

As expected, a majority of urban households purchased cereals while rural and nomadic households had a higher tendency to use subsistence cereals. This was true in all provinces except Northern, Nile and White Nile where there were no urban–rural differences. This is probably because urban households often owned nearby farms from which they obtained cereal.

Table 10.7. *Sources of cereal consumed in North Sudan (from Ministry of Health, 1988b)*

	Subsistence		Purchased		
Province	No.	%	No.	%	Total
Khartoum	2547	46.7	2911	53.3	5458
Northern	966	70.3	408	29.7	1374
Nile	386	22.0	1369	78.0	1755
Red Sea	801	41.8	1116	58.2	1917
Kassala	2687	68.3	1248	31.7	3935
Blue Nile	1938	59.5	1320	40.5	3258
Gezira	1778	33.3	3568	66.7	5346
White Nile	651	25.6	1896	74.4	2547
North Kordofan	2456	51.2	2344	48.8	4800
South Kordofan	2434	61.7	1514	38.3	3948
North Darfur	2304	52.9	2054	47.1	4358
South Darfur	3350	59.5	2277	40.5	5627
North Sudan					
Urban	3869	40.8	5625	59.2	9494
Rural	17624	53.0	15630	47.0	33254
Nomadic	805	51.1	770	48.9	1575
Total	22298	50.3	22025	49.7	44323

No. refers to number of households surveyed.
Significance of differences $p < 0.001$ among provinces and among locations.

Price of cereals

Over half of the households who bought cereal paid more than 150 piastres per malwa (US$ 0.75 per kg) of which one quarter paid over 250 piastres per malwa (Table 10.8). On a provincial basis many more households in Northern Province, than in any other province, paid a higher price for grain, indicating that the cereal being purchased was wheat. More households in South Darfur and, to a lesser extent, North Kordofan and North Darfur paid higher prices for grain compared with the remaining provinces. This is due to two factors. First millet consumption is relatively high in these three provinces and millet is more expensive than sorghum. Secondly cereal production in these three provinces and Khartoum is lower than in the others and demand is greater than supply. This, in addition to greater transport distances and costs, results in grain prices being higher in these three provinces than elsewhere.

Urban households, in general, paid more for cereals than nomads who in turn paid more than rural households, this reflects the relative ease of access to cereals.

Table 10.8. *Price (piastres) per malwa of cereal in North Sudan (from Ministry of Health, 1988b)*

Price (piastres)	<100		100–149		150–199		200–249		250–299		>300		
Province	No.	%	No.	%	No.	%	No.	%	No.	%	No.	%	Total
Khartoum	380	12.9	702	23.9	660	22.5	412	14.0	498	16.9	285	9.7	2937
Northern	0	0.0	2	0.5	9	2.2	22	5.4	66	16.2	309	75.7	408
Nile	14	1.0	50	3.6	1066	77.9	140	10.2	45	3.3	54	3.9	1369
Red Sea	35	3.1	185	16.6	571	51.2	187	16.8	95	8.5	43	3.9	1116
Kassala	264	21.1	490	39.3	368	29.6	39	3.1	38	3.0	48	3.8	1248
Blue Nile	284	21.6	924	70.2	75	5.7	24	1.8	3	0.2	7	0.5	1317
Gezira	252	7.1	2180	61.1	889	24.9	188	5.3	45	1.3	15	0.4	3569
White Nile	91	4.8	1136	59.9	335	17.7	197	10.4	112	5.9	25	1.3	1896
North Kordofan	50	2.1	520	22.2	833	35.6	505	21.6	196	8.4	240	10.2	2344
South Kordofan	158	10.4	700	46.2	408	26.9	121	8.0	55	3.6	72	4.8	1514
North Darfur	230	11.2	290	14.1	458	22.3	716	34.8	220	10.7	141	6.9	2055
South Darfur	121	5.3	383	16.8	500	21.9	632	27.7	343	15.0	300	13.2	2279
North Sudan													
Urban	275	4.9	1498	26.5	1398	24.8	1142	20.2	713	12.6	622	11.0	5645
Rural	1520	9.7	5938	38.0	4482	28.7	1883	12.0	941	6.0	860	5.6	15624
Nomadic	84	10.9	126	16.4	293	38.1	158	20.5	62	8.1	47	6.1	770
Total	1879	8.5	7562	34.3	6173	28.0	3183	14.4	1716	7.8	1539	7.0	22052

No. refers to number of households surveyed.
Significance of differences $p < 0.001$ among provinces and among locations.

Table 10.9. *Food aid July 1985 to July 1987 in North Sudan (from Ministry of Health, 1988b)*

	Received food aid				
	No		Yes		
Province	No.	%	No.	%	Total
Khartoum	5998	90.6	620	9.4	6618
Northern	1376	91.7	124	8.3	1500
Nile	1597	76.9	479	23.1	2076
Red Sea	1105	52.3	1008	47.7	2113
Kassala	1043	21.6	3792	78.4	4835
Blue Nile	462	15.1	2601	84.9	3063
Gezira	2654	49.6	2696	50.4	5350
White Nile	493	20.0	1967	80.0	2460
North Kordofan	314	6.0	4880	94.0	5194
South Kordofan	661	16.0	3468	84.0	4129
North Darfur	446	9.0	4521	91.0	4967
South Darfur	841	13.9	5219	86.1	6060
North Sudan					
Urban	7121	64.0	3998	36.0	11119
Rural	9340	26.3	26197	73.7	35537
Nomadic	529	30.9	1180	69.1	1709
Total	16990	35.1	31375	64.9	48365

No. refers to number of households surveyed.
Significance of differences $p<0.001$ among provinces and among locations.

There was a seasonal effect on the price paid for grain. Grain prices were lower during September–November 1986 as a result of the increased availability of cereal in the market because of the harvest. By January–March 1987 cereal prices had increased, they were even higher by May–June 1987. Indeed, prices were higher in May–June 1987 than they were during June–July 1986. This could have been, in part, due to the availability of food aid in 1986 and/or to farmers storing rather than selling cereal in 1987. This had the effect of exacerbating shortage and thus forcing up prices.

The relationship between the price of grain and the availability of grain per caput was not strong or evident in every province. Where a relationship did exist, the availability of grain per person tended to be greater where cereal was cheaper. This suggests that cereal prices *per se* do not always influence the availability of grain per person. This is perhaps not surprising, since, where the diet is monotonous, people have little choice except to eat the staple cereal. In other words, even though cereal prices fluctuate they do not greatly influence the availability of grain per caput per day.

Food aid

In addition to subsistence and market sources there was also a third source of cereal, namely food aid. Table 10.9 shows that, on average, about two thirds of households in the Ministry of Health study had received food aid since July 1985. However, there was considerable provincial variation. Over 90% of households in both North Kordofan and North Darfur received food aid. In all other provinces except Khartoum, Northern, Nile and Red Sea over half of the households received food aid.

Over two thirds of the households in rural and nomadic areas received food aid, while just over a third of those living in urban areas did. These location differences reflect that households in both the rural and nomadic areas lead a more marginal existence and require food assistance in times of stress, for example during and just after a drought.

Nutritional vulnerability in children

The nutritional status of some 80,000 children under five years old in north Sudan during four periods between May–July 1986 and May–July 1987 was determined using the NCHS (National Centre for Health Statistics) sex specific reference standards (WHO, 1979). The two indices used were weight-for-height and height-for-age.

Taken together these measurements indicate the extent of both long term, or chronic undernutrition, and short term, or acute undernutrition. In chronic undernutrition, which results in delayed growth in stature or shortness, height and weight are reduced, but weight is normal for height. In acute undernutrition, which leads to thinness, weight is low relative to height.

The overall mean *Z* wt/ht score (distance from the National Centre for Health Statistics median in NCHS standard deviation units), of children under five years in north Sudan, was −0.97 with a standard deviation 1.02. The same values for *Z* ht/age were -1.34 ± 1.43. Children below −2 SD *Z* wt/ht, i.e. thin, comprised 14.1%, of which 1.7% were below −3 SD, i.e. very thin. Also 32.1% were below −2 SD *Z* ht/age, i.e. short, of which 12.6% were below −3 SD, i.e. very short. This is to say one in seven children were at best thin and one in three short.

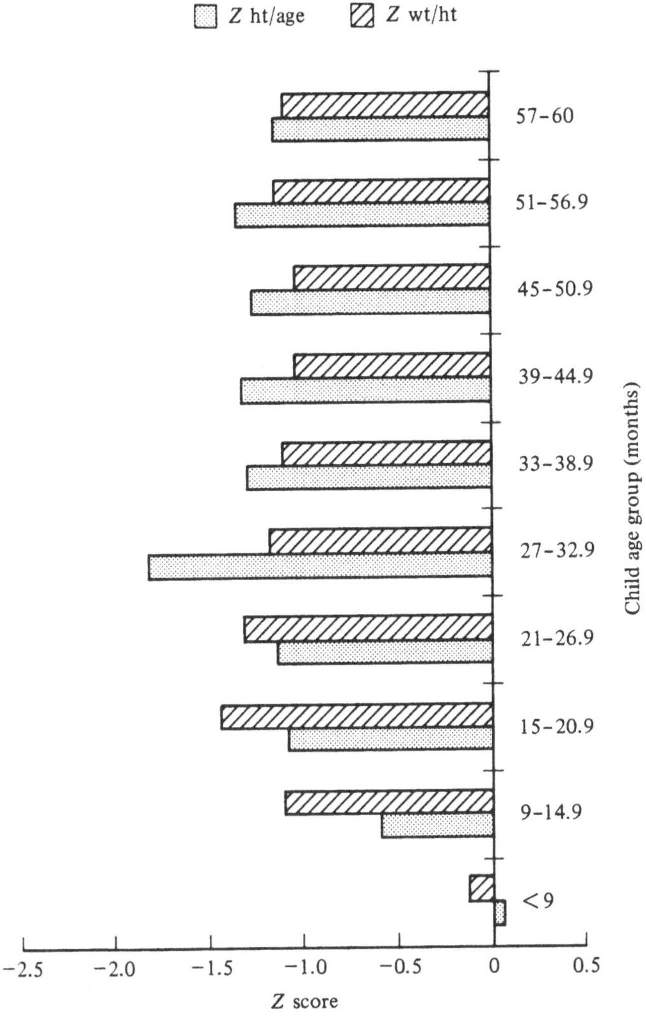

Fig. 10.5 Nutritional status by child age group. Significance of differences: chronic (ht/age) $p<0.001$, acute (wt/ht) $p<0.001$. (Source: Ministry of Health, 1988a).

Effect of age

The associations between child age and both thinness and shortness were large and highly significant. These data and the ones in the figures which follow show the residual effect of child age on nutritional status after 27 other social, economic, demographic and child variables were removed using multiple regression analyses.

Fig. 10.5 shows the cross-sectional growth pattern of children under five years in north Sudan. Children up to nine months old were only just

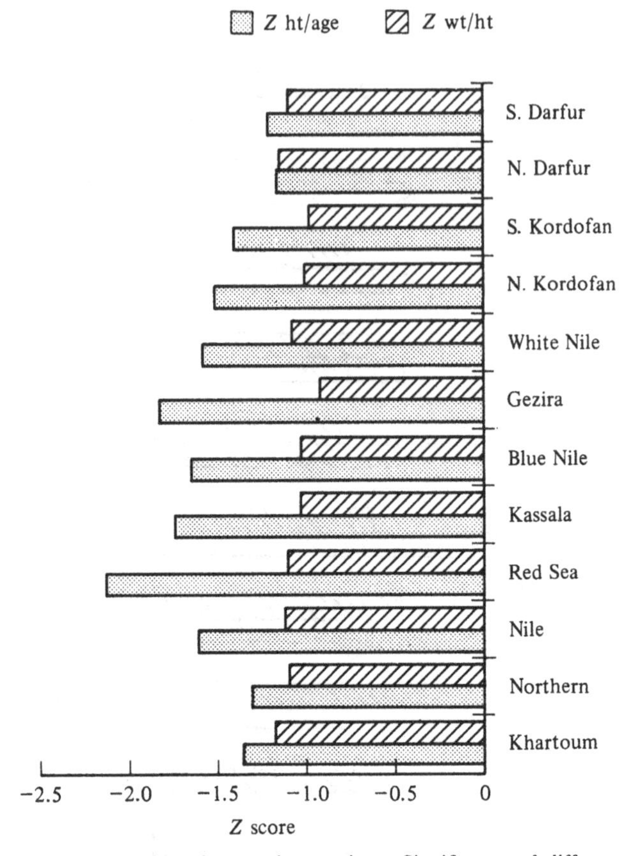

Fig. 10.6 Nutritional status by province. Significance of differences: chronic (ht/age) $p<0.001$, acute (wt/ht) $p<0.001$. (Source: Ministry of Health, 1988a).

below the norm in terms of weight-for-height but children aged 9 to 15 months old were noticeably thinner. The situation among the 15 to 21 month age group was the worst of all age groups. Thereafter there was some improvement up to the age of between 39 and 51 months whereupon weight stabilised, in relation to height. There was a further decline in the 51 to 57 month age group.

The same downward trend was observed for growth in height, although the turning point here was not until the children were in the 33 to 39 months old age group. Even so, between the age of 33 and 57 months the children were only able to maintain growth in height relative to age. By 57 months, however, it would appear that the children underwent a second small growth spurt. Although this data was cross-sectional, the findings were consistent in the four periods surveyed and are therefore assumed to reflect longitudinal growth patterns.

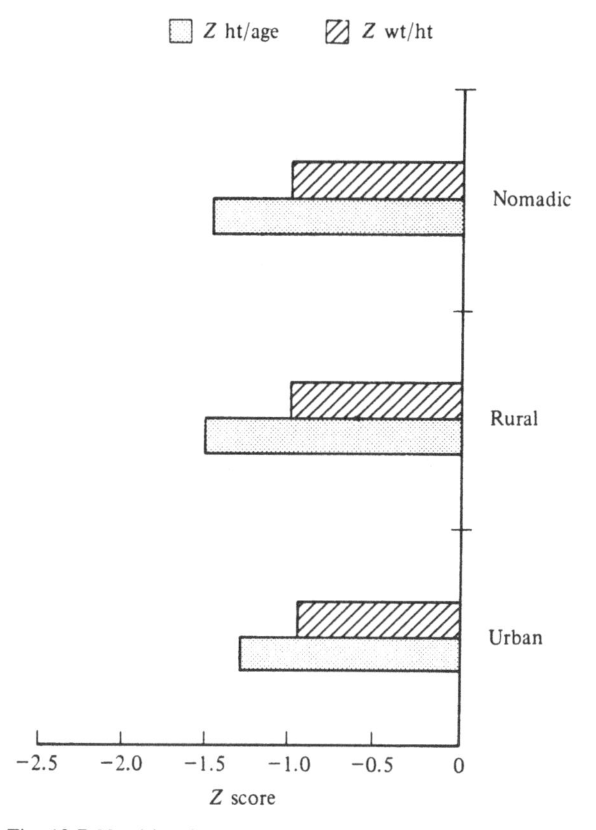

Fig. 10.7 Nutritional status by location. Significance of differences: chronic (ht/age) $p<0.001$, acute (wt/ht) $p<0.001$. (Source: Ministry of Health, 1988a).

Effects of province and location

Fig. 10.6 shows the levels of thinness and shortness by province. On average the thinnest under five year olds were in Khartoum, North Darfur and South Darfur. The heaviest children were in Gezira followed by South Kordofan, North Kordofan, Kassala and Blue Nile provinces. In contrast children in Red Sea, followed by Gezira, Kassala and Blue Nile were shorter, on average, while those in Khartoum, Northern, North Darfur and South Darfur were taller than those in other provinces, i.e. there is evidence of an inverse relationship between thinness and shortness. The importance of these data are that they put a historical perspective on the extent of undernutrition. The large provincial differences in levels of shortness compared with those for thinness shows that undernutrition has been a particularly chronic problem in Red Sea, Gezira, Kassala and Blue Nile provinces. The finding for Red Sea is not surprising given that it is not an agricultural area and there are few opportunities for work outside Port Sudan. In addition Red Sea province has never fully recovered from the 1972 drought. In other words, access to both income and subsistence food is limited, compared with other provinces.

The highest levels of chronic undernutrition are found in Gezira, Kassala, and Blue Nile, the three main agricultural provinces of Sudan. This probably reflects the fact that these provinces are dominated by large farms and major agricultural schemes. Many of the households are hired labourers, share-croppers or owners of uneconomically small farms, and have limited means of acquiring the surplus grain produced in these provinces.

In addition to a provincial effect, there was also a location effect on nutritional status (Fig. 10.7). Children in urban areas were both heavier and taller than their counterparts in rural and nomadic areas. This is attributed to the better access to a more balanced diet in urban areas than in rural or nomadic ones.

Effect of type of cereal

The ability to purchase cereal also influences nutritional status. There was a highly significant association between the type of cereal consumed and both acute and chronic undernutrition among under five year olds (Fig. 10.8). Children from households consuming wheat, which is generally purchased, were both heavier and taller than those consuming sorghum or millet.

Wheat is more expensive than millet which is in turn more expensive than sorghum. These data, therefore, show that the difference in nutritional status, both in the long and short term, by type of cereal can be correlated with wealth since wealthier households are more able to buy wheat than are poorer households.

The availability of cereal per caput per day was not, however, found to be associated with either chronic or acute undernutrition. This was because the type of cereal, which was related to price and hence to access, had already been taken into account in the regression analysis.

Effect of price of cereal

The associations between price of cereal and both acute and chronic undernutrition were significant (Fig. 10.9). Children from households paying more than 250 piastres per malwa (US$ 1.23/kg) of cereal were more likely to be heavier and taller than those who paid less, or depended on subsistence cereal.

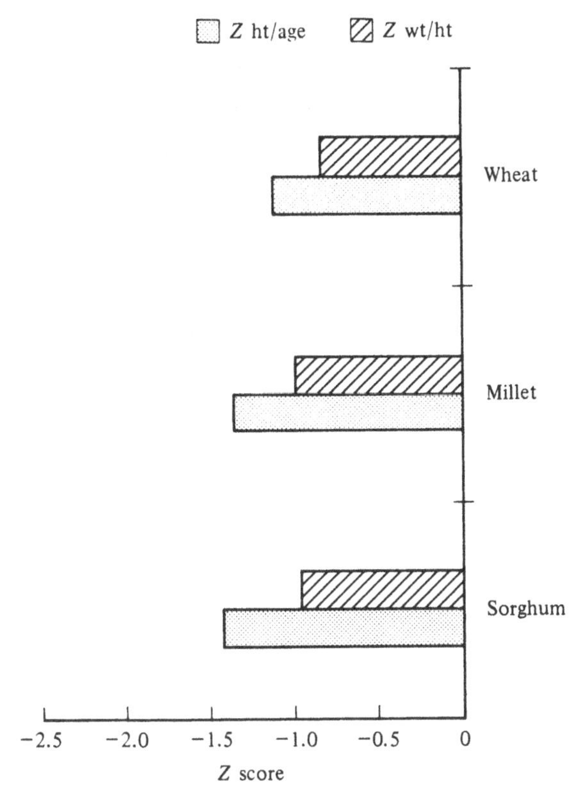

Fig. 10.8 Nutritional status by type of cereal consumed. Significance of differences: chronic (ht/age) $p<0.001$, acute (wt/ht) $p<0.001$. (Source: Ministry of Health, 1988b).

Given that the effect of type of grain, which was correlated with price of grain, had been removed, the results show that there is an independent effect of price of grain beyond that of type of grain. This may reflect different qualities of grain, in that adulteration of grain occurs and such grain is cheaper. The fact that children of households dependent on subsistence grain were less well nourished compared with those from households paying higher prices suggests that subsistence grain stocks were insufficient and, as a result, consumption was low.

Effect of food aid

Food aid was also associated with nutritional status. Fig. 10.10 shows that the association between months since food aid was last received and both acute and chronic undernutrition was highly significant. On

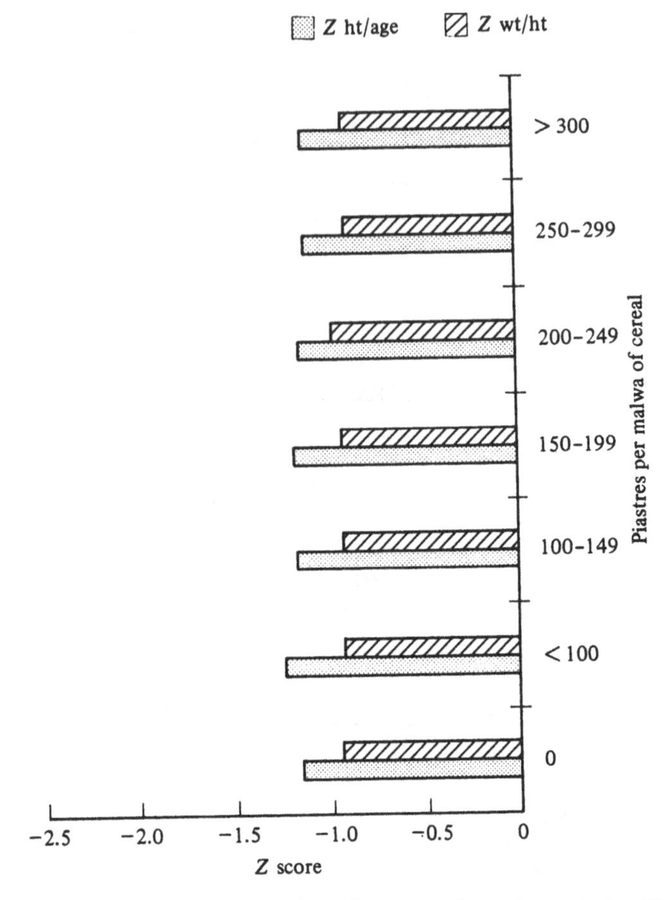

Fig. 10.9 Nutritional status by price per malwa of cereal. Significance of differences: chronic (ht/age) $p<0.001$, acute (wt/ht) $p<0.001$. (Source: Ministry of Health, 1988b).

average, children from households which had received food aid most recently, that is within the last three months, were likely to be heavier but not taller than those who had never (zero in Fig. 10.10) received food aid or had not received food aid within the last three months.

Effect of season

Seasonality, too, was a good predictor of nutritional status. Fig. 10.11 shows that, on average, children put on weight from May–July 1986 to January–March 1987 (*Z* wt/ht score deficit decreasing), but they were failing to gain weight in relation to height by May–July 1987. This was the

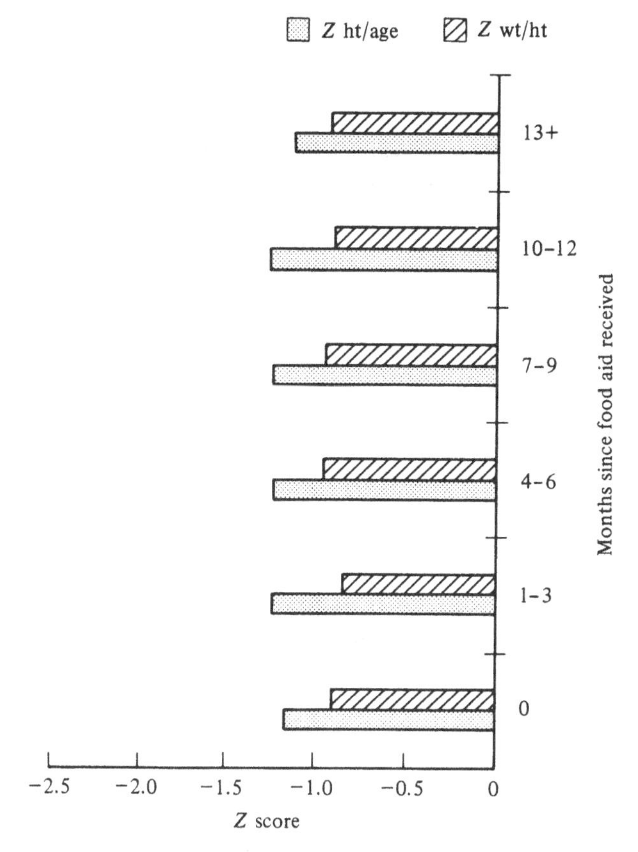

Fig. 10.10 Effects of food aid on nutritional status. Significance of differences: chronic (ht/age) $p<0.001$, acute (wt/ht) $p<0.001$. (Source: Ministry of Health, 1988b).

opposite to what was happening to growth in height. Between May–July 1986 and January–March 1987 children failed to grow in height as can be seen by the increase in shortness. By May–July 1987, however, the level of shortness had decreased indicating that growth in height was taking place.

The inverse relationship between weight gain and growth in height is not atypical and reflects that where food shortages exist children first gain weight at the expense of height. After this they will grow in height once food intake increases; this, of course, takes place after the harvesting season.

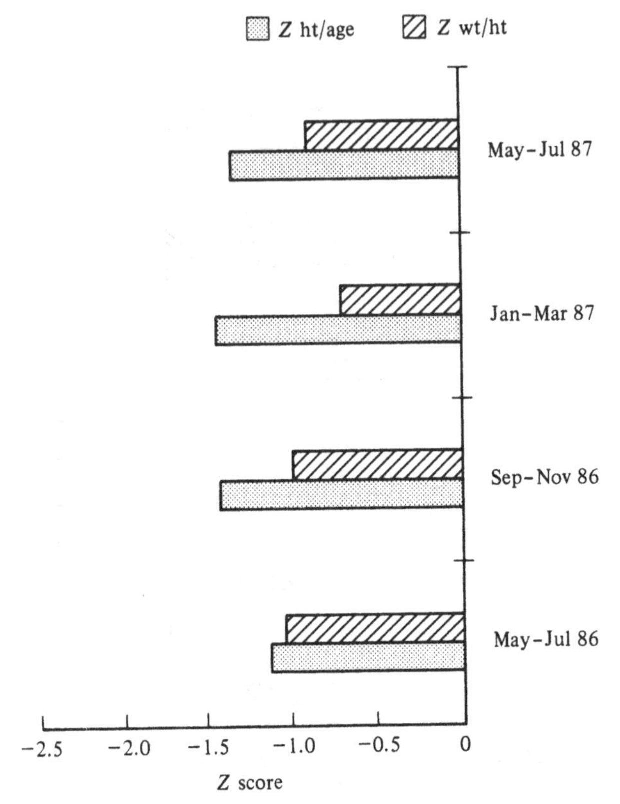

Fig. 10.11 Nutritional status by season. Significance of differences: chronic (ht/age) $p < 0.001$, acute (wt/ht) $p < 0.001$. (Source: Ministry of Health, 1988b).

Effect of feeding practices

In addition to the above variables, which relate access to food to nutritional status, there was also an independent effect of type of meal and number of meals on nutritional status. Fig. 10.12 shows that, on average, purely breast fed children were heaviest in relation to their height and all breast fed children were heavier, relative to their height, than children fed one or two meals a day, but they were not significantly different to children fed three or more meals a day. However, all breast fed children were shorter, relative to their age, compared with children already completely weaned.

These results indicate that breast fed children are more vulnerable to undernutrition in the long term. This is a reflection of inadequate feeding practices because mothers do not introduce supplementary food early enough.

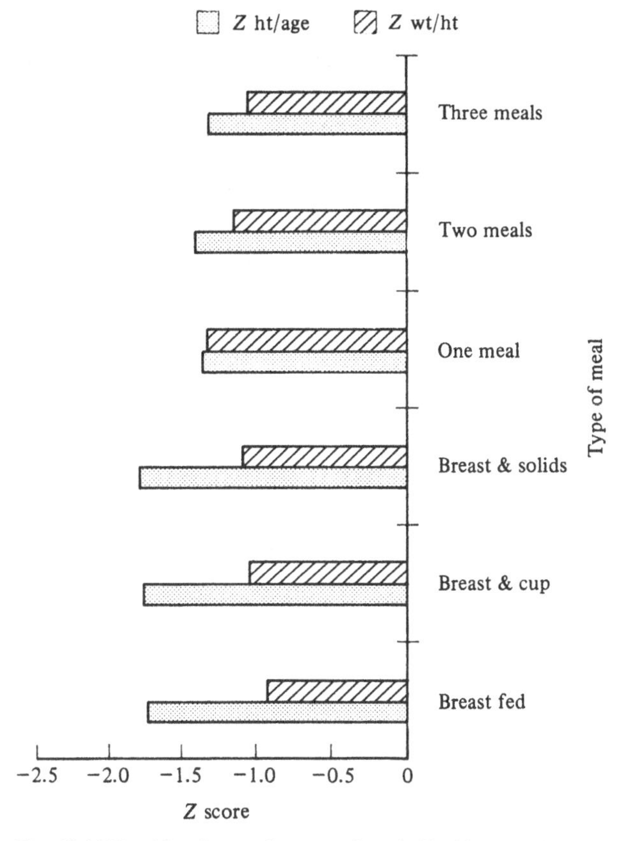

Fig. 10.12 Nutritional status by type of meal. Significance of differences: chronic (ht/age) $p<0.001$, acute (wt/ht) $p<0.001$. (Source: Ministry of Health, 1988a).

Summary

In Sudan agriculture provides 33% of the GDP and absorbs 80% of the labour force. Of the cultivated land, 78% is used to produce cereals, principally sorghum and millet. Wheat is produced on a small scale and imported through food and programmes.

At the national level Sudan usually produces a surplus of cereals but, if the past trends in demand and supply continue, this surplus will vanish before the end of this century. The area and yield of cereals vary widely from year to year and are markedly influenced by rainfall, the availability of inputs and the price paid for the previous harvest. When conditions are unattractive to farmers they diversify out of cereals into cash crops such as sesame. Such diversifications accentuate existing problems of unequal distribution of cereals and low intake levels, especially in those rural and nomadic peoples with low purchasing powers.

Amongst such populations undernutrition resulting from both chronic and transitory food insecurity is widespread. Chronic food insecurity is reflected in the fact that one in three children are short relative to their age. Transitory food insecurity is shown up by one in seven children being thin relative to their height.

There are a number of public policies which could be used to improve the nutritional status of the Sudanese. These include nutrition and pricing policies, provision of buffer stocks, developing marketing structures and food aid. Unless attention is given to developing appropriate policies there is little prospect of improving the nutritional status, since the underlying cause is poverty.

References

Agricultural Bank of Sudan (1987). *Annual Report*. Agricultural Bank of Sudan, Khartoum.

Central Bureau of Statistics (1987). *1983 Population Census*. Ministry of Finance & Economic Planning, Khartoum.

D'Silva, B. (1985). *Sudan: Policy Reforms and Prospects for Agricultural Recovery after Drought*. International Economics Division, Economic Research Service, USDA, Washington.

Maxwell, S. (1988). *Food Security Study: Phase I*. Institute of Development Studies, University of Sussex.

Ministry of Agriculture (1988a). *Situation and Outlook Annual Report 1986/1987*. Department of Agricultural Economics, Khartoum.

Ministry of Health (1988b). *Food in Relation to Nutritional Status*. Ministry of Health Nutrition Department, Khartoum.

Ministry of Health (1988). *Child Variables in Relation to Nutritional Status*. Ministry of Health Nutrition Department, Khartoum.

United Nations High Commission for Refugees (1988). *Annual Report*. UNHCR, Khartoum.

UNICEF (1988). *State of the World's Children*. UNICEF, New York.

WHO (1979). *Measuring Changes in Nutritional Status*. WHO, Geneva.

World Bank (1986). *World Development Report*, John Hopkins, Washington DC.

World Bank (1988). *World Development Report*. World Bank, Washington DC.

11 *Traditional economies and patterns of nutritional disease*

S.S. STRICKLAND

Introduction

Whichever criteria of nutritional health are adopted, the question of how traditional economic practices interact with nutritional experience can be posed in many ways. This paper discusses three factors. These are: the contribution of child labour to the health status of siblings; selective buffering of children against seasonal nutritional stress; and the functional consequences of small body size for physical work capacity and productivity.

The question could be approached in other terms, and a general argument linking traditional economies with nutritional disorders is that: first, the diets of 'primitive rural economies' are relatively balanced though perhaps quantitatively inadequate on a seasonal or permanent basis; and such diets depend on the relative population pressure of neighbouring areas. Second, diet deteriorates with increasing population pressure in slash-and-burn economies and among dry-land farmers as a result of reduced animal protein availability, falling crop yields from short fallow or eroded land, and seasonality in supplies (Whyte, 1974). This is summarized for monsoon Asia in Table 11.1.

This argument is familiar from other sources (Polunin, 1977). From an evolutionary perspective, it poses the problem of modelling how reproductive rate and environmental resources combine to determine the pattern of population growth and appropriate evolutionary strategies (Foley, 1987). From a contemporary viewpoint, it suggests that demographic imbalance between fertility and mortality underlies uniformly deleterious changes in nutritional outcome. In traditional economies there is limited evidence for secular trend in growth performance, but it is rarely possible to judge whether apparently deleterious decline results from this process (Høygaard, 1941; Bailey, 1962a, b; Tobias, 1975, 1985; Billewicz & McGregor, 1981, 1982). Without such changes, hunter-gatherer and other populations would probably manage a complex balance between food resources, experience of disease, fertility and mortality rates (Howell, 1976; Neel, 1977; Payne, 1985).

Table 11.1. *Rural economies and nutritional status in monsoon Asia (adapted from Whyte, 1974)*

Rural economies	Nutritional status
1 Hunters and collectors at low population density.	Relatively balanced diet. Short stature probably compensating for possible energy deficiency.
2 Hunters and collectors with swiddens at low population density.	Diversified diet. More carbohydrate from cultivated crops than wild sources.
3 Swidden farmers in secondary forest; greater population density.	Availability of animal protein dependent chiefly on fish resources.
4 Coastal or lakeside fishermen with some cultivation of plants.	Energy adequate or marginal but inadequate protein, vitamins and iron.
5 Swidden farmers of short fallow and decreasingly fertile land: excessive population pressure.	Probable energy and nutrient deficiency, with marked seasonal fluctuation.
6 Permanent with/without swidden dry land farmers on poor soils: excessive population density.	Marginal energy supplies, protein deficiency; also deficiencies of vitamins, especially retinol, and of minerals. Marked seasonality.
7 Mixed agro-pastoralists, migratory, nomadic or settled.	Probably adequate protein, vitamins and minerals for males, consistently inadequate for vulnerable groups.
8 Mixed irrigated and dry permanent field farmers with swiddens: excessive population pressure.	Marginal energy supplies; protein, vitamin and mineral deficiencies.
9 Irrigated rice monoculture; marketing of livestock produce. High to excessive population density.	Serious protein, thiamine, calcium, iron, retinol and riboflavin deficiencies.

In addition, this argument poses the problem of defining 'malnutrition' or 'nutritional health'. Several small population studies suggest that hunter-gatherers and horticulturalists traditionally present little evidence of nutritional stress by conventional biochemical and clinical criteria, though non-specific indices of growth performance may suggest otherwise (Neel, 1977; Sinnett, 1977; Truswell, 1977). It is not known how far contemporary hunter-gatherers resemble pre-agricultural Palaeolithic populations, but it has been suggested that with the emergence of agriculture in various places from about 10 000–5000 BC arose a number of adverse nutritional and epidemiological changes to which man

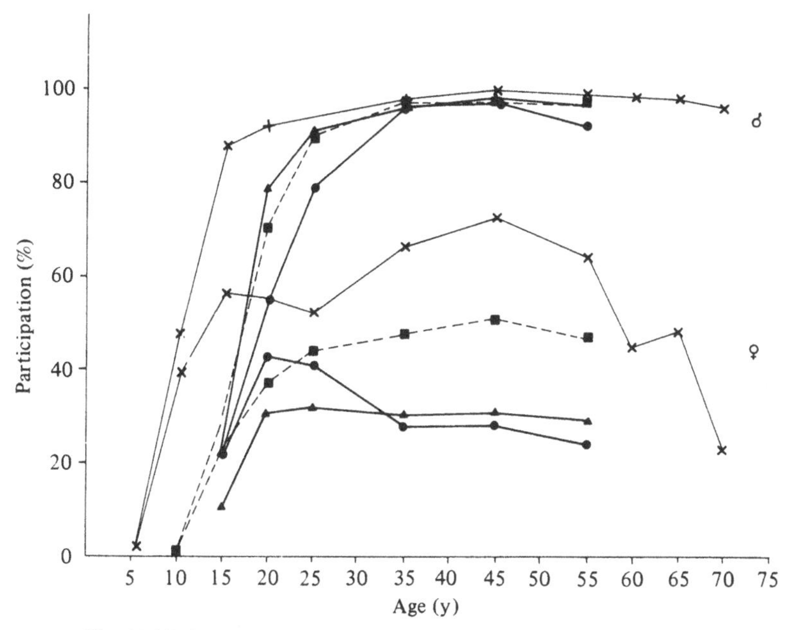

Fig. 11.1 Labour force participation rates (%) by age group and sex (drawn from data in Mueller, 1976). Crosses = rural Egypt, closed squares = rural India, closed triangles = less developed countries, closed circles = rural Taiwan.

is still in the process of adjusting (Cassidy, 1980; Eaton & Konner, 1985; Schoeninger, 1982; Kent, 1986; Gordon, 1987).

As an alternative to conventional indicators, absence of undesirable dysfunction, assessed by immuno-competence, life-expectancy or mortality rate, could be a criterion of adequate adjustment to environmental conditions (Payne, 1985). On the other hand, the impact of modernization on traditional economies has led to prevalence rates of diabetes mellitus, alcoholism, obesity, and clinical nutrient deficiencies which suggest that formerly such groups were relatively well-adjusted to their nutritional environment (Høygaard, 1941; Sinclair, 1953; Dugdale & Payne, 1977; Neel, 1977; Robson & Wadsworth, 1977; Fernandes-Cosat *et al.*, 1984; Koike *et al.*, 1984; Gracey, 1986).

The factors considered in this paper are therefore sources of variability within populations; in general anthropometric proxies of nutritional plane are used.

Child labour and its contributions to health

There is strong evidence for high levels of child participation in economic activities in traditional economies. Fig. 11.1 shows some data from surveys covering unpaid family work among other forms of labour. It has

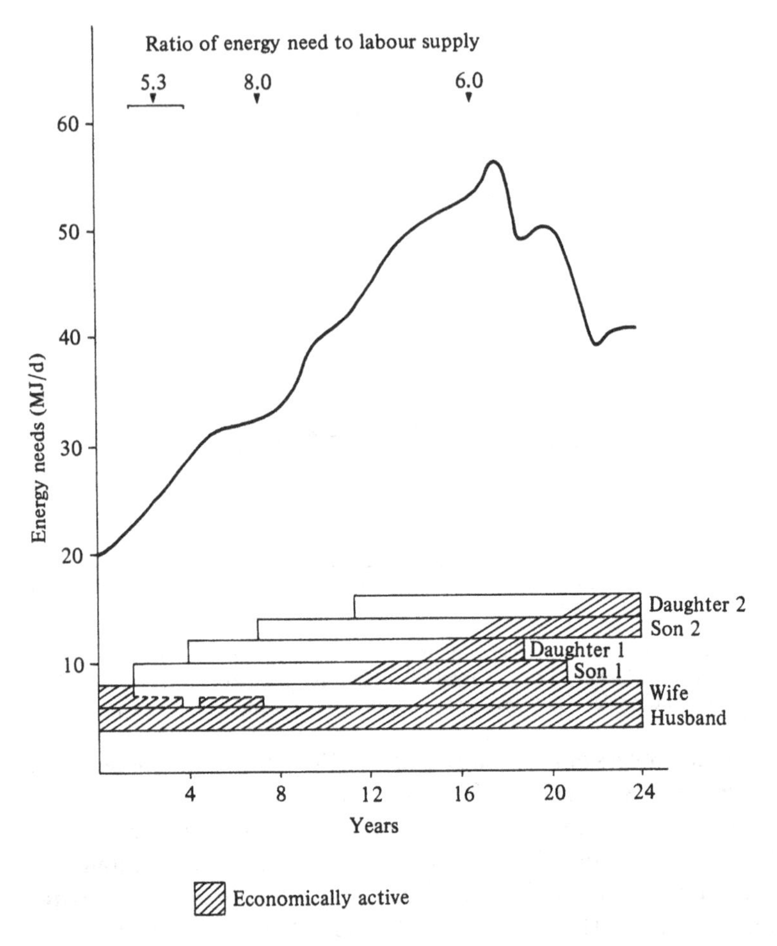

Fig. 11.2 Energy needs and labour supply for a household over 24 years (after Pacey & Payne, 1985; Payne 1985).

been argued speculatively that in rural Bangladesh male children become net producers by 12 years of age, repay their own accumulated consumption by 15, and cover for one sister's by the age of 22 (Cain, 1977). It has also been argued that children of individual couples consume more than they produce until they themselves become parents (Mueller, 1976); that where child labour is economically significant then it will promote high fertility (Kasarda, 1971); and that systems which do not penalize high fertility depend on the persistence of the extended family (Caldwell, 1977).

In many traditional economies the household forms the principal unit of food production and consumption. The household changes demographically with the life cycle, in its resource needs and in its capacity to

Table 11.2. *Effect of birth order on growth performance*

Population	Age	Variable	Effect	Parity affected	Reference
Papua New Guinea	0–maturity	*W*, *H*	–	2, 3, 4	Malcolm, 1970
Papua New Guinea	12–24 mo	*W*/*A*	–	2, 5, 8	Haravey & Heywood, 1983
Guatemala	3.0 y	*H*	?	1–7[a,c]	Russell, 1976
Colombia	0–71 mo	*W*/*A*	–	6+	Wray & Aguirre, 1969
Mali	0.3–3 y	*W*/*A*	–, +	3.7, 5.6[d]	Dettwyler, 1986
Urban Gambia	6–35 mo	*H*/*A*	–	8+	Tomkins *et al.*, 1986
Rural Gambia	0–18 mo	*W*/*A*	–	10+[c]	Prentice, Cole & Whitehead, 1987
Czechoslovakia	6–18 y	*H*	–	3+	Procopec, 1969
California	6–18 y	*H*, *W*	–	4+	Boas, 1895
California	5.0 y	*H*	–	2+[c]	Wingerd & Schoen, 1974
Wisconsin	Young adults	*H*	–	3+	Howells, 1948
U.K.	7–10 y	*H*	+	3, last born	Grant, 1964
U.K.	7.0 y	*H*	–	3+[b,c]	Goldstein, 1971
Germany	20–70 y	*H*	+	2+	Hermanussen, Hermanussen & Burmeister, 1988

W = weight
H = height
A = age
+/– = positive/negative effect; ? = equivocal
[a] = holding number of live siblings constant
[b] = holding number of younger siblings constant
[c] = definition of parity includes still births
[d] = average birth order of low and high *W*/*A* children

meet them. Fig. 11.2 shows a theoretical case to illustrate this over 24 years. The period likely to be of particular risk is when child care time and labour needs most conflict, where the second and third born children indicate that the ratio of overall energy needs to labour available is highest. Over this stage, the first born is not yet or just beginning to be economically active, perhaps by taking on some share in the burden of caring for younger siblings.

Few attempts have been made to assess the effect of child economic contributions on the health status of the household (Stinson, 1980). This could be examined by estimating the effect of birth order or variable dependency ratios on child growth (Russell, 1976). An effect of birth order alone would not differentiate pre-natal from post-natal influences, but evidence from birth weights suggests that the pre-natal environment improves with parity (Rajalakshmi, 1971; Rosenberg, 1988), while the post-natal environment has been argued to be less favourable to the later born (Goldstein, 1971). If child labour significantly affects health outcome of siblings then some effect could be expected from examining data by parity. Evidence supporting this proposition is diverse and conflicting, as shown in Table 11.2. Comparability is affected by inconsistent definitions of birth order and household, by an assumption that the effect of variable birth intervals does not introduce systematic bias, and by a tendency to ignore the confounding effect of family size.

Malcolm (1970) describes the Bundi evidence from Papua New Guinea as highly significant in its overall pattern, though the published statistics are incomplete. There was no effect of family size, nor of maternal height and weight, on children of mothers aged over 30 years grouped by parity. However, compared to the mean height and weight of children at all ages up to maturity, first born children and second born girls appeared to be of above average stature and body weight. This was followed by a decline to the third and fourth born, while subsequent children showed an increase to well above average. Malcolm argues that the observed pattern results from the privileged access to child care of the first born, and of the fifth and subsequent children. Their relative welfare is partly explained by the presence of older siblings helping to forage for useful protein resources such as insects, frogs, rats and mice in an area in which dietary protein is argued to have been limiting linear growth. The intervening second, third and perhaps fourth born are less privileged in that sense, and tend to compete with each other.

The strongest evidence to date is from longitudinal data for pre-school children under five years of age from two rural maize farming communities in Guatemala (Russell, 1976). Fig. 11.3 shows the mean heights at age three for both sexes combined plotted against birth rank, number of

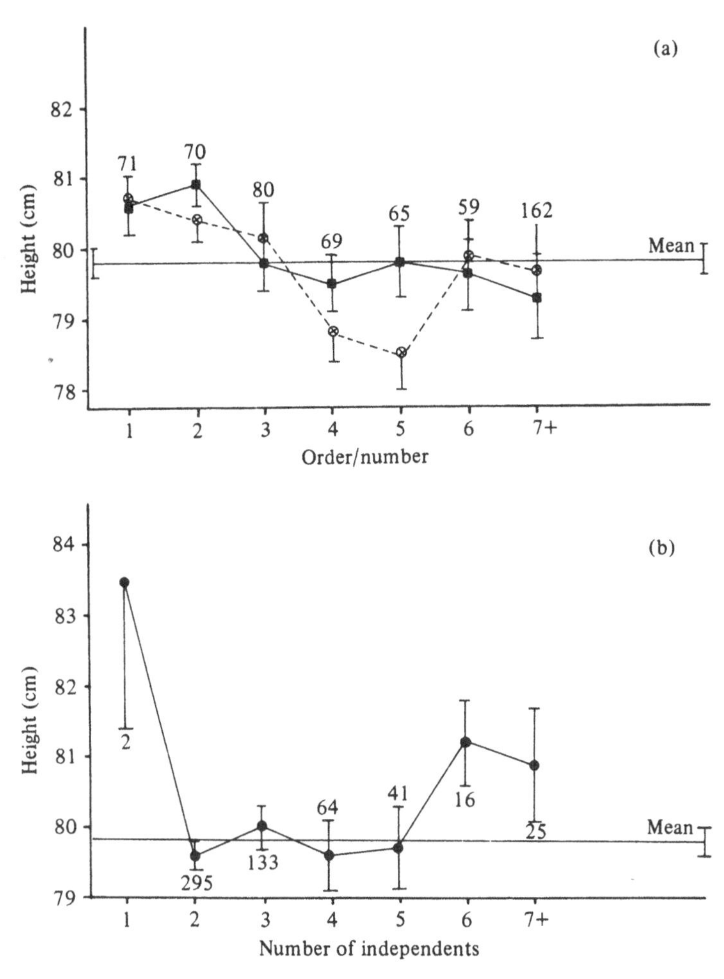

Fig. 11.3 Height at age three (a) by birth order (closed squares) and number of live siblings (crossed circles) and (b) by number of independents aged over 10½ at the birth of the three year old. (Mean ± SE, number of observations for each point given on graph. Data from Guatemala after Russell, 1976).

live siblings, and number of economically significant independents. Birth rank and number of live siblings negatively correlated with height, but the number of economically significant independents, defined as those aged over 10½ at the birth of the index child, was positively associated with height: those with fewer than six independents were significantly shorter ($p<0.05$) than those with six or more. The ratio of independents to dependents also correlated positively with height ($r=0.15$, $p<0.001$).

The implication is that with extended dependency the contribution of older siblings to the welfare of younger ones will tend to lessen. For

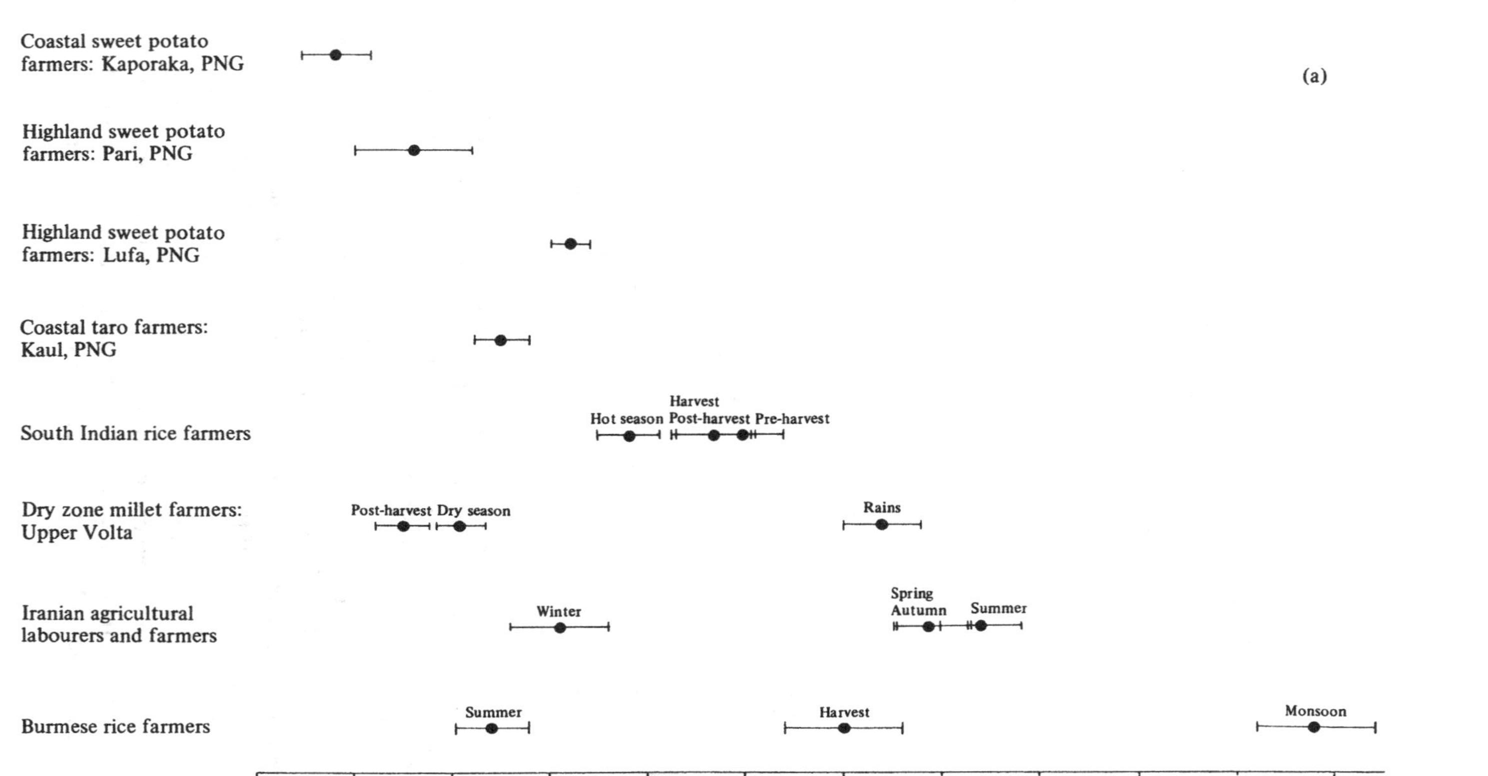

Fig. 11.4 Daily energy expenditure (mean ± SE) for a range of subsistence patterns of (a) men and (b) non-pregnant women. (PNG = Papua New Guinea.)

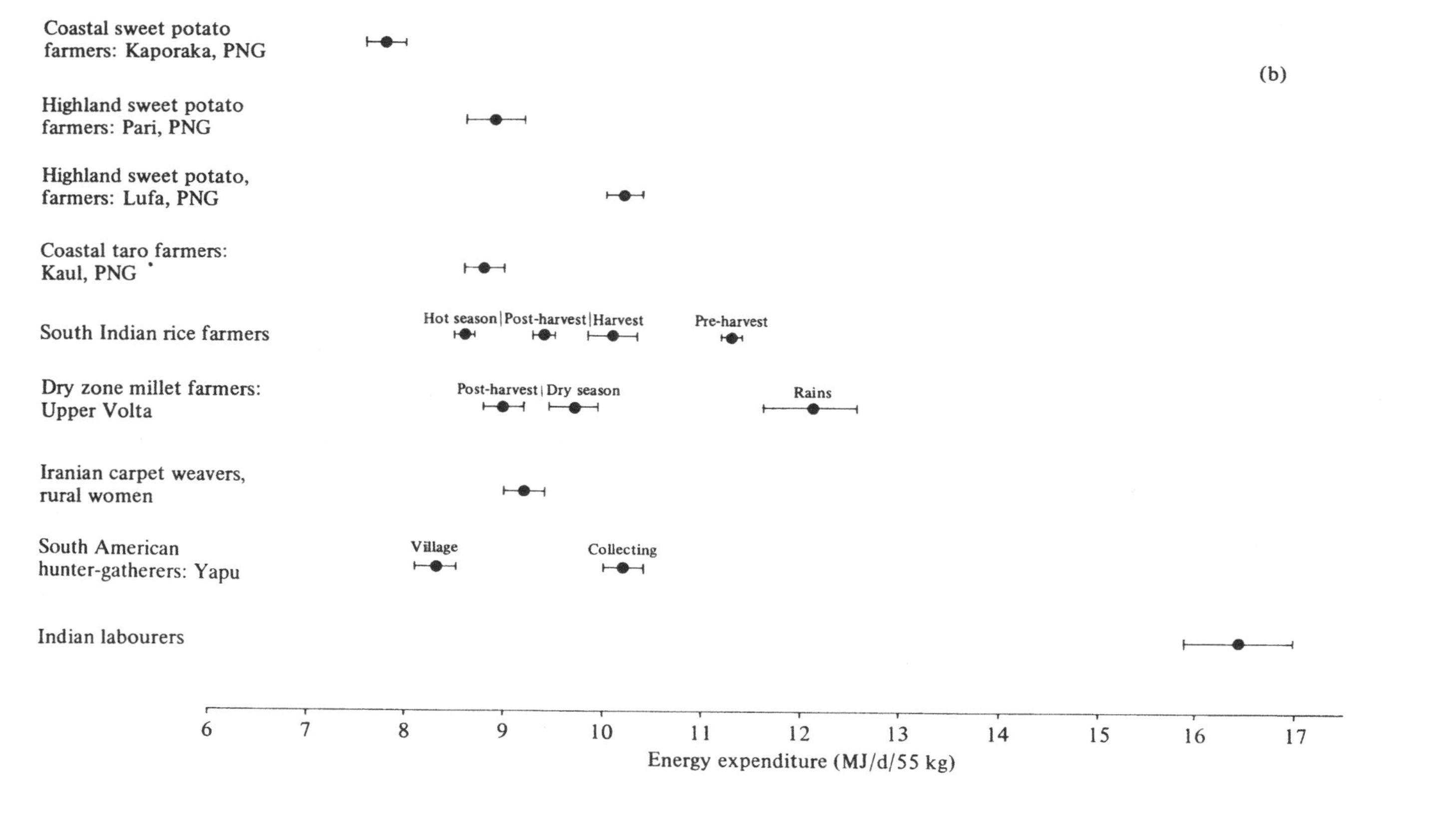

(b)
Coastal sweet potato farmers: Kaporaka, PNG
Highland sweet potato farmers: Pari, PNG
Highland sweet potato, farmers: Lufa, PNG
Coastal taro farmers: Kaul, PNG
South Indian rice farmers
Hot season | Post-harvest | Harvest
Pre-harvest
Dry zone millet farmers: Upper Volta
Post-harvest | Dry season
Rains
Iranian carpet weavers, rural women
South American hunter-gatherers: Yapu
Village
Collecting
Indian labourers
6
7
8
9
10
11
12
13
14
15
16
17
Energy expenditure (MJ/d/55 kg)

Table 11.3. *Mean daily energy expenditure in a range of subsistence patterns*

People	No.	Age (y)	Wt. (kg)	Season	kJ/kg/d	Reference
Male adults						
Kaporaka	8	27.6	60.7	non-seasonal	150	Hipsley & Kirk, 1965
PNG[g] Pari	5	25–40	58.3	non-seasonal	160	Hipsley & Kirk, 1965
PNG Lufa	40	18–49	57.8	non-seasonal	190	Norgan *et al.*, 1974
PNG Kaul	42	18–48	56.3	non-seasonal	170	Norgan *et al.*, 1974
S. India	159	20–49	50.8	pre-harvest	220	McNeill, Payne & Rivers, 1988[a]
	199	20–49	51.2	harvest	210	
	119	20–49	51.3	post-harvest	210	
	129	20–49	50.3	hot	200	
Burma	10	38.4	53.6[b]	monsoon	310	Tin-May-Than & Ba-Aye, 1985
				harvest	230	
				summer	170	
Iran	10	38.6	55.7	spring	250	Brun, Bleiberg & Goihman, 1979
	14	38.3	56.1	summer	260	
	9	32.1	58.8	autumn	250	
	12	37.6	59.1	winter	180	
Volta	23	36.6	58.5[c]	dry	170	Brun *et al.*, 1981
	16	36.6	58.5[c]	wet	240	
	11	28–58	56.5	post-harvest	160	Bleiberg *et al.*, 1981
Machiguenga	15	30	53.9[d]	dry	230	Montgomery & Johnson, 1977
				wet	270	

Female adults						
Kaporaka	5	26.2	50.3	non-seasonal	150	Hipsley & Kirk, 1965
PNG Pari	17	21–35	51.3	non-seasonal	160	Hipsley & Kirk, 1965
PNG Lufa	38	18–49	50.5	non-seasonal	190	Norgan *et al.*, 1974
PNG Kaul	40	18–48	48.1	non-seasonal	160	Norgan *et al.*, 1974
S. India	159	20–49	43.8	pre-harvest	210	McNeill *et al.*, 1988[a]
	220	20–49	43.6	harvest	180	
	166	20–49	43.5	post-harvest	170	
	147	20–49	43.8	hot	160	
Iran	13	28.8	50.6	non-seasonal[e]	170	Geissler *et al.*, 1981
Volta	12	20–40	50.6[f]	dry	180	Bleiberg *et al.*, 1980
				wet	220	
	14	19–48	49.9	post-harvest	160	Bleiberg *et al.*, 1981
India	7	18	43.5	non-seasonal[e]	300	Devadas, Anuradha & Rani, 1975
Yapu	11	34.0	49.6	in village	150	Dufour, 1984
				foraging[e]	190	
Machiguenga	20	16–41	44.5	non-seasonal[e]	180	Montgomery & Johnson, 1977

[a]Data for this study were derived by 24-hour activity recall and the energy cost of activities using multiples of BMR according to FAO/WHO/UNU (1985) recommendation. The BMR predicted from body weight was corrected by − 12.1%. All other data in this table derive from studies using indirect calorimetry and time budget diaries.

[b]Mean for the whole year.

[c]Mean for group of 30 from whom subjects were drawn for energy expenditure estimates.

[d]No seasonal change in weight is described in this report.

[e]The work pattern is not reported to be seasonal.

[f]Mean dry season body weight for a sample of 15 women studied.

[g]PNG = Papua New Guinea.

women in traditional economies, the demands of child care and food procurement are likely to conflict. Solutions to this vary. Hunter-gatherer groups like the !Kung, the Eskimos, and the Ache differ in women's contribution to food procurement and in the emphasis placed on high-risk, high-return strategies adopted by men (Høygaard, 1941; Hill *et al.*, 1985; Hurtado *et al.*, 1985). Microenvironmental factors influence the costs and benefits of specific methods of resource use (for the Pygmies; Milton, 1985). Among sedentary populations, different farming systems have different implications for the need to look beyond the immediate family for labour at certain seasons, as indicated for several areas (Roberts *et al.*, 1982; Nabarro, 1984; Richards, 1986). Those who are poorly cared for in the rainy season will also be at worst risk from the combined effects of nutritional deficiency and infection. This can pose difficult decisions for the farming household faced with incurring debt, compromising next year's harvest or current family health. In such circumstances it is not surprising that child labour contributions could be perceived to be desirable for the health of a household. As yet, however, there appears to be little substantive evidence supporting the conjecture that households pass through predictable periods of risk with lasting consequences for health, and that child contributions to labour significantly alleviate this risk.

Adult energy expenditure and seasonality

Levels of physical exertion differ among subsistence patterns and in some there is significant seasonal variation. Fig. 11.4 shows the mean daily energy expenditure of adult men and women at distinct seasons for a variety of economies, and for the whole year in economies with insignificant seasonal variation in activity patterns. These data are derived from studies using different methods, so that the comparisons are imprecise and should be treated cautiously.

When these values are expressed per kilogram body weight (see also Table 11.3), the extent of seasonal disparity among men is greatest in the Burmese rice farmers (ratio of peak to trough 1.8), intermediate for the W. African farmers and Iranian agricultural labourers (1.4–1.5) and lowest for the farming villagers of S. India (1.1). The exertion required for the non-seasonal subsistence patterns, represented by examples from Papua New Guinea, approximates to that of the off-season of the other economies.

Among women, the seasonal disparity appears somewhat less than in men. The seasonality of W. African agriculture shows the highest ratio (1.4). The S. Indian ratio is about 1.3. The non-seasonal economies again

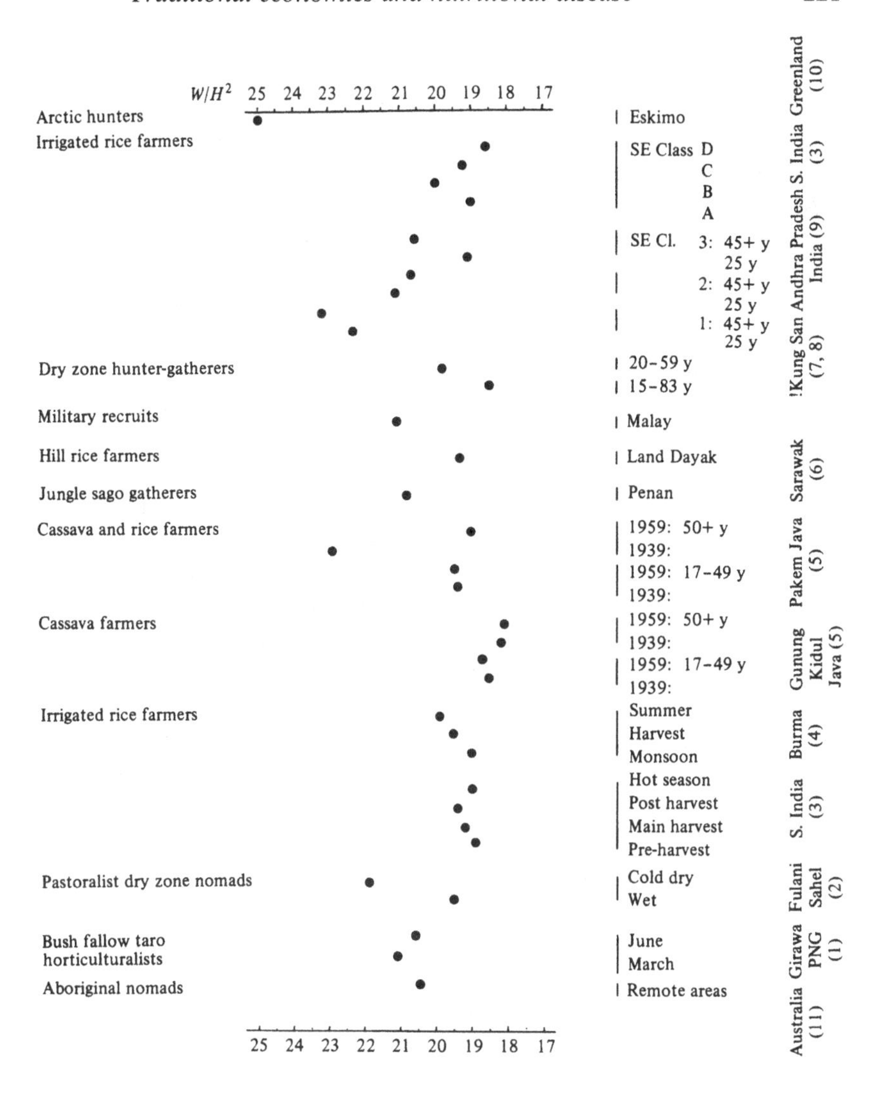

Fig. 11.5 Mean Quetelet index (W/H^2) for adult males from a variety of populations and subsistence patterns. References: (1) Spencer & Heywood, 1983, (2) Benefice, Chevassus-Agnes & Barral, 1984, (3) McNeill *et al.*, 1988, (4) Tin-May-Than & Ba-Aye, 1985, (5) Bailey, 1962a, (6) Anderson, 1979, (7) Truswell & Hansen, 1976, (8) Lee, 1979, (9) Rao & Satyanarayana, 1976, (10) Høygaard, 1941, (11) Gracey, 1986.

exhibit magnitudes which are similar to those of off-season activity in the others. A report for Turkana pastoralists indicates a ratio of about 1.1 for women; and a low level of physical exertion (130–150 kJ/kg/d) relative to all the other examples given here (Galvin, 1985).

As for the relative exertion of men and women, the data suggest that for each season or instance the ratios are low and range between 1.1 and 0.9. The S. Indian evidence suggests a ratio varying from 1.0 to 1.3 over the seasons. The highest ratio of 1.5 is reported for Machiguenga hunter-gatherers.

Some seasonal changes in adult energy balance have been observed. These changes seem rarely to exceed about 2% of the mean body weight for the year (Høygaard, 1941; Galvin, 1985; Dugdale & Payne, 1987). Fig. 11.5 shows average male Quetelet Index (QI) values (W/H^2) as a measure of current energy balance for a range of distinct subsistence patterns, some of which show seasonal variation. Most of these values fall below the recent FAO/WHO/UNU (1985) Report's suggested cut-off point of 20.1, either seasonally in the case of the Sahelian Fulani, or throughout the year for which the data apply in the case of the S. Indian and Burmese rice farmers, Javanese cassava growers, Sarawaki hill rice shifting cultivators, and Kalahari bushmen. The Indian examples shown by social economic class suggest that a similar range can be observed within a given population. There are limited data on the relative risk of morbidity or mortality for populations differing in this Index, apart from Waaler's Norwegian survey (1984). Thus it is not clear how deviations below the cut-off point are to be interpreted.

There is limited evidence that the seasonal changes which occur in this way are consistent with an attempt to buffer non-adults from seasonal nutritional stress, as suggested for !Kung Bushmen (Lee, 1979). In a study of the E. African Turkana, Galvin found that adolescent and adult energy intakes and body weights showed significant seasonal variation while those of children were apparently protected (Galvin, 1985). Thus in these cases there was little to support the hypothesis that, under normal conditions, those who perform the most essential productive tasks will have access to foodstuffs at the expense of young dependents. There is evidence elsewhere that nutritionally vulnerable children tend to be favoured when access to resources is increased, either by nutritional intervention programmes as in Guatemala (Martorell *et al.*, 1979), or by raised incomes as in urban Indian (Payne, 1985). A recent study on the allocation of food within families in Bangladesh suggested that family members over age five would share seasonally reduced intakes in proportion to activity level and body size, but the share of children aged one to four was lower than their theoretical requirements.

The share of young girls increased when food was least available and work least strenuous (Abdullah & Wheeler, 1985). This has also been noted for South India, though the reverse was found in Japan (Wheeler, 1988). There are no such studies for most traditional economies, for good logistical reasons, but a review of the scanty literature, mostly from South Asia, suggests that the youngest age group of one to three year olds consistently appears underfed in energy terms, while adult women suffer a relative lack of micronutrients notably iron (Wheeler, 1988).

It could be argued that there is 'buffering' of small children if infanticide of the newborn occurs where there are difficulties with early weaning, as suggested by Townsend (1971) for some New Guinea sago-gatherers. From the woman's viewpoint, this is arguably an alternative to the policy of an abrupt weaning when she becomes pregnant again, but for most types of subsistence pattern the question of effective selective buffering cannot be answered adequately on present nutritional evidence.

Adult nutritional status and physical work capacity

Traditional economies can be defined as those dependent largely on human energy expenditure for the procurement of food resources, making little use of fossil fuel based energy subsidies (Pimentel & Pimentel, 1979; Bayliss-Smith, 1984). Optimal foraging theory suggests that human groups tend to exploit resources according to the energy and nutrient returns to time investment or energy expenditure required, though with non-nutritional requirements playing a highly significant role (Hill, 1982; Keene, 1985). It has been suggested that calorie returns to energy expenditure of foraging women will significantly influence fertility through effects on energy balance and duration of post-partum amenorrhoea (Howell, 1976; Bentley, 1985), but a convincing biological basis for this argument has not been presented.

Factors impairing physical work capacity may have consequences for household nutritional status on the model of general linkages presented in Fig. 11.6. Since differing subsistence patterns will make varying demands on physical exertion, the extent to which factors affecting aerobic capacity limit productivity and income will likewise be variable.

Fig. 11.7 shows $\dot{V}_{O_2}$ max estimates derived by a mixture of direct and indirect methods for various populations and plotted against average QI values. Comparison of western with non-western populations suggests that differences in $\dot{V}_{O_2}$ max largely disappear when corrected for body weight (Ferro-Luzzi, 1985). They are likely to be functions of lean body mass or muscle cell mass, partly also of other differences in oxygen

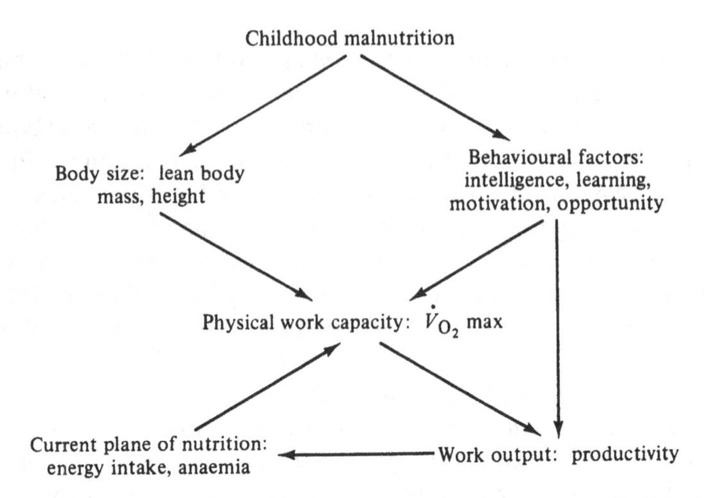

Fig. 11.6 General relationships between nutrition, fitness and productivity (after Martorell & Arroyave, 1988).

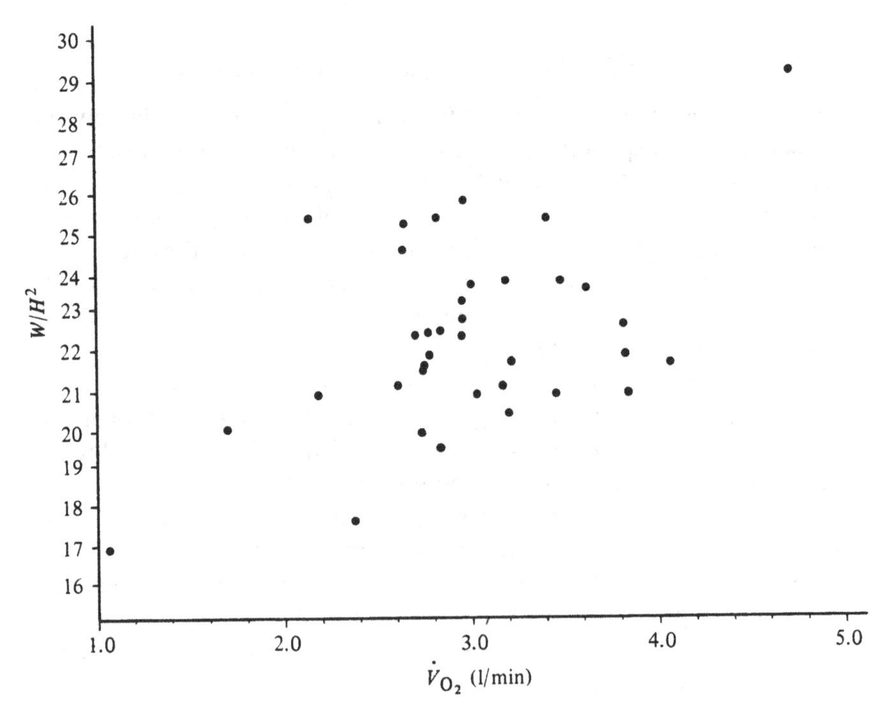

Fig. 11.7 Mean W/H^2 by $\dot{V}_{O_2}$ max of adult males of various occupations and economies (drawn from data in Ferro-Luzzi, 1985).

transport capacity. The plot suggests that small body size in these terms would be associated with reduced physical work capacity to a significant extent (by linear regression, $r^2 = 0.199$; $p < 0.01$). If it could be shown that productivity, household income and nutritional outcomes were significantly affected by small body size, then the functional significance of low QI values could be interpreted more readily.

Small body size has sometimes been argued to be an advantageous adaptation to poor living conditions (Seckler, 1984). If the evidence for advantage is broadened from morbidity risk to reproductive success, then there is further support on grounds of larger family size and survival among shorter fathers (Russell, 1976; Sinnett, 1977). On the other hand, if the cost of an activity can be estimated in terms of the $\dot{V}_{O_2}$ uptake required to perform it, independently of body weight, then this will represent a variable proportion of the aerobic capacity of individuals differing in physical work capacity. Since about 35–40% $\dot{V}_{O_2}$ max is approximately the upper limit sustainable over an eight hour working period, the amount of the task completed will be limited by those factors which influence aerobic capacity (Spurr, 1984).

For sugar cane cutters there is evidence for a significant effect of $\dot{V}_{O_2}$ max and maximum aerobic power, height, weight and lean body mass on productivity at work, at least for a severely malnourished group of QI about 17.0 (Spurr, 1988). While cane cutting is continuous and moderately heavy work, evidence suggesting an effect on productivity of several indices of physical fitness was also observed for people engaged in rather intermittent work of cane loaders (Spurr, Maksud & Barac-Nieto, 1977). Immink *et al.* (1984) found a consistent association between lean body mass, coffee picking and weeding, cane loading and cane cutting performance. These are consistent with Viteri's observations (1971). On the other hand, Satyanarayana *et al.*'s (1977) study of Indian industrial workers observed that productivity per day at lighter work loads was related to weight, height and lean body mass.

It would appear that the percentage of $\dot{V}_{O_2}$ max sustainable and absolute aerobic capacity may be a constraint to productivity and income in certain types of regimented and vigorous work where there is a clear relation between immediate work output and economic gain; but that body size may, for reasons that are unclear, also affect productivity in other ways.

It is unclear how far household productivity in traditional economies is adversely affected by adult nutritional status in terms of this mechanism. One hypothesis derives from an argument advanced by Weiner (1972) in a discussion of tropical slash-and-burn agricultural systems. Where there are thin soils and high rainfall, plant residues are the principal source of crop nutrients, and a common characteristic of slash-and-burn systems is an extended bush fallow cycle with relatively low population densities. If, as is often the case, governments wish to protect primary forest areas for commercial or ecological reasons, several responses to population growth can occur other than expansion into otherwise virgin territory and use of fossil fuel energy supports to production. Most attention has been paid to social changes rather than to physiological, biological or behavioural adjustments. The first includes migration or migrant labouring; a variety of conservative farming practices; changes in local land tenure custom; reduced age-specific fertility rates; and greater seasonal and total work inputs to agricultural enterprise (Padoch, 1982). Furthermore, change in cropping patterns, with greater reliance on cassava or other 'famine' crops, has been argued to occur as just such a response in Java (Bailey, 1962b).

Descriptive models of the capacity for land to support populations engaged in shifting agriculture include several variables. These represent the percentage of accessible territory that is cultivatable with the technology at hand (C_p); the area necessary to support one person per year at the level observed (C_a); the number of plots needed to permit an adequate fallow rest period (L, which $= (r/u) + 1$, where $r =$ years fallow, and $u =$ years of successive use) (Brush, 1975). For example, the formula Weiner gives for the overall area per caput required by a given population is:

$$K = 100 \times (C_a \cdot L)/C_p$$

Thus, if the observed values for a given population show that it is broadly self-sufficient in food production, these values will be used to represent the variables in the model, which therefore assumes a steady state. Such models pose problems in the specification of 'adequate' fallow periods; and the need to indicate the yield per unit area required to support one person per annum raises the question of whether to adopt officially recommended estimates of nutritional requirements, or to adopt those values which represent intakes locally judged to be sufficient. Human capacity to tolerate and to adjust to variable planes of nutrition and infectious disease loads then becomes a factor significant to the model.

Weiner's argument takes account of this factor by incorporating physical work capacity as an element. He postulates that a family of five persons would require about 12,000 kcal per day. To work the necessary acreage would need about 5,000 kcal per day, entailing an extraction rate of about 1:2.5. The most active worker, who would contribute about half the labour energy required, would expend about 300 kcal/h or a rate of

Table 11.4. *Rice farming: human energy expenditure estimates for two systems (after Bayliss-Smith, 1981; Payne, 1985)*

Shifting cultivation			Paddy farming		
Stage	man-d/ha	kJ/kg/h	Stage	man-d/ha	kJ/kg/h
Clearing	18	16	Irrigation + nurseries + first plough	47	27
Burning	4	16	Ploughing + manuring + transplanting	38	25
Tillage + sowing + fencing	26	13	Weeding + fertilizing	21	12
Weeding	35	10	Weeding + pesticides	19	12
Harvest	43	8	Harvest + transport	26	17
Threshing + winnowing	25[a]	19	Threshing + winnowing	23	19
Total input	158	692 MJ/ha[b]		174	1354 MJ/ha[c]
Yield[d]	1012 kg/ha	14890 MJ/ha		1604 kg/ha	23600 MJ/ha
Extractive ratio		22:1			18:1

[a]Estimate from unpublished field data from Sarawak for six fields.
[b]Average of male and female adult body weights 46 kg derived from Anderson (1979) from Sarawak Land Dayak and assumed here, with 8 h/day.
[c]Average of male and female adult body weights 47 kg derived from McNeill *et al.* (1988) S. Indian data assumed here, with 8 h/day.
[d]From Bayliss-Smith (1981); assumes metabolizable energy value of 14.7 MJ/kg rice, converted from Platt's (1962) estimate.

O_2 uptake of about 1 l/min. Since it is expected that work output at 35–40% of $\dot{V}_{O_2}$ max could be sustained for about eight hours, those principal workers whose maximal aerobic capacity fell below about 3.0 l O_2 would be unable to achieve such a level of work. Thus, if anaemia, food shortage or other factors could be shown to impair work output through their effects on physical work capacity, these would justify incorporating this variable (3.0 l O_2/min $= W_c$ of 1) into the carrying capacity model. Weiner's model therefore is as follows:

$$K = 100 \times (C_a \cdot L)/(C_p \cdot W_c)$$

Is there any empirical support for this model?

First, it can be noted that if literature values for energy costs of work in shifting agriculture (Table 11.4) are applied to observed body weights of shifting cultivators of hill rice, the resulting levels of work output are considerably less than the 300 kcal/h suggested by Weiner. This is the case even where the tasks involved are significantly constrained by time limitations. This is partly because the body weight values are low: Sarawaki group average values (adults, sexes combined) fall around 46 kg (Anderson, 1979); while those for a recent Philippine sample aged 18 and over are, for males 50.6 (se $=0.85$, $n=41$) and females 44.3 (se $=0.97$, $n=51$) (unpublished data from FAO (Rome) and University of the Philippines, Los Baños).

Second, comparison between the labour time spent on short as against long fallow plots (Strickland, 1986), and whether within or between communities (Padoch, 1982), suggests that weeding is the task most likely to be modified if labour needs increase. This is a task of moderate energy cost, estimated in FAO/WHO/UNU (1985) to be about $2.9 \times$ BMR for women, and ranging between 2.5 and $5.0 \times$ BMR for men. It may not require much lifting of the body mass, but rather skilled and vigorous use of the upper part of the body. It seems unlikely that capacity to sustain a high fixed level of energy expenditure, at a proportion of $\dot{V}_{O_2}$ max varying with fitness, would limit the productivity of weeding per unit area.

For the severely malnourished adults considered by Spurr (1988), with QI of about 17.0 and body weight 43 kg, the most strenuous agricultural task listed in Table 11.4 as 16 kJ/kg/h would amount to about 0.55 l O_2/min or about 52% of the $\dot{V}_{O_2}$ max observed (1.05 l/min). The cost of weeding would amount to about 33% $\dot{V}_{O_2}$ max. Thus, even on Spurr's argument, the major variable component in slash-and-burn agricultural method would seem unlikely to be out of reach for the severely malnourished adult. Since the extent of weeding tends to be greater the shorter the fallow, and the shorter the fallow the less time required to fell large trees rather than clearing scrub, there is some flexibility in being able to choose how hard to work.

It could be argued that only time-intensive tasks would be affected by reduced physical working capacity. The tasks of sowing and harvesting would, for Spurr's severely malnourished subjects, require them to work at about 42 and 26% of $\dot{V}_{O_2}$ max respectively. However, there are various means of coping with the need to complete the sowing and harvesting of crops in short periods of time, notably the arrangements in which households pool their labour resources to complete work on each other's fields successively. This, indeed, is one way of spreading the variable effects of different working skill and capacity throughout a community, and one reason why communities seem to require a minimum size to subsist in some areas.

Moreover, even if it could be shown that factors impairing physical work capacity set limits upon productivity at work, it is not clear how far this would translate into reduced food energy output. Slash-and-burn farming systems tend to be diverse in the range of crops planted together on a single plot, and wide ranging in the combinations of wild, protected and cultivated resources exploited. Inter-household variability in yields of a single crop can be remarkably high, for example with a coefficient of variation for rice yields of 164% in one Bornean study (Dove, 1982). On the other hand, there are few published estimates for year-on-year variation of individual households' yields (Dugdale & Payne, 1988). The shifting cultivators work a variety of options, each of which may seem to have low productivity, but which overall can provide a broadly adequate subsistence pattern. Such diverse patterns constitute long-term, risk-spreading strategies rather than attempts to maximize immediate gain.

Whereas on these grounds there seems little evidence to support Weiner's model, it is clear that for paddy farmers (Table 11.4) the first and second stages are particularly strenuous and approach or exceed 300 kcal/h. Overall, paddy farming would seem more labour intensive, although the total days' work involved is not dissimilar; and with relatively high yields (in the mid 1970s) the extractive ratios are comparable between the two systems. It seems that factors limiting $\dot{V}_{O_2}$ max are more likely to have some impact on farm productivity in this type of system than under shifting agriculture. Thus even though such systems are conventionally described as 'land limited' rather than 'labour limited', there is probably a sense in which they are limited by labour power.

Although the model assumes a steady state, there are factors other than food energy output which affect physical work capacity. There is good evidence for adverse effects of anaemia on physical work capacity,

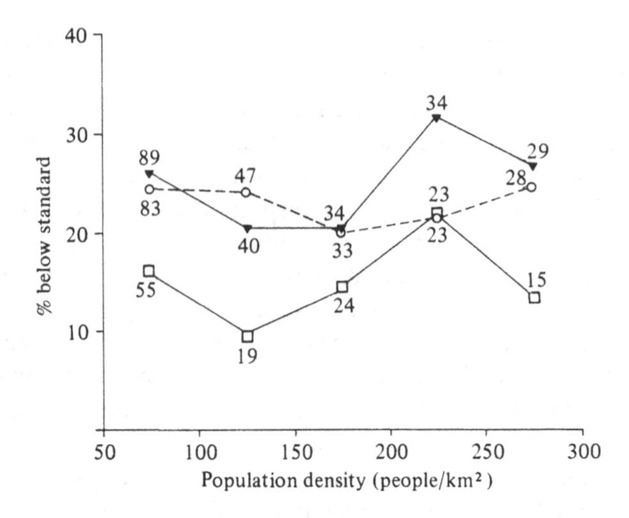

Fig. 11.8 Nutritional status by population density: children aged 6–30 mo in Simbu, Papua New Guinea (after Harvey & Heywood, 1983). Triangles represent <80% *W/A*, open circles <90% *L/A* and open squares <90% *W/L*. Differences among population densities were not significant ($p > 0.05$). Numbers of observations making up each point are given on the graph.

but recent experimental and field results suggest that compensatory mechanisms may reduce the functional effects of mild anaemia at submaximal exercise levels (Collins, 1982; Collins, Abdel-Rahaman & Awad El Karim, 1988). There appear to be no studies linking prevalence of parasitic infections, anaemia, physical work capacity and nutritional outcome for slash-and-burn farmers, but Polunin (1977) argued that shifting cultivation in Malaysia may promote risk of malaria, since by admitting sunlight removal of forest cover would favour breeding of *Anopheles maculatus*. He also noted that periodic mobility and small susceptible population size reduce risk of parasitic infection by breaking the life-cycle. Shifting cultivators, however, usually move their fields from year to year rather than their settlements, although there are exceptions like the Batak (Cadeliña, 1985).

In terms of outcome, an attempt has been made to relate prevalence of child malnutrition to an agricultural stress index in an area of highland Papua New Guinea (Harvey & Heywood, 1983). This index was devised from the number of plantings during cultivation periods and the average length of fallow. It was based on small numbers of plots in seven areas, but it incorporates elements of the carrying capacity model. While the coefficient of correlation between this index and proportion of children below 90% *L/A* (*L* = length) in seven areas was 0.76 ($p<0.05$), there was no significant variation in numbers of children below the cut-off point.

Table 11.5. *Criteria of social economic disparity associated significantly with nutritional status indicators in a range of studies*

	W/A	*W/H*	*H/A*	*W/H*^{2}	X	B	Hb	SF	AC	DW	DH	DQ
House rating	o a g j q	a	o a g q	a	a	a			g			
Income	b p l	b l	b l									
Parental occupation	j r n		n r		s			r		n	n	
Parental education	e b c r		o e i r		s		e c	r	e			
Land availability	e m j	f	e f						e			
% food expenditure	p											
Food exp. (gross or per caput)	p i											
Family size					s							
Birth order	p		q k									
Number of siblings	p j		k									
Caste	e b		e					e				
Crop production	e		e				e		e			
Assets profile	g h i j		g q					h	g h			
Water supply			q		s							
Sanitation			q									
Working mother	o	o	o									o

X = Indices of severity of xerophthalmia
B = Vitamin B complex deficiency signs
Hb = Haemoglobin
SF = Skinfold thickness(es)
AC = Mid upper arm circumference
DW = Velocity of growth in weight
DH = Velocity of growth in height
DQ = Developmental quotient

References

	Country	Rural/Urban	Source
a	India	R	Rao & Satyanarayana, 1976
b	India	R	Levinson, 1974
c	India	R	Devadas, Rajalakshmi & Kaveri, 1980
d	India	R	McNeill *et al.*, 1988
e	Nepal	R	Martorell, Leslie & Moock, 1984
f	Nepal	R	Nabarro, 1984
g	Honduras	R	Martorell, Mendoza & Castillo, 1988
h	Mexico	R	Malina *et al.*, 1985
i	Haiti	R	Smith *et al.*, 1983
j	Costa Rica	R	Rawson & Valverde, 1976
k	Guatemala	R	Russell, 1976
l	Guatemala	R	Valverde *et al.*, 1981
m	Guatemala	R	Valverde *et al.*, 1977
n	Guatemala	U	Bogin & MacVean, 1978
o	Jamaica	U	Powell & Grantham-McGregor, 1985
p	Colombia	R	Wray & Aguirre, 1969
q	The Gambia	U	Tomkins *et al.*, 1986; Tomkins, Dunn & Hayes, 1989
r	Bolivia	R	Stinson, 1983
s	Indonesia	R	Sommer, 1982

Similarly difficult to interpret was an attempt to relate nutritional status to population density (Fig. 11.8). Variation in nutritional status was significantly related to absenteeism from the village at population densities exceeding 150/km^2. Absenteeism would be expected to be one response to increased population under such subsistence patterns (Padoch, 1982). This suggests a more complex picture than that proposed by the argument linking traditional economies and nutritional disorders, as summarized in Table 11.1 at the outset of this paper. It also suggests the importance of collecting adult nutritional status data in such populations in future.

Conclusions

Much evidence exists for associations between social economic disparity and various indices of nutritional outcome. For a number of examples this is summarized in Table 11.5. Such data presented in static form pose questions about the emergence and distribution of nutritional disorders over time. While detailed discussion of this problem falls beyond the scope of this paper, it has been useful to consider three mechanisms whereby observed patterns of nutritional status could come about.

Within the household, the evidence of relationships between growth performance, birth rank and dependency ratio suggests that there is scope for more investigation into the health implications of child labour. In particular, these may partly explain persisting levels of high fertility in traditional economies. In addition, there is limited evidence for selective buffering of non-adults against seasonal nutritional stress, but such evidence is available for few populations. There is scope for further research in this area.

While adult nutritional status may affect physical work capacity and productivity under strict working conditions, it seems that in traditional economies this is more likely to occur under irrigated rice farming than under slash-and-burn subsistence patterns. Studies of factors affecting physical work capacity among the economically active and nutritional outcome among dependents in both types of economies are required to test this hypothesis.

Acknowledgements

Grateful thanks for comments on an earlier draft of this paper are due to P.R. Payne, G.B. Spurr, A.H. Strickland, and S.J. Ulijaszek. The author thanks FAO (Rome) and the University of the Philippines at Los Baños for permission to cite unpublished anthropometric statistics collected under investigation number TCP/PHI/6652.

References

Abdullah, M. & Wheeler, E.F. (1985). Seasonal variations, and the intra-household distribution of food in a Bangladeshi village. *American Journal of Clinical Nutrition*, **41**, 1305–13.

Anderson, A.J.U. (1979). Subsistence of the Penan in the Mulu area of Sarawak. *The Sarawak Gazette*, 30th November, 204–16.

Bailey, K.V. (1962a). Rural nutrition studies in Indonesia. X. Weight and height of Gunung Kidul adults. *Tropical and Geographical Medicine*, **14**, 230–7.

Bailey, K.V. (1962b). Rural nutrition studies in Indonesia. XI. The Gunung Kidul problem in perspective. *Tropical and Geographical Medicine*, **14**, 238–58.

Bayliss-Smith, T.P. (1981). Seasonality and labour in the rural energy balance. In: *Seasonal Dimensions to Rural Poverty*, ed. R. Chambers, R. Longhurst & A. Pacey, pp. 30–8. Frances Pinter, London.

Bayliss-Smith, T.P. (1984). Energy flows and agrarian change in Karnataka: the Green Revolution at micro-scale. In: *Understanding Green Revolutions*, ed. T.P. Bayliss-Smith & S. Wanmali, pp. 153–72. Cambridge University Press, Cambridge.

Benefice, E., Chevassus-Agnes, S. & Barral, H. (1984). Nutritional situation and seasonal variations for pastoralist populations of the Sahel (Senegalese Ferlo). *Ecology of Food and Nutrition*, **14**, 229–47.

Bentley, G.R. (1985). Hunter-gatherer energetics and fertility: a reassessment of the !Kung San. *Human Ecology*, **13**, 1, 79–109.

Berkes, F. & Farkas, C.S. (1978). Eastern James Bay Cree Indians: changing patterns of wild food use and nutrition. *Ecology of Food and Nutrition*, **7**, 155–72.

Billewicz, W.Z. & McGregor, I.A. (1981). The demography of two West African (Gambian) villages, 1951–75. *Journal of Biosocial Science*, **13**, 219–40.

Billewicz, W.Z. & McGregor, I.A. (1982). A birth-to-maturity study of heights and weights in two West African (Gambian) villages, 1951–75. *Annals of Human Biology*, **9**, 4, 309–20.

Bleiberg, F.M., Brun, T.A., Goihman, S. & Gouba, E. (1980). Duration of activities and energy expenditure of female farmers in dry and rainy seasons in Upper-Volta. *British Journal of Nutrition*, **43**, 71–82.

Bleiberg, F.M., Brun, T.A., Goihman, S. & Lippman, D. (1981). Food intake and energy expenditure of male and female farmers from Upper-Volta. *British Journal of Nutrition*, **45**, 505–15.

Boas, F. (1895). The growth of first-born children. *Science*, **1**, 15, 402–4.

Bogin, B.A. & MacVean, R.B. (1978). Growth in height and weight of urban Guatemalan primary school children of low and high socioeconomic class. *Human Biology*, **50**, 4, 477–87.

Brun, T.A., Geissler, C.A., Mirbagheri, I., Hormozdiary, H., Bastani, J. & Hedayat, H. (1979). The energy expenditure of Iranian agricultural workers. *American Journal of Clinical Nutrition*, **32**, 2154–61.

Brun, T.A., Bleiberg, F. & Goihman, S. (1981). Energy expenditure of male farmers in dry and rainy seasons in Upper-Volta. *British Journal of Nutrition*, **45**, 67–75.

Brush, S.B. (1975). The concept of carrying capacity for systems of shifting cultivation. *American Anthropologist*, **77**, 799–811.

Cadeliña, R.V. (1985). *In time of want and plenty: the Batak experience*. Silliman University, Dumaguete City.
Cain, M.T. (1977). The economic activities of children in a village in Bangladesh. *Population and Development Review*, **3**, 201–27.
Caldwell, J.C. (1977). The economic rationality of high fertility: an investigation illustrated with Nigerian survey data. *Population Studies*, **31**, 5–27.
Cassidy, C.M. (1980). Nutrition and health in agriculturalists and hunter-gatherers: a case study of two prehistoric populations. In: *Nutritional Anthropology*, ed. N.W. Jerome, R.F. Kandel & G.H. Pelto, pp. 117–45. Redgrave, New York.
Collins, K.J. (1982). Energy expenditure, productivity and endemic disease. In: *Energy and Effort*, ed. G.A. Harrison, pp. 65–84. Taylor & Francis, London.
Collins, K.J., Abdel-Rahaman, T.A. & Awad El Karim, M.A. (1988). Schistosomiasis: field studies of energy expenditure in agricultural workers in the Sudan. In: *Capacity for Work in the Tropics*, ed. K.J. Collins & D.F. Roberts, pp. 235–47. Cambridge University Press, Cambridge.
Dettwyler, K.A. (1986). Infant feeding in Mali, West Africa: variations in belief and practice. *Social Science and Medicine*, **23**, 7, 651–64.
Devadas, R.P., Anuradha, V. & Rani, A.J. (1975). Energy intake and expenditure of selected manual labourers. *Indian Journal of Nutrition and Dietetics*, **12**, 279–84.
Devadas, R.P., Rajalakshmi, R. & Kaveri, R. (1980). Influence of family income and parents' education on the nutritional status of preschool children. *Indian Journal of Nutrition and Dietetics*, **17**, 237–44.
Dove, M.R. (1982). The myth of the 'communal' longhouse in rural development: the Kantu' of Kalimantan. In: *Too Rapid Rural Development*, ed. C. MacAndrews & Chia Lin Sien, pp. 14–78. Ohio University Press, Athens, Ohio.
Dufour, D.L. (1984). The time and energy expenditure of indigenous women horticulturalists in the Northwest Amazon. *American Journal of Physical Anthropology*, **65**, 37–46.
Dugdale, A.E. & Payne, P.R. (1977). Pattern of lean and fat deposition in adults. *Nature*, **266**, 349–51.
Dugdale, A.E. & Payne, P.R. (1987). A model of seasonal changes in energy balance. *Ecology of Food and Nutrition*, **19**, 231–45.
Dugdale, A.E. & Payne, P.R. (1988). Variability in crop yields as a cause of failure among peasant farmers. *Ecology of Food and Nutrition*, **22**, 117–23.
Eaton, S.B. & Konner, M. (1985). Palaeolithic nutrition. A consideration of its nature and current implications. *New England Journal of Medicine*, **312**, 5, 283–9.
FAO/WHO/UNU (1985). *Energy and Protein Requirements*. World Health Organization, Geneva.
Fernandes-Costa, F.J., Marshall, J., Ritchie, C., Tonder, S.V. van, Dunn, D.S., Jenkins, T. & Metz, J. (1984). Transition from a hunter-gatherer to a settled lifestyle in the !Kung San: effect on iron, folate, and vitamin B12 nutrition. *American Journal of Clinical Nutrition*, **40**, 1295–303.
Ferro-Luzzi, A. (1985). Work capacity and productivity in long-term adaptation to low energy intakes. In: *Nutritional Adaptation in Man*, ed. K.L. Blaxter & J.C. Waterlow, pp. 61–9. John Libbey, London.

Foley, R. (1987). *Another Unique Species*. Longmans, London.
Galvin, K. (1985). Food procurement, diet, activities and nutrition of Ngisonyoka, Turkana pastoralists in an ecological and social context. PhD Thesis. State University of New York.
Geissler, C.A., Brun, T.A., Mirbagheri, I., Soheli, A., Naghibi, A. & Hedayat, H. (1981). The energy expenditure of female carpet weavers and rural women in Iran. *American Journal of Clinical Nutrition*, **34**, 2776–83.
Goldstein, H. (1971). Factors influencing the height of seven year old children – results from the national child development study. *Human Biology*, **43**, 92–111.
Gordon, K.D. (1987). Evolutionary perspectives on human diet. In: *Nutritional Anthropology*, ed. F.E. Johnston, pp. 3–39. A.R. Liss, New York.
Gracey, M. (1986). The nutrition of Australian Aborigines. *Recent Advances in Clinical Nutrition*, **2**, 57–68.
Grant, M.W. (1964). Rate of growth in relation to birth rank and family size. *British Journal of Preventative and Social Medicine*, **18**, 35–42.
Harvey, P.W.J. & Heywood, P.F. (1983). Nutrition and growth in Simbu. From: *Research Report of the Simbu Land Use Project*, vol. 4. Simbu Provincial Government, Simbu, Papua New Guinea.
Hermanussen, M., Hermanussen, B. & Burmeister, J. (1988). The association between birth order and adult stature. *Annals of Human Biology*, **15**, 2, 161–5.
Hill, K. (1982). Hunting and human evolution. *Journal of Human Evolution*, **11**, 521–44.
Hill, K., Kaplan, H., Hawkes, K. & Hurtado, A.M. (1985). Men's time allocation to subsistence work among the Ache of Eastern Paraguay. *Human Ecology*, **13**, 1, 29–47.
Hipsley, E.H. & Kirk, N.E. (1965). *Studies of Dietary Intake and the Expenditure of Energy by New Guineans*. South Pacific Commission, Noumea, New Caledonia.
Howell, N. (1976). Towards a uniformitarian theory of human palaeodemography. In: *The Demographic Evolution of Human Populations*, ed. R.H. Ward & K.M. Weiss, pp. 25–40. Academic Press, London.
Howells, W.W. (1948). Birth order and body size. *American Journal of Physical Anthropology*, **6**, 449–60.
Høygaard, A. (1941). *Studies on the Nutrition and Physiopathology of Eskimos*. I Kommisjon Hos Jacob Dybwad, Oslo.
Hurtado, A.M., Hawkes, K., Hill, K. & Kaplan, H. (1985). Female subsistence strategies among Ache hunter-gatherers of Eastern Paraguay. *Human Ecology*, **13**, 1, 1–28.
Immink, M.D.C., Viteri, F.E., Flores, R. & Torun, B. (1984). Microeconomic consequences of energy deficiency in rural populations in developing countries. In: *Energy Intake and Activity*, ed. E. Pollitt & P. Amante, pp. 355–76. A.R. Liss, New York.
Kasarda, J.D. (1971). Economic structure and fertility: a comparative analysis. *Demography*, **8**, 3, 307–17.
Keene, A.S. (1985). Nutrition and economy: models for the study of prehistoric diet. In: *The Analysis of Prehistoric Diets*, ed. R.I. Gilbert & J.H. Mielke, pp. 155–90. Academic Press, London.
Kent, S. (1986). The influence of sedentism and aggregation on porotic hyperostosis and anaemia: a case study. *Man*, **21**, 605–36.

Koike, G., Yokono, O., Iino, S., Adachi, M., Yamamoto, T., Puloka, T. & Suzuki, M. (1984). Medical and nutritional surveys in the Kingdom of Tonga; comparison of physiological and nutritional status of adult Tongans in urbanized (Kolofo-ou) and rural (Uiha) areas. *Journal of Nutritional Science and Vitaminology*, **30**, 341–56.

Lee, R.B. (1979). *The !Kung San: Men, Women and Work in a Foraging Society*. Cambridge University Press, Cambridge.

Levinson, F.J. (1974). *Morinda: an Economic Analysis of Malnutrition among Young Children in Rural India*. Cornell-MIT international nutrition policy series, Cambridge, MA.

Malcolm, L.A. (1970). *Growth and Development in New Guinea – a Study of the Bundi People of the Madang District*. Institute of Human Biology, Madang.

Malina, R.M., Little, B.B., Buschang, P.H., Demoss, J. & Selby, H.A. (1985). Socioeconomic variation in the growth status of children in a subsistence agricultural community. *American Journal of Physical Anthropology*, **68**, 385–91.

Martorell, R. & Arroyave, G. (1988). Malnutrition, work output and energy needs. In: *Capacity for Work in the Tropics*, ed. K.J. Collins & D.F. Roberts, pp. 57–75. Cambridge University Press, Cambridge.

Martorell, R., Leslie, J. & Moock, P.R. (1984). Characteristics and determinants of child nutritional status in Nepal. *American Journal of Clinical Nutrition*, **39**, 74–86.

Martorell, R., Mendoza, F. & Castillo, R. (1988). Poverty and stature in children. In: *Linear Growth Retardation in Less Developed Countries*, ed. J.C. Waterlow, pp. 57–73. Vevey/Raven Press, New York.

Martorell, R., Valverde, V., Mejia-Pivaral, V., Klein, R.E., Elias, L.G. & Bressani, R. (1979). Protein–energy intakes in a malnourished population after increasing the supply of dietary staples. *Ecology of Food and Nutrition*, **8**, 163–8.

McNeill, G., Payne, P.R. & Rivers, J.P.W. (1988). *Patterns of Adult Energy Nutrition in a South Indian Village*. Department of Human Nutrition, London School of Hygiene and Tropical Medicine, Occasional Paper no. 11, London.

Milton, K. (1985). Ecological foundations for subsistence strategies among the Mbuti Pygmies. *Human Ecology*, **13**, 1, 71–8.

Montgomery, E. & Johnson, A. (1977). Machiguenga energy expenditure. *Ecology of Food and Nutrition*, **6**, 97–105.

Mueller, E. (1976). The economic value of children in peasant agriculture. In: *Population and Development: the Search for Selective Interventions*, ed. R.G. Ridker, pp. 98–153. The Johns Hopkins University Press, Baltimore.

Nabarro, D. (1984). Social, economic, health, and environmental determinants of nutritional status. *Food and Nutrition Bulletin*, **6**, 1, 18–32.

Neel, J.V. (1977). Health and disease in unacculturated Amerindian populations. In: *Health and Disease in Tribal Societies*. Ciba Foundation Symposium 49 (new series), pp. 155–77. Elsevier, Amsterdam.

Norgan, N.G., Ferro-Luzzi, A. & Durnin, J.V.G.A. (1974). The energy and nutrient intake and the energy expenditure of 204 New Guinean adults. *Philosophical Transactions of the Royal Society of London, B*, **268**, 309–48.

Pacey, A. & Payne, P.R. (1985). *Agricultural Development and Nutrition*. Hutchinson and Co. (for FAO and UNICEF), London.

Padoch, C. (1982). *Migration and its Alternatives among the Iban of Sarawak*. Martinus Nijhoff, The Hague.

Payne, P.R. (1985). Nutritional adaptation in man: social adjustments and their nutritional implications. In: *Nutritional Adaptation in Man*, ed. K.L. Blaxter & J.C. Waterlow, pp. 71–88. John Libbey, London.

Pimentel, D. & Pimentel, M. (1979). *Food, Energy and Society*. Edward Arnold, London.

Platt, B.S. (1962). *Tables of Representative Values of Foods Commonly Used in Tropical Countries*. HMSO, London.

Polunin, I. (1977). Some characteristics of tribal peoples. In: *Health and Disease in Tribal Societies*. Ciba Foundation Symposium 49 (new series), pp. 5–24. Elsevier, Amsterdam.

Powell, C.A. & Grantham-McGregor, S. (1985). The ecology of nutritional status and development in young children in Kingston, Jamaica. *American Journal of Clinical Nutrition*, **41**, 1322–31.

Prentice, A., Cole, T.J. & Whitehead, R.G. (1987). Impaired growth in infants born to mothers of very high parity. *Human Nutrition: Clinical Nutrition*, **41C**, 319–25.

Procopec, M. (1969). The effect of birth order of children in the family on their physical development (height). *Anthropologie*, **VII/2**, 27–32.

Rajalakshmi, R. (1971). Reproductive performance of poor Indian women on a low plane of nutrition. *Tropical and Geographical Medicine*, **23**, 117–25.

Rao, D.H. & Satyanarayana, K. (1976). Nutritional status of people of different socio-economic groups in a rural area with special reference to pre-school children. *Ecology of Food and Nutrition*, **4**, 237–42.

Rawson, I.G. & Valverde, V. (1976). The aetiology of malnutrition among preschool children in rural Costa Rica. *Journal of Tropical Pediatrics*, **22**, 1, 12–17.

Richards, P. (1986). *Coping with Hunger: Hazard and Experiment in an African Rice-Farming System*. Allen & Unwin, London.

Roberts, S.B., Paul, A.A., Cole, T.J. & Whitehead, R.G. (1982). Seasonal changes in activity, birth weight and lactational performance in rural Gambian women. *Transactions of the Royal Society of Tropical Medicine and Hygiene*, **76**, 5, 668–78.

Robson, J.R.K. & Wadsworth, G.R. (1977). The health and nutritional status of primitive populations. *Ecology of Food and Nutrition*, **6**, 187–202.

Rosenberg, M. (1988). Birth weights in three Norwegian cities, 1860–1984. Secular trends and influencing factors. *Annals of Human Biology*, **15**, 4, 275–88.

Russell, M. (1976). The relationship of family size and spacing to the growth of preschool Mayan children in Guatemala. *American Journal of Public Health*, **66**, 12, 1165–72.

Satyanarayana, K., Naidu, A.N., Chatterjee, B. & Narasinga Rao, B.S. (1977). Body size and work output. *American Journal of Clinical Nutrition*, **30**, 322–5.

Schoeninger, M.J. (1982). Diet and the evolution of modern human form in the Middle East. *American Journal of Physical Anthropology*, **58**, 37–52.

Seckler, D. (1984). Malnutrition: an intellectual odyssey. In: *Interfaces Between Agriculture, Nutrition, and Food Science*, ed. K.T. Achaya, pp. 195–206. United Nations University, Tokyo.

Sinclair, H.M. (1953). The diet of Canadian Indians and Eskimos. *Proceedings of the Nutrition Society*, **12**, 69–82.

Sinnett, P.F. (1977). Nutritional adaptation among the Enga. In: *Subsistence and Survival*, ed. R.G. Feachem & T.P. Bayliss-Smith, pp. 63–90. Academic Press, London.

Smith, M.F., Paulsen, S.K., Fougere, W. & Ritchey, S.J. (1983). Socio-economic, education and health factors influencing growth of rural Haitian children. *Ecology of Food and Nutrition*, **13**, 99–108.

Sommer, A. (1982). *Nutritional Blindness: Xerophthalmia and Keratomalacia*. Oxford University Press, Oxford.

Spencer, T. & Heywood, P. (1983). Seasonality, subsistence agriculture and nutrition in a lowlands community of Papua New Guinea. *Ecology of Food and Nutrition*, **13**, 221–9.

Spurr, G.B. (1984). Physical activity, nutritional status and physical work capacity in relation to agricultural productivity. In: *Energy Intake and Activity*, ed. E. Pollitt & P. Amante, pp. 207–61. A.R. Liss, New York.

Spurr, G.B. (1988). Marginal malnutrition in childhood: implications for adult work capacity and productivity. In: *Capacity for Work in the Tropics*, ed. K.J. Collins & D.F. Roberts, pp. 107–40. Cambridge University Press, Cambridge.

Spurr, G.B., Maksud, M.G. & Barac-Nieto, M. (1977). Energy expenditure, productivity, and physical work capacity of sugar cane loaders. *American Journal of Clinical Nutrition*, **30**, 1740–6.

Stinson, S. (1980). Child growth and the economic value of children in rural Bolivia. *Human Ecology*, **8**, 2, 89–103.

Stinson, S. (1983). Socioeconomic status and child growth in rural Bolivia. *Ecology of Food and Nutrition*, **13**, 179–87.

Strickland, S.S. (1986). Long term development of Kejaman subsistence: an ecological study. *Sarawak Museum Journal*, **XXXVI**, 117–71.

Tin-May-Than & Ba-Aye (1985). Energy intake and energy output of Burmese farmers at different seasons. *Human Nutrition: Clinical Nutrition*, **39C**, 7–15.

Tobias, P.V. (1975). Stature and secular trend among Southern African Negroes and San (Bushmen). *South African Journal of Medical Sciences*, **40**, 4, 145–64.

Tobias, P.V. (1985). The negative secular trend. *Journal of Human Evolution*, **14**, 347–56.

Tomkins, A.M., Dunn, D.T. & Hayes, R.J. (1989). Nutritional status and risk of morbidity among young Gambian children allowing for social and environmental factors. *Transactions of the Royal Society of Tropical Medicine and Hygiene*, **83**, 282–7.

Tomkins, A.M., Hayes, R.J., Dunn, D.T. & Pickering, H. (1986). Socio-economic factors associated with child growth in two seasons in an urban Gambian community. *Ecology of Food and Nutrition*, **18**, 107–16.

Townsend, P.K. (1971). New Guinea sago gatherers: a study of demography in relation to subsistence. *Ecology of Food and Nutrition*, **1**, 19–24.

Truswell, A.S. (1977). Diet and nutrition of hunter-gatherers. In: *Health and Disease in Tribal Societies*. Ciba Foundation Symposium 49 (new series), pp. 213–26. Elsevier, Amsterdam.

Truswell, A.S. & Hansen, J.D.L. (1976). Medical research among the !Kung. In: *Kalahari Hunter-Gatherers: Studies of the !Kung San and their Neighbours*, ed. R.B. Lee & I. DeVore, pp. 166–94. Harvard University Press, Cambridge, MA.
Valverde, V., Martorell, R., Mejia-Pivaral, V., Delgado, H., Lechtig, A., Teller, C. & Klein, R.E. (1977). Relationship between family land availability and nutritional status. *Ecology of Food and Nutrition*, **6**, 1–7.
Valverde, V., Mejia-Pivaral, V., Delgado, H., Belizan, J., Klein, R.E. & Martorell, R. (1981). Income and growth retardation in poor families with similar living conditions in rural Guatemala. *Ecology of Food and Nutrition*, **10**, 241–8.
Viteri, F.E. (1971). Considerations on the effect of nutrition on the body composition and physical working capacity of young Guatemalan adults. In: *Amino Acid Fortification of Protein Foods*, ed. N.S. Scrimshaw & A.M. Altschul, pp. 350–75. MIT Press, Cambridge, MA.
Waaler, H.T. (1984). Height, weight and mortality: the Norwegian experience. *Acta Medica Scandinavia*, suppl. 679. Gruppe for Helsetjemesteforskining, Oslo.
Weiner, J.S. (1972). Tropical ecology and population structure. In: *The Structure of Human Populations*, ed. G.A. Harrison & A.J. Boyce, pp. 393–410. Clarendon Press, Oxford.
Wheeler, E.F. (1988). *Intra-Household Food Allocation: a Review of Evidence*. Department of Human Nutrition, London School of Hygiene and Tropical Medicine, Occasional Paper no. 12, London.
Whyte, R.O. (1974). *Rural Nutrition in Monsoon Asia*. Oxford University Press, Kuala Lumpur.
Wingerd, J. & Schoen, E.J. (1974). Factors influencing length at birth and height at five years. *Pediatrics*, **53**, 5, 737–41.
Wray, J.D. & Aguirre, A. (1969). Protein–calorie malnutrition in Candelaria, Colombia. 1. Prevalence; social and demographic causal factors. *Journal of Tropical Pediatrics*, September, 76–98.

12 *Social adaptation to season and uncertainty in food supply*

IGOR DE GARINE AND SJORS KOPPERT

Sociocultural factors

In this paper we shall deal with the sociocultural aspects of seasonal food shortage, considering culture as 'the residue of social knowledge passed on through social transmissions . . . which differs from the processes of biology and is embodied in artifacts and technology as well as in ideas, beliefs and attached values.' (Linton, 1936; Kroeber & Kluckhohn, 1952). Since the material, technological, etic aspects have often been dealt with, we will concentrate more on the non-material, ideational, emic aspects of the problem (Geertz, 1966). How do people visualize the seasonal shortage and, by extension, the field of nutrition in the framework of the total cultural system? How does such a foreseeable shortage affect behaviour? What is the cultural meaning of seasonal food shortage? We will also attempt to show the influence of monetization and westernization on the traditional system of coping. Examples will be drawn mostly from our own field work in Senegal, Chad and Cameroon (mainly among the Massa and the Mussey), and from that of the team to which we belong (Research Team 263 – 'Differential Food Anthropology' of the French National Scientific Research Centre) in tropical and equatorial Africa and in Nepal.

Seasonal food uncertainty

In relation to the disasters which have occurred during these last decades, the focus has been given by most observers to what Watts (1988, p. 287) called 'Famine as a large-scale calorie deprivation associated with the terminal points of a famine cycle' (Currey, 1984). Famines trigger off whole sets of behaviour and, according to many authors, jeopardize the very existence of society and the principles of reciprocity on which it is based (Turnbull, 1972; Laughlin & Brady, ed., 1978).

Clearly, seasonal hunger and famine are on the same continuum and, as we have seen recently, repeated droughts can cause famines, but the centre of our debate will be on the seasonal reduction in dietary intake which is a permanent recurrent feature of many traditional food systems: not a disaster but rather 'routine hazards to which a population must be adapted if it is to survive' (Turton in Torry, 1979).

Nutritional and health aspects

In the nutritional field Platt (1954) was one of the first to formulate the generalized concept of seasonal hunger as it is understood today by most nutritionists, anthropobiologists and economists: 'Throughout tropical Africa there is a hungry season: towards the end of the agricultural year the stock of grain runs out and this occurs when extra physical effort is expended in getting the new season's crop into the ground' (p. 101). This statement has induced many studies since Fox (1953) and has led to emphasis on the adverse effect of the period in utilitarian terms, pointing out the vicious circle between low food availability and low energy resources for securing a new crop. Later, the FAO (1958) seized hold of the idea and built it up as a main issue. Three years later, Miracle (1961) criticized seasonal hunger as a 'vague concept and an unexplored problem' and questioned its existence on the grounds of 'the paucity and the unreliability' of the data available, pointing out that 'food consumption surveys made in tropical Africa (have) been painfully few and incomplete inasmuch as seasonal variations have often not been adequately covered (p. 279). Today there is an abundance of studies, mostly anthropometrical, however, demonstrating the existence of such a preharvest shortage period (Teokul, Payne & Dugdale, 1986). What can be drawn from available data is that every situation has its originality in the nature, timing and extent of the shortage as well as in its biological effects, the environmental factors involved and the cultural responses made to it. Efforts have been made to correlate it with signs of under- and malnutrition among the total population and especially the vulnerable groups: pregnant and nursing mothers, and children (weight at birth and growth rate (MacGregor *et al.*, 1968)).

The Massa (and the Mussey) are similar to other savanna populations of Africa: during good years, e.g. 1976, they lose about 2 kg between the dry and the wet season and catch up rapidly. In bad years, e.g. 1980 and 1985, the difference is about 7 kg for men and 5 kg for women (Tables 12.1 and 12.2). These figures are slightly higher than those recorded by Brun, Ancey & Bonny (1979) in Burkina Faso (ex Upper Volta). Severe food shortage correlates with low and untimely rainfall as well as flooding from the Logone River and its tributaries during the preceding years.

Many authors emphasize the combination during the wet season of malnutrition, morbidity and mortality. Seasonal food shortage seems to

Table 12.1. *Variations in anthropometric measurement and rainfall among the Massa (Kogoyna village) in different years, (a) men and (b) women. Completed from Garine & Koppert (1988, p. 229)*

		Dry season April 1976	Rainy season Aug 1976	Crop season Dec 1976	Rainy season Sep 1980	Crop season Dec 1980	Rainy season Aug 1985
Annual rainfall (mm)		1975 = 821	1976 = 732	1979 = 552	1980 = 932	1984 = 606	
(a) No. men 15 y and older		23	23	23	26	26	71
Height (cm)		173.0	173.0	173.0	172.8	172.8	173.7
		4.9	4.9	4.9	4.5	4.5	5.0
Weight (kg)	m	61.4	58.0	60.9	53.1	58.3	54.5
	se	6.5	6.0	5.8	6.9	7.0	6.0
	sl	***	***		***		
Arm circumference (cm)	m	27.1	26.3	27.1	25.2	26.3	25.0
	se	1.8	1.9	1.5	2.5	2.4	2.1
	sl	***	**		***		
Triceps skinfold (mm)	m	4.9	4.4	4.5	3.7	4.3	3.7
	se	1.2	0.9	1.0	0.6	1.0	0.6
	sl	*	ns		***		
Haemoglobin (g/100 ml)	m				11.8	14.2	
	se				1.9	1.5	
	sl				***		
Haematocrit (%)	m				35	38	
	se				5	4	
	sl				***		

(b) No. women 15 y and older		35		35		35	50		50	90
Height (cm)	m	161.4		161.4		161.4	160.7		160.7	161.3
	se	5.7		5.7		5.7	6.3		6.3	7.0
Weight (kg)	m	49.7		47.6		48.8	43.5		47.8	45.0
	se	8.7		6.3		6.9	6.5		6.7	7.0
	sl		***		*			***		
Arm circumference (cm)	m	25.6		24.7		25.2	23.2		24.6	23.7
	se	2.5		2.0		2.1	2.3		2.0	2.0
	sl		***		*			***		
Triceps skinfold (mm)	m	8.5		7.0		7.4	5.3		6.5	4.8
	se	3.4		3.0		3.2	2.0		2.4	1.8
	sl		***		*			***		
Haemoglobin (g/100 ml)	m						10.9		13.0	
	se						1.2		1.5	
	sl							***		
Haematocrit (%)	m						32		35	
	se						3		3	
	sl							***		

m Mean.
se Standard error.
sl Significance level between means: ns, not significant; * $p<0.1$; ** $p<0.01$; *** $p<0.001$ (Student's *t*-test for paired observations).

Table 12.2. *Nutritional value of the diet in different seasons in Kogoyna, Massa (per capita) 1976–80 (Garine & Koppert, 1988, p. 224)*

Nutrient	Units		Annual average 1976		Dry season (Feb–May) 1976		Rainy season (Jun–Sep) 1976		Crop season (Sep–Dec) 1976		Crop season dry vs 1980		significance level of differences: rainy vs rainy season	dry vs crop season	crop season
Food energy	kcal (MJ)	m	2544	(10.4)	2362	(9.9)	2347	(9.9)	2924	(12.3)	2540	(10.4)	ns	***	***
		se	636	(2.6)	400	(1.7)	460	(1.9)	773	(3.2)	480	(2.0)			
Animal protein	g	m	37		44		24		45		32		***	***	ns
		se	16		15		9		15		20				
Vegetable protein	g	m	60		55		56		69		60		ns	***	***
		se	15		10		11		19		11				
Fat	g	m	30		28		27		35		32		ns	**	**
		se	11		6		9		14		11				
Carbohydrate	g	m	507		461		481		579		511		ns	***	***
		se	120		80		92		147		96				
Calcium	mg	m	950		950		720		1180		970		ns	***	ns
		se	510		570		360		510		400				
Iron	mg	m	35		37		30		38		33		**	***	ns
		se	9		10		6		10		5				
Retinol	μg	m	520		580		440		550		480		***	**	ns
Beta carotene	μg	m	1900		430		3600		1700		2200		***	**	ns
		se	1800		190		1500		1600		2300				
Thiamine	mg	m	2.5		2.3		2.3		2.8		2.5		ns	***	***
		se	0.6		0.4		0.4		0.8		0.5				
Riboflavin	mg	m	1.2		1.1		1.3		1.3		1.3		***	ns	***
		se	0.3		0.2		0.4		0.4		0.4				
Niacin	mg	m	27		26		24		31		27		ns	**	**
		se	7		6		5		9		5				
Ascorbic acid	mg	m	28		5		49		29		39		***	**	***
		se	36		3		25		22		41				

m Mean.
se Standard error.
sl Significance level between means: ns, not significant; * $p<0.1$; ** $p<0.01$; *** $p<0.001$ (Student's t-test for paired observations).

correlate with a peak in the incidence of endemic disease (Cantrelle *et al.*, 1967; Chambers, 1982; Harrison, 1988).

Cultural perception of seasonal food shortage

This does not mean that the people themselves visualize the shortage period and its hazards in the same way as Western scientists do. This is an important part of the picture for basic, as well as for applied, research.

Is the seasonal shortage period conceived of as a stress by the populations involved? Being aware of these biological variations, which are often minimal (a loss of weight generally less than 10% of total weight, as remarked by Hussain (1985) and Hunter (1967) implies precise quantified observations and the use of an array of the sophisticated techniques available to scientists. Few of the indicators in question are acknowledged by the people involved, especially in societies where malnutrition is endemic and, if they are, they are not necessarily traced to lack of food (Garine, 1984).

How do the Massa and the Mussey of Northern Cameroon acknowledge the seasonal food shortage? Like most traditional societies, they have the concept of a cyclical time, divided into a number of lunar months and seasons, according to meteorological observations (rainfall, temperature, wind, heat). Months are named according to many criteria (references to historical events, rituals) but mostly to the dominant type of activity undertaken (agriculture, fishing, herding, house building) and the crop available at the time. As a matter of fact, the cycle of the main staple, the rainy season red sorghum (*Sorghum caudatum*), is the backbone of the ritual calendar.

The Massa and the Mussey distinguish two main seasons: the dry season (*walla*)*, and the rainy season (*dolla*). This does not mean that there is a bad and a good season, feelings about each of them are always mixed. Each one is divided in two: dry and cold, with sand mist (*simetna*), dry and hot with burning sun (*fata*), rainy and cold, the agricultural work period (*til ma baratna*), rainy and hot, the crop season (*fena*). It is interesting to note that the peak of the shortage period (July–August) is called *til yan ansu*, the month 'Did they call you?', referring to the fact that people listen attentively to any noise which might be interpreted as an invitation to share food. The same month is also sometimes named *til mayra*, 'the hungry month' (Table 12.3).

Many traditional societies distinguish and sometimes name a regular lean month (Annegers, 1973), which does not necessarily correspond to the rainy period. In the case of the Bushmen, the lean period would be at

* We are using here the Mussey terminology, the Massa is similar.

Table 12.3. *Calendar of the Jarao clan (Mussey)*

	Translation of name	Comments in relation to the elements	'Seasons'	Stresses and epidemics	Overall conditions
Dec	Month of the Domo clan festival	Beginning of the dry and idle period	Dry season (*walla*) Cold winds (*simetna*)	Cold, colds, pneumonia	Good
Jan	Month of the Gunu clan	Middle of the dry and idle period	Mist (*kutna*) Dusty winds	Whooping cough, measles	Bad
Feb	Month of the Jarao clan protector	Sun!			Average
Mar	Month of the Jarao clan hunt	Sun!	Very hot (*zamalla*)	Cerebro-spinal meningitis	Very bad
Apr	Month of the Es clan hunt	Sizzling!		Smallpox, rabies	Very bad
May	Month of the Es clan fishing	The Rain God begins to cry	Rainy season (*dolla*) Begins to get cooler (*bu heppa*)		Average
Jun	Month of the Leo clan fishing	Rain falls a little	Months of cultivation (*til baratna*)	Too much work	Average
Jul	Month of flooding Month of 'Did you call me for food?'	Water God begins the floods		Diarrhoea Food shortage	Bad
Aug	Month of the 'furtive' crop of finger millet	Rain falls on the sorghum stems	Cool	Malaria	Average
Sep	Month of the red finger millet	Rain God goes back up to his home	Drier and hotter		Good
Oct	Month of the Bogodi clan festival	Rain falls a little more on the late sorghum	Month of recreation (*til luta*)		Good
Nov	Month of the Gunu clan festival	Rain God still cries a little	Beginning of the cold season (*vun simetna*)		Good

the hottest and driest time of the year (October and November when the distance between food and water appears to reach an annual maximum (Lee, 1979; Silberbauer, 1981). This does not necessarily mean that the people themselves acknowledge the matter and this is important in terms of stress or perception of well-being.

Wet season – not necessarily associated with disease

To the Massa and the Mussey, who live in a very hot part of the world, the wet season is acknowledged as having positive aspects (coolness). Although it is recognized as a period when fevers (malaria) and diarrhoeal diseases increase, it does not appear as a time characterized by the highest morbidity and mortality. The dangerous periods of the year are the first months of the dry season which bring the cold (*simetna*), when dust mists due to the Harmattan wind hang around and are considered to bring all sorts of epidemics, e.g. measles, chicken pox. It is also the period when wide temperature variations and cold nights are responsible for respiratory ailments and death in populations which have few clothes and blankets at their disposal. (The Bushmen experience the same cold stress.) The sun – *fatta* – is also to be feared in these climates as it is held responsible for many causes of death. It is associated with the Death God: the wild bush, bush fires and the colour red (skin diseases, leprosy, and the main epidemic infection: cerebro-spinal meningitis). The ailments which reach their peak during the wet season, such as malaria or diarrhoea, are not easily distinguished as specific sicknesses and appear to be permanent endemic diseases. As Lee (1979) writes, 'it is by no means clear that the maximum of ill health coincides with the season of food scarcity.' As a matter of fact, it is interesting to notice that most populations, including those having a bimodal rainfall system, distinguish 'shortage periods'. This is what appears among the Yassa of Southern Cameroon in June, when fishing has become difficult and cassava is not yet available. Although there is little seasonal variability in the food, this also occurs among the Aka Pygmies of the Central African Republic in October, at the end of the rains when the men have deserted the camps to go hunting large animals and little food is available to the women in the settlements (Bahuchet, 1985). The same applies to the Oto and the Twa of the Zaire forest, where Pagezy (1988) recorded some seasonal fluctuations in nutritional status and small variations in food availability (especially animal food). This is enough to arouse among the villagers a kind of 'meat hunger'.

Food shortage and agricultural work

It seems that in many societies the relation between food intake and work is perceived during the field tilling and weeding period of the year and that the villagers are trying not to experience what Fox (1953) perceived as a negative energy balance. While the notion of permanent nutritional needs does not appear very explicitly, many traditional societies are

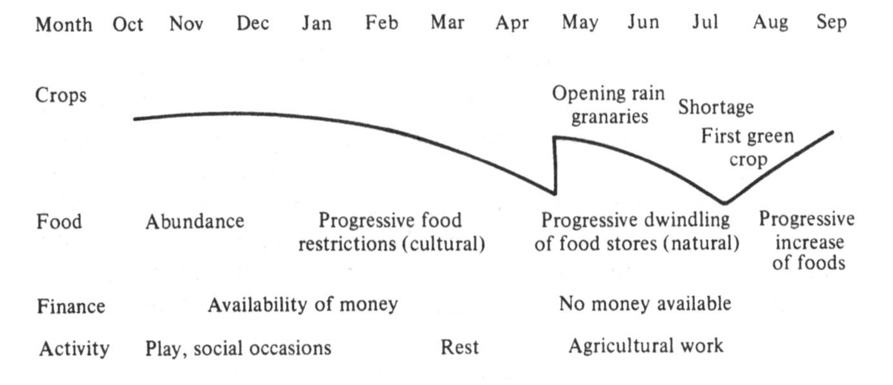

Fig. 12.1 The food year among the Massa and Mussey (from Garine & Koppert, 1988, p. 235).

aware of the bottleneck in terms of agricultural tasks and the need to provide food for work. The Massa and the Mussey name this period *til ma baratna*, 'the month of the agricultural toil'. The Guéré of Ivory Coast call April the 'weariness month', when trees have to be felled in order to clear land for cultivation (Schwartz, 1968). In the dry tropics, field clearing and weeding, which takes place during the wet season, appear as the critical period for which food has to be set aside. This is one of the few areas where traditional customs and nutritionists' concerns coincide, but the villagers are responding to a work stress rather than to a food one. We have documented this type of strategy among the Serer of Senegal (Garine, 1962, 1981). The villagers keep millet stores for many months in order to have fuel available for the agricultural work needed to obtain a new crop. At the same time they reduce their energy expenditure and food consumption during the hot months of the year (February to April) when no work is needed in the fields (Fig. 12.1). A similar strategy exists among other groups as well as the Massa, who add that, anyway, heat is not an incentive to heavy eating. This strategy generates a culturally controlled lean period which will allow food supplies to be attuned to the busy period, after which the true, uncontrolled shortage period may occur.

Spatial distribution of food resources and seasonality

Traditional societies also attempt to diminish seasonal uncertainty by making the best of spatial differentiation of food resources. Many years ago Mauss & Beuchat (1904), when dealing with the social morphology of the Eskimos, showed that moving the community according to the spatial availability of resources is an efficient way of combatting seasonal

Table 12.4. *Seasonal variations in daily energy and animal protein intakes between village and fishing camps among the Ba-Oto and the Ba-Twa (from Pagezy, 1988)*

		Dry season 1		Rainy season		Dry season 2		Dry season 3	
		Ba-Oto	Ba-Twa	Ba-Oto	Ba-Twa	Ba-Oto	Ba-Twa	Ba-Oto	Ba-Twa
Village									
Animal protein	(g)	175	107	96	71	162	157	174	103
Energy	(kcal)	2168	1891	1970	1744	1953	1529	2101	1764
Camps (Ba-Oto fishermen)									
Animal protein	(g)			324		463		375	
Energy	(kcal)			2086		2081		2159	

Table 12.5. *Protein consumption (g/d) in two sub-populations of Salmé (Tamang, Nepal) (from Koppert, 1988)*

Category	Vegetable proteins	Animal proteins	Total	% Animal origin
Per capita average	46.8	2.9	49.7	5.8
Under 7 years	16.8	2.7	19.5	13.8
7–15 years	40.7	2.3	43.0	5.3
Men	58.1	3.8	61.9	6.1
Women	54.3	2.6	56.9	4.6
Village	47.8	0.9	48.7	1.8
Herders (temporary shelters)	46.2	2.9	49.1	5.9
March–May	43.5	1.4	44.9	3.1
June–August	51.9	3.9	55.8	6.9
September–November	47.1	2.2	49.3	4.5
December–February	44.8	4.2	49.0	8.6

Table 12.6. *Diet of the gurna as compared to the adult male villager in Kogoyna (Massa) in 1976*

February–May			June–August			September–October		
Energy (kcal)	Animal protein (g)	*n*	Energy (kcal)	Animal protein (g)	*n*	Energy (kcal)	Animal protein (g)	*n*
Gurna								
3828	63.3	17	3224	74	22	5325	110.0	19
Villagers								
3090	57.6	27	2970	30	27	3090	47.3	27
Gurna bonus								
738	5.7	—	254	44	—	2235	63.6	—

food shortage. This is the way hunter-gatherers, herdsmen and in general all nomadic people solve the problem. This is also the case for sedentary groups who, according to environmental resources and known techniques, make permanent use of specific food availability (Minnis, 1985, p. 39). This is how the agricultural Oto take advantage of the existence of Lake Tumba; by maintaining fishing hamlets constantly visited by the village (Pagezy, 1988) (Table 12.4). The same applies to the Tamang (Nepal), where the village population constantly visits the herding area located at a higher altitude (Koppert, 1988) (Table 12.5). This allows spreading of the risk by permanently maintaining a broad range of food

Table 12.7. *Anthropometry of the Massa men, women and gurna (Kogoyna) in 1980*

	Men	Women	Gurna
September, end of shortage period			
Number of observations	24	50	12
Height (cm)	172.8	160.7	178.7
Weight (kg)	53.1	43.5	67.0
Arm circumference (cm)	25.2	23.2	27.5
Tricipital skinfold (mm)	2.7	2.3	3.0
Subscapular skinfold (mm)	6.2	7.9	7.6
December, full crop period			
Number of observations	24	50	10
Height (cm)	173.1	161.0	179.7
Weight (kg)	58.3	47.8	69.2
Arm circumference (cm)	26.3	24.6	28.1
Tricipital skinfold (mm)	2.9	2.6	3.4
Subscapular skinfold (mm)	6.5	8.4	8.3

procuring activities. This does not mean that all biological groups benefit from it, but rather mostly the adult men, which leads us to another point. Sexual division of labour, as well as ideas about food distribution favour adult men and create a differential seasonal access to food among different age/sex classes.

During the lean period among the Aka, women are left in the main camp and may perform rituals so that the meat becomes plentiful again, while the Pygmy male hunters, away in the forest, are already gorging on food (Bahuchet, 1985). Among the Massa, the men participating in the *guru** fattening sessions are those who benefit from the high milk yield of the cattle precisely during the shortage period (Garine, 1960, 1964, 1980, 1984) (Tables 12.6 and 12.7). There are groups within the population which are worse off than others during the seasonal food shortage and they are most often the same ones which are permanently disadvantaged (Table 12.8). These are vulnerable groups which are less mobile and have to hang around the main village: pregnant and lactating women, and children who ought to receive a higher protein intake during the seasonal shortage. In many societies, like the Massa, mild seasonal shortage may be increasing structural, permanent insufficiencies in the diet although this is not acknowledged by the society and not clearly reflected in the diet per capita.

Temporary migrations are another efficient device for coping with seasonal shortage. They take place according to food or work avail-

* We shall use the term *guru* to designate the institution and *gurna* the participants.

Table 12.8. *Percentage satisfaction of energy needs by sex and age (Kogoyna) (Koppert), 1980*

		February–May		June–September		October–December	
Age categories	Energy (MJ)[a]	% available	*n*	% available	*n*	% available	*n*
Children							
1–3	5.7	78	(9)	84	(8)	99	(9)
4–6	7.6	88	(13)	94	(15)	110	(11)
7–9	9.2	86	(11)	87	(8)	106	(11)
Men							
10–15	10.9–12.1	86	(5)	111	(4)	104	(5)
16–39	12.6–12.8	103	(25)	99	(15)	103	(15)
40+	10.8–10.4	103	(16)	105	(13)	137	(11)
Women							
10–15	9.8–10.4	85	(6)	77	(8)	104	(8)
16–39	9.2– 9.7	95	(33)	102	(30)	129	(31)
40+	7.9–8.7	108	(13)	113	(13)	141	(12)

[a]According to FAO/OMS (1974).

ability, but also appear as opportunities to break with the daily routine, see the world and acquire prestige, rather than as a way of saving food or money for the benefit of the whole household. The Massa used to go on fishing parties down the Logone River from March to the end of April in order to take advantage of the fish migration and secure fish and money. Nowadays the Massa and the Mussey leave their villages in the middle of the rainy season, from July to October, to work in the sorghum (*muskwari*) fields of the Fulani, a couple of hundred kilometers away. The money they earn (15 600 CFA francs per capita) is used to buy clothes (18%), pay taxes and debts, etc. (19%), 26% is spent on alcoholic drinks and 15% on other stimulants, and only 15% goes on food (Table 12.9). It usually comes too late to ease the current rainy season shortage, though it may be used partly to buy sorghum for the following year. In both cases these migrations have an important side effect in diminishing the number of consumers sharing the family's stores of food.

Degrees of seasonal food uncertainty

All traditional societies are aware of the cyclical availability of food and identify periods of normal food shortage which are usually easily coped with and which, even if they are accompanied by what nutritionists would identify as signs of mild malnutrition, will not give rise to exceptional measures.

Table 12.9. *Cash income and expenses in Kogoyna between 22 August and 21 November 1980*

Expenses	(%)		Receipts	(%)	
Tobacco	6	56%	Tobacco	1	75%
Kola nuts	1				
Drinks (alcoholic)	26		Drinks	24	
Tea and sugar	8				
Staple and relish	11		Fish	42	
Snacks and meals out	4		Agricultural production	1	
			Poultry	2	
			Goats, sheep	2	
Clothes	18	44%			
Other manufactured items	1				
Cosmetics	2				
Travel	4		Handicrafts	1	25%
Taxes	17		Loans, debts	10	
'Gifts'	2		Gifts	14	
Per capita Fr.CFA 6909 ± 4258 (about 23 US dollars)			Per capita Fr.CFA 11317 ± 8333[a] (about 38 US dollars)		

[a]Income from cotton is not included here. It is about Fr.CFA 15000 per capita (about 50 US dollars).

The analysis of the local vocabulary provides clues as to the physiological as well as the psychological stress perceived by the subjects. This aspect is sometimes overlooked by observers who stick to 'objective' indicators. What are the terms used to signify hunger? The Massa and the Mussey call *mayra* the normal feeling of hunger leading to feeding episodes. *Mayra canu*, 'hunger is killing me', merely means 'I am ready for a meal'. The same term designates tolerable, normal seasonal hunger triggering off normal seasonal responses: using household stores, taking advantage of kinship, friendship and neighbourhood ties, resorting to bush food and selling sheep and goats. Children are told not to wander in the village and solicit food since this would tarnish the family's image. Then comes *baknarda* – 'hunger with threat' – 'because it is really too long' and life is at stake. It is also called *may haana*, 'the white hunger', because people lose their pride, stop washing themselves and remain white with dust. Dogs, who are no longer fed, begin to die, people who have food hide themselves away to eat it, children cry constantly, people stop moving around, then children and elderly people begin to die. This stage causes shame, aggression, the selling of children for food (in previous times), eating dangerous bush tubers or, as remembered by very old villagers, eating even the residue of one's own faeces.

At a moderate level, the evaluation of hunger and satiety is to a large extent cultural (Bellisle, 1979). The Massa feel 'hungry' when they are unable to ingest at a meal two pounds of thick sorghum gruel. They value plumpness and the men indulge in fattening sessions (*guru*) where they consume large quantities of milk and sorghum gruel. Conversely the Gurage of Ethiopia, studied by Shack (1977), eat sparingly, despise over-eating and fatness and display (p. 269) anxiety over the prospect of unsatiated daily hunger, 'although they permanently keep large stores of their staple, the "false banana" (*Ensete ventricosum*)'. In this example, as the author mentions, 'environmental factors pose no limitations on food supply but sociocultural factors constitute a formidable impediment against satisfying nutritional requirements.'

'Cultural' food stress

At the same time, a minor qualitative change in the diet may arouse disproportionate emotional reactions. In her recent study of the food and nutrition system of two rain forest populations of Zaire, Pagezy (1988) recorded the existence of a mild decrease in energy content of the diet during the rainy season, accompanied by a more noticeable decrease in the availability of animal proteins, which boils down to consuming animal foods every two days instead of every day (Table 12.4). (There is no significant increase in morbidity or the workload, and only a small decrease in weight and fatness – hardly one kilogram – as measured by skinfolds.) The biological effects of the shortage and the wet season are minimal and have no statistical significance, but are experienced as a very dramatic period. 'The population complains about hunger and even famine (*nzala*). The village is filled with the sound of children crying, motivated by the monotony of the diet and by the lack of the animal protein relish to which they are accustomed.' What Pagezy calls 'seasonal meat hunger' is reported elsewhere. For instance, Bahuchet (1985) mentions that fear of meat hunger (*sené*) is a constant preoccupation among the Aka and is at the origin of specific rituals in October; the most difficult period. The author stresses the lack of anxiety in relation to vegetable foods, which can be obtained permanently, in comparison with meat which is available more irregularly. Both authors suggest that the food stress experienced seasonally, which has mild biological consequences, is not mainly linked to a decrease in the quantity of food ingested or its nutritional value, but to monotony and to the fact that the available diet does not satisfy the cultural norms related to what is identified as good and comforting food. It was to similar phenomena that Miracle (1961) referred in his highly criticized article when he mentioned 'the transformation of staples into beer (an inefficient conversion calorie-wise) as an attempt to fight the monotony of the diet'. Richards & Widdowson (1936, p. 179) referred to the same aspect when they quoted the 'irrational' conduct of the Bemba, who trade grain for fish on very unfavourable terms in order to relieve the boredom of their diet. The Mussey, and especially the Massa, have a very monotonous diet and do not complain about it.

Food stress in seasonal shortage may well be due to lack of access to normally valued foods: what Jelliffe (1967) called the 'cultural superfood', usually the staple (see also Ogbu, 1973). The Massa illustrate this point. During the 1985 hunger period, they sold goats and sheep at a very low price in order to buy their sorghum staple. It occurred many times that the amount of carbohydrate calories obtained from the money they received was smaller than the amount which would have been provided by the animal if they had eaten it instead of selling it. In the same way the Yassa, sea fishermen of Southern Cameroon, complain about 'hunger' when cassava, which is their staple, is missing although other tubers, breadfruit and fish, which provide similar nutritional value in the diet, are available.

Biological consequences of mental stress

'Ordinary' food shortage seems to induce variable physical and mental stresses. Many cultures have mentioned the emotional unrest during the seasonal shortage period. Pagezy (1988), referring to Gardner (1972) and Guedeney (1986), suggested the possibility of the emotional stress influencing the growth and nutritional status of children among the Oto and the Twa. We are unable to venture far on this ground, but if seasonal food shortage does not materially impair the fitness of individuals very much it does not appear to be a period when people enjoy a feeling of well-being. Nevertheless, such shortage periods do not monitor large-scale response at the level of the food system as a whole.

Most mammals react in a rather straightforward way to heat, cold or hunger. In human culture, 'collective conscience', as Durkheim (1893) named it, plays a part in acknowledging, increasing, deferring, decreasing or denying the non-material and maybe material effects of what we call aggression, angst and discomfort. For example, death may cause, within a whole community, mourning, sadness and despair, leading to the collective consumption of 'bitter foods' and a general feeling of ill-being over a socially determined period. Among the Massa, the death of an elder triggers off a very important festive display. Premature death, especially of a man in his youth or a pregnant woman, is apprehended as a social catastrophe and puts the whole community in a state of grief, fragility and lack of dynamism which perhaps amplifies the effects of material risks. Conversely, festive events or institutions like the *guru* result in dynamic and positive reactions to material discomfort, as in the case of a mild food shortage.

Among the Massa, although there are few stocks of food left during the rainy season, collective field work and weeding, accompanied by food, beer and girls, are (or rather were) an incentive to heavy work and high performance, even in a period of possible energy imbalance.

Seasonal hunger as a cultural choice

It appears that, in some cases, people such as the Massa prefer to undergo an expected lean period rather than to modify in a noticeable way their general food strategy. As remarked by Bennett (1953, p. 203), this option may be a 'mode of life, a peculiar way of ingesting calories'. They feel equipped to manage a recurrent level of seasonal food uncertainty. The situation is radically different in the case of severe shortage or famine, where physical survival depends on resorting to solutions beyond their normal scope.

Material culture

What, then, is the origin of this normal, foreseeable shortage? A whole range of specialists have abundantly commented on these aspects, but most of them, including the first anthropologists to have dealt with the problem, have put forward environmental factors and the 'etic' – material–technical aspects of culture (Heywood & Nurse, 1980; Ogbu, 1973). Under the influence of the human ecology school, traditional societies which often suffer the influence of adverse and unpredictable environmental factors are depicted as having most often devised the best possible solution to fill their stomachs and fulfil their biological needs. As Friedman (1979) mentioned, this slightly Rousseauist view should be moderated. We have given elsewhere a detailed account of the food techniques and strategies used by the Massa and the Mussey in coping with seasonal uncertainty (Garine & Koppert, 1988).

Agriculture

The Massa and the Mussey enjoy a mixed kind of food economy. They benefit from agriculture, herding, fishing and, to a certain extent, from small game hunting and gathering. Although these activities are complementary, there is a gap of about four weeks in food availability

Table 12.10. *Grain resources (kg) in Kogoyna (Massa) in November 1–76 and October 1980*

Households	No.[a]	25.11.76	8.10.80	Active consumers[b] & non-active	Availability per capita 1976	Availability per capita 1980
Bunrayna	1	1097	612	3 A, 1 N = 3.5	313	175
Dadira	2	602	681	3 A, 2 N = 4	150	170
Digidim	4	1600	1810	6 A, 4 N = 8	200	226
Fiwlinsu	5	2457	1110	3 A = 3	819	370
Fuyawsu	6	896	1225	5 A, 6 N = 8	112	153
Gamasna	7	3380	1378	4 A, 2 N = 5	676	276
Helsu	9	3152	1302	5 A, 4 N = 7	450	186
Jibetsu	11	1376	715	2 A, 1 N = 2.5	550	286
Likusidi	14	1448	1235	6 A, 2 N = 7	206	176
Molgyna	15	1196	665	4 A, 2 N = 5	170	133
Nakutsia	16	767	1897	3 A, 2 N = 4	191	474
Rayna	18	2415	1015	6 A, 4 N = 8	302	126
Samasu	20	3016	736	5 A, 5 N = 7.5	402	98
Sampulla	22	971	490	4 A = 4	242	122
Suburssu	23	626	302	3 A, 4 N = 5	125	60

[a]Refers to the same numbers as in Table 12.14 and Fig. 12.3.
[b]Active members (A) are counted as 1 consumer.
Non-active members (N) are counted as ½ consumer.

between July and September, when the stocks of grain are depleted and water is too high in the river to allow for efficient fishing. As far as agriculture is concerned, they practise, to a certain extent, mixed farming and favour early maturing crops (sorghum for the Massa; sorghum, pulses and finger millet for the Mussey). Both have recently taken to cultivating a late maturing pricked sorghum (*Sorghum durra*), which ripens closer to the food shortage period. The Massa do not cultivate large areas and, although they own fertile riverine land, they are not very prone to doing so. Among the Massa, the cultivated acreage per active adult worker was, in 1985, 0.6 hectares. The yield of sorghum, the main crop, is approximately 700 kg per hectare, but it is very irregular (Table 12.10). At the normal daily ration of 600 grams per capita, about 220 kg are needed every year. We have been able to measure the food reserves in 15 households of Kogoyna in 1976 and 1980, about one month after the harvest. Although 1976 was a good year, five households were unable to meet their needs. In 1980, an average year, ten out of 15 were in the same situation, some of which had only four months of reserves. Our picture is optimistic as it does not take into account storage and grain processing losses (amounting to a total of 10%) and assumes that all the cereals will be consumed as food (a sizeable amount is actually used to make beer and alcohol). It is reasonable to consider that in many households about two months of grain (40 kg) are missing. Crop success varies among the households from one year to the other. The Massa rely on their sales of fish to reach an equilibrium.

The Mussey, who do their best to extend their acreage, dispose of less fertile land, which is often exhausted by overcropping, and have difficulties in balancing the areas they dedicate to cotton and to food crops. Tubers like cassava, which could at least alleviate the seasonal shortage, are not cultivated. Wild tubers as well as bush leaves, fruits, berries and wild cereals (especially *Dactyloctenium aegyptium*) play an important role in easing hunger during difficult times. In order to minimize the shortage period, the Massa and the Mussey consume sorghum cobs, raw or toasted, before they are totally ripe. They have devised a special oven (*jojok'a*) to dry the green grain and allow it to be converted into flour. Small game, such as rodents or monitor lizards (*Varanus griseus*), are collected from the bush during the shortage period.

Other production activities

The Massa are equipped for fishing during times of low water and during the increase and decrease of flooding, but not during the high waters which correspond to the seasonal shortage period (July, August). During the last decade, over-fishing has diminished the fishermen's efficiency.

Cattle are used for the bride price, fattening sessions and prestige lending (*golla*). Sheep and goats are the main capital assets and can be sold to obtain food. Cattle are rarely slaughtered and only on ritual occasions, and milk is mostly used by the privileged group of Massa taking part in the fattening sessions of the *guru*. Animals are reluctantly sold in emergencies (including food shortages), but their main function is in extending the network of social ties, through marriage and prestige lending, enabling the creation of a clientele on which one will be able to draw in case of food shortage, as well as on relatives, affines (related by marriage) and friends.

Storage

As is the case in many traditional societies, storage is not especially efficient and losses in food, grain and pulses are quite noticeable. Among the Massa and the Mussey, the head of the household is in charge of the main granary, where the product of the collective work of the domestic group is stored. It is the 'rain granary', used to provide food on special occasions and during agricultural work. Each of the mothers within the family owns a number of granaries in which she stores the product of her own cereal and pulse fields and of her garden. She is supposed to feed herself and her children daily, and to provide a dish for the adult males of the compound. The household head and the housewives have divergent ideas regarding the management of the stores; the family head wants the women's granaries to be used first and to keep the main granary for the shortage period and celebrations. The housewives wish the rain granary to be opened as soon as possible in order to dispose more freely of their own stocks. This opposition is not conducive to dealing smoothly with the seasonal food shortage.

Cash utilization

The Massa and the Mussey have been cultivating cash crops for about thirty years; rice in the north and cotton in the south of their region. The cash available from them is not systematically used to buy food stores in order to cope with the foreseeable seasonal food gap. As demonstrated by direct observation and by the questionnaire we used – 'If you had more money, what would you buy?' – money does not appear to be spent primarily on such a trivial item as food; this is the case only in emergency. Money is used to improve housing, buy clothes, manufactured goods and, most of all, cattle to provide bridewealth, in order to increase the number of wives, thereby raising one's social and economic status, and fulfilling 'the socially desirable activity in the home and community.' These aspects have been taken into account by the recent WHO/FAO panel of experts in their estimation of the recommended energy requirements (WHO, 1986, p. 51).

The simultaneous availability of food crops and money from cash crops, as well as the festive atmosphere, tend to minimize the need for buying cereals when they are at a low price to cope with what appears as a remote shortage period, even if there is a correct evaluation of its magnitude (Fig. 12.1).

Household budgets show that in most developing countries a high percentage of the income is spent on food. This is true because of the small size of the income, the disappearance of self-sufficiency in food production and the conversion by statisticians of self-produced food into the amount of money it represents. It is not the case among the Massa and the Mussey, who still attempt to spend as little money as possible on the food they produce or believe they should be able to produce. However, as remarked by Spitz (1982, p. 24) 'the forces to retain the food for self provisioning are still higher than to sell it for cash'. As written by Watts (1988, p. 273) 'householders are sensitive to the seasonal fluctuations in millet prices and the debilitating effect of wet season purchases when prices are perhaps 50 to 100% above normal'. But, having spent their money during the festive harvest period according to social priorities (bridewealth, debts, entertaining various kinds of relatives and friends), they are compelled to sell some of their grain and their sheep or goats at a low price in order to find money for compulsory expenses such as taxes. In many cases, during the hunger season, they will again have to buy back – at twice the price – the very grain they have sold. This mechanism is one of the main levers of local trading. It is quite obvious that, although they have the awareness and the economic means to do so, the Massa and the Mussey in normal times do not accord a high priority to putting aside buffer stocks for the seasonal shortage period. It is time that researchers got rid of what Cook (1973) called 'nutritional reductionism' and Brookfield (1972) 'caloric obsession'. Most human groups are not out to ensure the best possible and permanent nutritional equilibrium according to the latest Western scientific findings. They use their food potentialities according to their specific demands and somehow an adequate nutritional balance is achieved. However, some groups may reach a high level of biological fitness and others a lower one.

Awareness of the nutritional value of foods and dishes is no more widespread than in our society. In traditional groups, people also eat for pleasure, prestige and other aims which are not attuned to the most efficient fulfilment of biological needs as defined by Western scientists.

Wastage

If we look into the food consumption system of the Massa and the Mussey, it is obvious that these populations could consume their food in a more economical way. In each polygynous family, each wife has to cook daily to feed herself and her children, and to provide one dish for her husband and the other adult males of the compound. The latter are provided daily with several dishes which they partake in and share with neighbours and visitors in a very hospitable way. There is no doubt that a more economical (and stingy, according to local values) solution could result in the self-sufficiency of households in the wet season, but food hospitality, and even wastage, during the festive period is status-enhancing and helps in establishing a vast network of social ties, which can be drawn upon in case of emergency.

Another example of food wastage is the lavish use of grain resources to brew beer and distill alcohol. The former may have some nutritional interest – not the latter. However, being drunk is good for prestige in traditional societies and, as we shall see, alcohol brewing is an efficient way for women to obtain cash.

Daily food consumption

Cutting down on food consumption is an obvious way of coping with the seasonality of resources. Among the groups we are dealing with, during the harvest season rations are abundant. During the dry season people try to eat as much as possible outside the household. Then comes the cultural restriction period to which we have referred, after which the housewives progressively diminish the number of meals and the amount of food contained in each meal. At the peak of the shortage period, only the evening meal is maintained. In normal times, the nutritional value of the diet does not vary much from one season to another, although its composition changes noticeably.

Non-material culture

Ceremonial uses of food

There is another way of diminishing food consumption during the rainy season. On the occasion of a funeral, the Massa, and especially the Mussey, slaughter large numbers of cattle, sheep and goats in order to honour the deceased and to demonstrate the affluence of his paternal and maternal kin and affines. When death occurs during the lean period, the funeral feast is deferred until the next dry season. Many authors have reported a decrease in the number of weddings and of bridewealth size during lean periods (Torry, 1979; Minnis, 1985; Watts, 1988). Among the Mussey, on the contrary, marrying a daughter to a still affluent villager is a well-used device to bring in a little wealth and food during the seasonal emergency.

Education

There are other ways of coping with the seasonal food uncertainty, without dealing with material assets. One is education. Among the Massa and the Mussey, the attitude towards eating (etiquette) and food sharing (hospitality) is clear, and eating abundantly and becoming fat are valued. However, individuals are taught to behave in a restrained manner when eating, especially when little food is available. During the seasonal shortage and hunger period of 1985, it was very striking to see that the people, including children, did not throw themselves on food and saw that everyone received his share, however small. The only example of selfish behaviour during a dearth was displayed by an elderly man who was rather antisocial at any time. The same general attitude was recorded by Shack among the Gurage and Ethiopian christians.

Although joking about food and adopting a gluttonous attitude goes on among peers during the *guru*, it is not considered well-behaved when food is scarce. Youths and adults are supposed to be able to undergo ordinary lean periods without complaint.

Oral literature and myths

If we turn towards oral literature, we see that food themes and stories about dearths or hunger are numerous and, among other things, that seasonal food shortage is acknowledged. The Serer have a rather explicit myth about early maturing crops: 'The main millet staple and the early millet are up in the sky with God (*Rog*) and they are looking at the men below. The main staple (*matye*) says to the other one, "I have to go down, my people are suffering". The early ripening variety (*mbot*) replies, "No, don't! I shall go first so that they can eat, but you will go later and fill up their granaries".'

Among the Massa, the main character in tales is a fat, selfish glutton, who always has misadventures, and even reaches death, because of his defects. He goes as far as eating his own children's fingers when they give him food.

Many of the anthropologists dealing with the traditional personality have perceived a high food anxiety component, for instance, among the Marquisans (Kardiner, 1939, p. 94). It also appears that this insecurity may find its expression in myths and the oral literature, and does not necessarily trigger off appropriate material measures to secure food in ordinary life.

Ritual action

Traditional societies in all parts of the world have to cope with seasonal climatic hazards in order to secure their subsistence. The Massa and the Mussey are able to take practical measures after they have assessed, at harvest period, the amount of grain available to them. As the crops now yield over several months, thanks to the use of pricked sorghum which matures later, some hope is left that the last one may be successful. This may preclude taking radical measures for provisioning ahead of time. They have no material way of dealing with environmental uncertainties. Even in Western societies there is little to be done to secure rain, or to

Table 12.11. *Rituals concerning the food cycle of the Mussey*

	Jan	Feb	Mar	Apr	May	Jun	Jul	Aug	Sep	Oct	Nov	Dec
Sacrifice to the Rain God			●									
Sowing finger millet				●	●							
Sacrifice for the rains					●	●						
Sacrifice to fix the limits of flooding					●	●						
Imprisonment of the spirits during the growth of the crops						●						
Sacrifices for the repose of Mother Earth							●					
Ritual for the growth of red sorghum and millet							●					
First fruits to Mother Earth of finger millet, sorghum and cucumbers								●				
First fruits to the ancestors of finger millet, sorghum and cucumbers									●			
First fruits of Bambara groundnuts									●			
First fruits of bulrush millet										●		
Rituals to the guardian spirits of the clans and to Mother Earth	●	●	●									
First fruits of the late varieties of sorghum											●	●
Sacrifice for the cotton market												●
Sacrifice for the rice market	●											
Sacrifice for the safety of food stocks	●	●										
Rituals for the building of granaries and houses		●	●									
Fishing rituals	●	●	●	●	●							
Hunting rituals	●	●	●									

prevent hail or flooding. Prayers and offerings to God are universal and the first fruits of the new crop are offered in the framework of catholicism as well as in that of ancestors worship. Among traditional groups, who used to be believers, it appeared as an efficient technique. The Mussey even have a ritual to delineate the expansion of the flood waters from the tributaries of the Logone river; at the end of the seasonal cycle, they actually call the new year into being. These recourses, as irrational as they may appear to the cynical Western technician, relieve the anxiety and comfort the community in the feeling that they are not helpless in the face of food shortage; that they have a responsibility in relation to it. (See also Grivetti on the Tlokwa, 1981, p. 533) (Table 12.11). These attempts should appear as a humanization and a socialization of the inescapable seasonal uncertainty which seems to be a punishment, but in which humanity has its own responsibility. In this context, religion and magic appear as efficient techniques to bring the uncertainties of the natural environment into the framework of culture.

Among the Mussey, a bad year may appear to be the result of incest displeasing Mother Earth. It may also be thought to be the result of witchcraft performed by an angry villager who has buried a rain stone in a hollow tree trunk. Among the Tupuri, the main Rain Priest actually owns the first rains, and physically releases them on the land in a ceremony we were lucky enough to attend. Traditional societies used to have faith in the supernatural and integrated at least reasonable hardships into the normal destiny of the community. The ritual activities were there to relieve the tension.

Other techniques

Other techniques overlap the border between material efficiency and symbolic action. As we have already mentioned, the Massa value plumpness and abundant eating and have developed an institution, the *guru*, in which most men participate several times in the course of their lives. During the sessions, which can last from two to six months successively, the participants enjoy a diet of milk mixed with sorghum porridge, averaging an extra 1000 kcal per day as compared to the villagers' regime. They are slightly fat. Harrison (1988, p. 28) remarked: 'there is no better example of a true body store than the store of energy as fat', but it goes further: they permanently display a feeling of well-being and an easy-going attitude which pervades the whole village. The women do not participate in the same institution, they are not the trend setters but they are usually proud of the beautiful shape of their husbands, brothers or sons. It creates, even in the rainy season, an optimistic atmosphere.

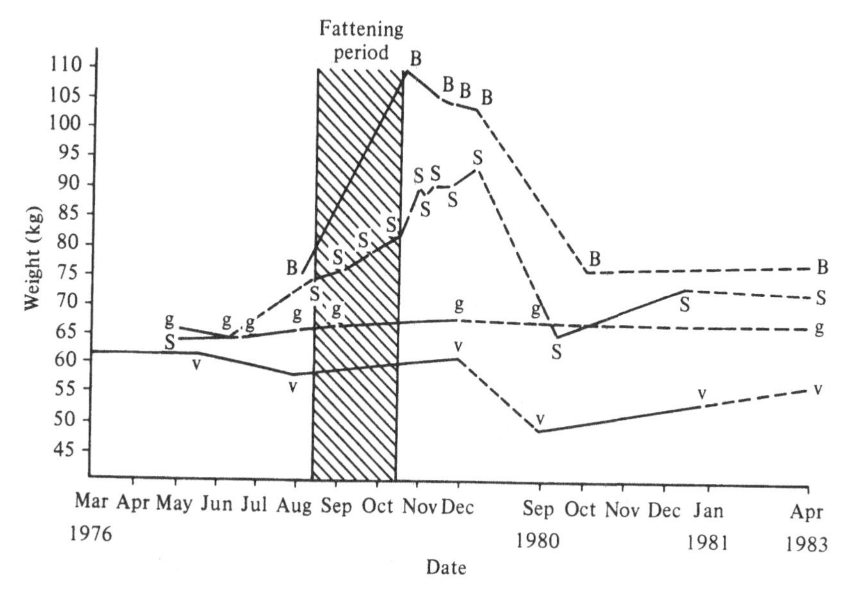

Fig. 12.2 Weight variations of two *guru walla* (B: Bungmasu, S: Sirina) as compared to ordinary *gurna* (g) and adult male villagers (v).

One step further, the Massa display a very singular institution, the *guru walla*. This consists in a wealthy man inviting a young, popular fellow (usually one of his paternal or maternal kin) to gorge on milk and sorghum, undergoing a very stern and authoritative session of fattening during which he must ingest, over a two-month period, a diet of about 12 000 calories per day while being secluded indoors: he usually puts on around 30 kg (Fig. 12.2). These sessions take place during the rainy season, when cows have abundant fodder and milk and when the weather is sufficiently cool to permit the ingestion of large quantities of food indoors without sweating. It also takes place at precisely the period when there is a seasonal shortage of sorghum and, as such, appears as a typical example of conspicuous consumption, demonstrating that the host is wealthy, and that the participant is cherished by the supernatural powers because he undergoes the session without becoming sick or dying. On a collective basis, it attempts to show that the seasonal food shortage does not exist and that the total community to which he belongs is prosperous, well-loved by the supernatural powers and protected against bad death. (The death of a young man, considered to be too young to die, immediately interrupts the *guru* session.) Nowadays, the influence of christianity, islam and rational thinking precludes resorting to this kind of belief to relieve anxiety. It is not necessarily replaced by efficient material techniques to suppress seasonal hunger and anxiety.

As a matter of fact, since the investigations of Kardiner (1939) and Cora du Bois (1960), food anxiety, rather than sexual frustration, has been said to be a component of the basic personality in traditional society. This may be substantiated by the frequency of food themes in daily conversation. Among the Bushmen, Wiessner (1981, p. 651) recorded that 59% of the conversations were on this theme and one gets the same feeling from Shostak's life stories on the same group (1976). The Massa and the Mussey display a large amount of interest in food in their daily conversations. Food themes are also very common in oral literature, and we have already referred to the selfish glutton in their literature. The most conspicuous example of food anxiety may be found in the Mussey initiation (*fulisu bagayna*). At the first stage, the novices take part in a totally masculine cuisine. At the second degree, it is revealed to them that the offerings left for the initiation spirits are actually eaten by the full initiates. Thus they are given the privilege of being fed after their death, while the ordinary people's ghosts have to go hungry for a long time before becoming guardian spirits. The initiate, after his death, goes straight away into the magic circle of the initiation, where he will meet his initiate mates and where food is ritually offered according to the religious calendar; he is thus relieved of the fear of starving after death.

Sociological aspects

Social ties and reciprocity

Many authors have remarked on the intertwining of social and economic ties among traditional societies, especially in food dealing, the privileged framework of which is the household and the local community, 'using everyday kinship, friendship and neighbourly relations . . . regarding which an open stipulation of return would be unthinkable and unsociable' (Sahlins, 1974, p. 191).

Hunter-gatherers illustrate this kind of 'generalized reciprocity' at any time, including the seasonal shortage period. Among the Aka Pygmies, the successful hunters do not consume their quarry, but instead distribute it, initiating a kind of risk-pooling which will allow them to be fed when they come back empty-handed (Bahuchet, 1985). Things are not quite so serene in other societies.

A number of scholars have attempted to devise models relating the intensity of food shortage and the responses it arouses, to social and territorial proximity of the groups involved, as well as the reversibility of the solutions adopted. Watts (1988, p. 276) suggested a low to high

Table 12.12. *Solutions to food shortage according to Torry (1979), Minnis (1985) and Watts (1988)*

Torry	Minnis	Watts
Homeostatically-defined systems	Diversification of food production	*Low order responses*
Dispersal	Improvement of storage	Planting changes
Exodus	Transmission of knowledge on famine foods	Restructuring activities to maximise food production
Displacement	Conversion of surplus into valuables to be stored and traded for food	Borrowing food (kin)
Change in identity	Social relationships outside the region	Minimising unnecesssary allocations of resources (ceremonial expenditure)
Inter-ethnic economic exchange	Taking advantage of regional differences	Wage labouring
Retrenchment of social activities	Fasting, reducing intake	*Slower and deeper*
Shortening fallow periods	Changing food preparation	Selling livestock
Sharper definition of cultivation rights	Intensification of food-acquiring activities	Grain loans
Reduction of bridewealth	Developing craft production to obtain food	Dispersing goods
Ritual regulation of weather	Raiding and warfare	Borrowing, mortgaging, liquidation of assets or pledging
Inter-tribal raiding	Cannibalism	*Terminal*
Developmentally-defined systems	Diminishing cost of rituals and bridewealth	Sale of farm
Reforms from alien economic and political pressure	Ritual regulation of uncertainty	Migration
Improving administration		
Improving decision apparatus		
Transformation of land-use protection		

Table 12.13. *Solutions to seasonal food shortage (Massa and Mussey)*

↓ Increasing intensity	Individuals	Household	Neighbourhood	Paternal kin	Maternal kin	Affines	Outside 'friends'	Market traders	Temporary migration hosts (Fulani)	Government and towns
	Increasing distance →									
Producing enough food	×	×	× +		+		+			
Earning more money	×	×					+	+	+	
Better storage	×	×								
Consuming more sparingly		×								
Buying grain outside		×	+				+	+		
Eating outside	×		+							
Working little and fasting	×	×								
Temporary migrations	×				+		+	+	+	+
Receiving food gifts	×		+	+	+	+	+			
Selling goats and sheep	×	×	+				+	+		
Gathering/hunting famine foods	×	×								
Asking for help	×		+	+	+		+	+		
Selling cattle	×	×	+				+	+		
Stealing	×		+				+	+		
Beseeching help	×			+	+	+			+	+
Begging	×								+	+
Migrating permanently	×	×			+		+	+	+	+
Changing activity	×							+	+	+

× = decision-making unit
\+ = group concerned

commitment of domestic resources, with an increased risk of irreversibility, i.e. at first, a few unimportant domestic resources are committed, then increasingly important items are drawn upon. Laughlin & Brady (1978, p .334), as well as Sahlins (1974), showed that, with the intensity of the ecological stress, the level of social cooperation first increases, then one notices 'a progressive decrement toward zero'. Minnis (1985, p. 20) contended that: 'In the absence of unrestricted mobility, social groups faced with food provisioning problems will have to enlarge their social/economic network so as to have access to a more reliable food supply'. With increasing stress severity, that should be what he called 'increasing social inclusiveness of the response used'. These solutions apply to more urgent situations than those we witnessed among the Massa and the Mussey (Tables 12.12 and 12.13). In normal shortage periods, these two populations are able to draw on their paternal and maternal relatives as well as on their acquaintances. Between paternal kinsmen (especially the groups participating in the same bridewealth system) the food solidarity is total and one can theoretically expect food help in any circumstances. Among the maternal kin, mutual help is also the rule, but less compulsory.

Where in-laws are concerned, reciprocal gifts are more strongly formalized and entertaining a mother-in-law or a son-in-law is an important social event. The prestige of the receiving group is involved and returns are expected. Of course, in the case of a food shortage, families may informally help their daughters who are married outside the community, but the relations with the sons-in-law and the kinship group are usually strained, as can be expected where a very heavy bridewealth system maintains mutual debts and tensions. Food is usually remitted to the in-law as a compulsory gift made by a son-in-law to the kinship group of his wife at the funeral of one of her elderly family members.

The bonds of friendship appear as an essential element of reciprocity, the *bananwalla*, 'the great friend', one of the main partners in life, does not have to be a relative or an affine. The *golla* (cattle lending system whereby friends are entrusted with lactating cows 'so that they can drink their milk'), may reinforce these friendship ties and demonstrate the total trust between the two partners, who can call on each other in case of emergency. However, there is nothing among the Massa and Mussey to compare with the Hxaro system of the Bushmen described by Wiessner (1981) in creating strong, reciprocal bonds and excuses for lasting visits to far-away communities, especially during lean periods. The exogamy rules stress the need for seeking spouses away from the territorial limits of the clan, which may be interpreted as a means of diversifying the territorial characteristics of potential partners in case of food hazards.

Importance of the local group

The Massa and the Mussey resort in most cases to individuals located in their close neighbourhood with whom informal general reciprocity is the rule. Among the Massa, this would be the *farana* (quarter) occupied by households, usually kinsmen, who keep their cattle in a single herd. The same neighbourhood is involved in cooperating in domestic life and food producing activities which have, in turn, an influence in making it possible to face the lean period. It applies in the constant sharing at all periods of the year. In 1976 we recorded the composition of the food consuming group. The number of participants from outside the compound was not affected by seasonality and it consisted mostly of neighbours (Table 12.14).

Table 12.14. *Consumers' attendance among the Massa (total of three one-week periods); Kogoyna village – 1976 survey*

Household	No.	Permanent eaters	Total guests	Neighbours friends	Paternal kins	Maternal kins	Affines	Lovers	Others	Dry season (Jan–May)	Rains – shortage (Jun–Aug)	Crop season (Sep–Dec)
Bunrayna	1	3	13	6	5	1		1		4	3	6
Dara	3	2	19	19						6	3	10
Digidim	4	9	19	14	3		2			3	14	2
Fiwlingsu	5	4	25	9	12		2		2	13	7	5
Fuyawsu	6	11	19	15	2		2			11	5	3
Gamasna	7	5	14	7	6	1				6	6	2
Haramma	8	5	13	10	2		1			3	5	5
Helsu	9	8	19	5	3		7	4		3	6	10
Jibetsu	11	3	8	6	2					1	4	3
Laò	13	5	9	7	1		1			1	4	4
Likusidi	14	7	19	15			4			4		15
Moleyna	15	7	25	20	1		3	1		13	7	5
Nakutsia	16	4	23	19			3		1	13	3	7
Ṅrañam	17	6	20	9	5		6			7	10	3
Reyna	18	8	10	4	3		2	1		3	2	5
Samata	19	6	21	16	3		1	1		9	9	3
Samasu	20	12	9	5	2			1	1	5	4	0
Samhindi	21	9	10	6		1	3			4	1	5
Sampulla	22	4	14	10	2	1	1			6	2	6
Subursu	23	4	17	11	2		4			6	4	7
Yawna-Mutsumuna	24	6	13	8	2		1	1	1	5	7	1
Total		128	339	222	56	4	43	9	5	128	106	105

In normal times food commensality is the rule, and only etiquette slows down the food hospitality during the month of 'Did you call me for food?' The Massa and the Mussey stick to the principle of 'out of sight, out of mind', and reserve their best treatment for relatives residing close by and with whom they are better acquainted. In case of emergency, they do their best to draw on their neighbouring relatives for one main reason; they will not have to pay interest in case of a loan. This solution, however, does not allow them to benefit from territorial and climatic diversification of food risks.

Outside ties

Like all traditional groups, the Massa and the Mussey are obliged to reach outside their face-to-face group and resort to traders on the local market to cope with food shortages. This is far from the reciprocal ties between trader and peanut growers witnessed in Senegal (Garine, 1962, p. 257), but, for trading in ordinary times as well as during difficult times, the Massa and the Mussey solicit 'their trader', the one with whom they attempt to establish a friendly, reciprocal relationship outside business matters, even if they have to cope with a 200% increase in prices at the peak of the hunger season.

During the 1985 seasonal shortage, which was close to a famine, the Massa and the Mussey also warmly acknowledged the rather stingy governmental help, amounting to two cups of rice and two of sorghum, a total for the season of about 3 kg of grain per capita. The last resort was to go to town to draw on salaried relatives, affines or friends less affected by the seasonal shortage.

Emphasis on the household

If we look closely, even in 1985, the Massa and the Mussey did their best to find solutions at the household level by selling goats, sheep and even cattle and buying grain with the money obtained. They also resorted to gathering wild food, cereals, tubers and leaves, and to a lesser extent, small game hunting.

This tendency towards seeking household and even individual solutions to food problems in most circumstances is a cultural characteristic and may also suggest that the traditional pattern of kinship solidarity is breaking down.

Social control

Food solidarity, as well as accepting the daily fare, are based upon beliefs about common ancestry and blood; acceptance of the daily fare stems from the belief that supernatural powers control the universe and access of food. These powers should be dealt with through prescribed channels. They are, among other things, held responsible for any kind of uncertainty at the community level. Dealing with the microcossm as well as the macrocosm implies respecting a precise set of rights and duties. These operate under the control of: (i) the traditional Earth Priest and other magico-religious leaders (*bum nagata, sa mbasta*) leaders of possession groups and initiation chiefs; (ii) the elderly men of the community; (iii) the older as opposed to the younger men; (iv) men over women. Today, due to the influence of outside patterns of life, the system is breaking down. The administrative chiefs no longer feel, as in the past, compelled to feed the needy villagers in case of food shortage. Actually, the 'Chef de Canton' appears as a very negative link between the central government and the villagers and uses every available opportunity to profit from them. The police are unable to enforce the system of laws which appears to be alien to the people, while they have lost most of their faith about the traditional system. This has a heavy impact on the food system and on the feeling of uncertainty.

Magico-religious beliefs

As Richards & Widdowson remarked as long ago as 1936 (p. 196), 'The breakdown of the kinship . . . without corresponding adoption of European concepts of foresight and individual savings leads to serious hardships . . . ' The people are between two faiths and systems of rules and have a tendency to cheat both. Minnis (1985, p. 41) observed that, confronted with outside models of life, religion and ways of thinking, traditional people of today no longer consider the social environment as part of the natural one. Among the Massa and the Mussey the discipline maintained by the Earth Priest in relation to land rights and the organization of food crops is breaking down. Traditionally, land is the property of the clan; it cannot be sold. Its members are allotted an acreage corresponding to the number of people they have to feed. Some of it can be loaned to outsiders, usually maternals or affines, with the consent of the community. Nowadays, land can be rented to outsiders and fierce competition is under way for access to the black, moisture-retaining soils suitable for late pricked sorghum. Being backed by authorities is now more important than retaining traditional rights.

Agricultural tasks are undertaken more haphazardly than in the past and people do not respect any discipline in harvesting the crops. Sometimes sorghum crops are severely depleted before they are ripe. Many authors have remarked that, in traditional societies, food appeared as a privileged commodity to be handled with respect, to be shared and not to be stolen (Mauss, 1950; Sahlins, 1974), especially in case of community emergency. Among the Massa and the Mussey, many magical devices were supposed to protect the crops in the fields and in the granary. Under the influence of outside beliefs and rational thinking, they are totally obsolete nowadays.

Theft

Among the Massa and the Mussey, stealing food is no longer an emergency strategy reserved for famine situations, it has become a daily activity in normal conditions, and especially when times are getting a little harder. It is appalling during shortage time to hear each night the cries of distress of the villagers discovering that their crops have been stolen from the fields, or from their very granary. In the past, stealing cattle, goats or sheep from outside communities was a young man's sport; today it has become a permanent activity, very often undertaken by the younger generation at the expense of the elders. Village youngsters may go as far as hiding in their own homes some of their outside friends, who will wait until nightfall to steal some of the cattle belonging to the head of the household.

Collapse of traditional authority

The lack of respect between the generations, together with a very demanding bridewealth system which still favours the elderly, causes constant conflicts within the household. Younger brothers and elder sons are repeatedly challenging the authority of the household head with regard to food production as well as to strategies in coping with seasonal variations. As a matter of fact, the goals of the two generations are contradictory. The older generation are trying, by all possible means, to enhance their social influence by accumulating cattle and increasing their numbers of wives and children, and by drawing as much profit as possible from the labour of the young men and the production of the women in the compound. The younger men want cattle, in order to get married, and money to gain access to cattle, manufactured objects, fun and drinking. They are not readily willing to share their agricultural production with the rest of the household. As a consequence, men are no longer certain of being surrounded by respect, or even being appropriately fed and sheltered during the rainy season, when they reach old age, especially if, like King Lear, they have distributed their cattle riches to their offspring. Nowadays, they share the lot of old widows, supposed to become tired of living as soon as possible. As demonstrated by Ancey (1975) in 14 West African populations, goal and decision making is of a more conflicting nature than in the past and does not necessarily stress self-sufficiency in food until the next crop as a priority.

The women, who often get married half-heartedly, no longer accept as readily as in the past working mainly for the benefit of their husband. They want to dispose freely of the crop and of the money they have earned with their labour. Within the household, there is a tendency for each wife to constitute a separate economic unit, relying on her own parents and lovers. Brown (1983) has documented the conflicting goals of the Nar women in a population living close to the Massa, who hesitate between favouring their 'husbands' or their brothers. This tendency is clearly illustrated by the depletion by the women of sorghum and millet stores for beer brewing and alcohol distillation. The same thing was recorded by Richards & Widdowson among the Bemba back in 1936 (p. 181). These activities are the best way for women to earn money and have access to the manufactured goods of which they have been deprived for a long time. In some cases, the use of the crop to make alcohol is made with the blessing of the husbands, in others this may not be the case and can result in breaking up the family, the wife establishing herself as a freelance bar tender – and occasional prostitute.

In any case, alcohol making is a constant drain on the family grain supplies. It is not likely to disappear, as getting drunk and 'behaving like a lion' are considered to be prestigious and an efficient method of temporarily relieving social anxiety (Table 12.15).

Resorting to alcohol as a means of reducing psychological uncertainty in changing traditional societies (e.g. Heath, 1987, p. 39), abashed by their economic backwardness, having lost faith in their traditional values and having little hope of improving their future, is characteristic of the Massa and the Mussey as well as other non-muslim and non-protestant groups in Northern Cameroon. It has a noticeable impact on the family budget and food resources – in Kogoyna, 26% of the total income. As in many parts of the world, drinking has shifted from the traditional collective drinking which accompanies most festive occasions to more individual bouts, the frequency of which is related to money availability, linked to the fact that it displays economic wealth and has a prestigious connotation. As in other regions of the world, it is becoming one of the main public health and economic issues (WHO, 1980, p. 19). With the

Table 12.15. *Alcohol production in Kogoyna in 1976, June to November; from Koppert (1981) (in 75 cl. bottles) (A noticeable increase in alcohol consumption is observed in November, after the harvest of the main food and cash crops.)*

June	July	August	September	October	November
144	118	109	138	165	239

disappearance of ritual festive occasions, it may soon stand out as one of the main ways of combatting the increasing boredom of village life.

Change in food preferences

One last aspect to be considered is the change in food preferences occurring under the impact of outside influences and in a quest for a new identity. We have shown elsewhere (Garine, 1980) some of the specific trends of social change among the Massa and the Mussey. Besides the Western and the African urban influences, they favour, for historical reasons, a Fulani islamic style of life and have, to a large extent, lost faith in their own traditional one. These various influences are apparent in the field of food. If the Massa and the Mussey continue to consume their red sorghum staple, it is to a large extent because they are not yet able technically and financially to move away from it. They have lost their taboo on white pricked sorghum. The interviews and questionnaires we have been using since 1976 show that there is a growing preference for white sorghum, bulrush millet and rice (all of which produce white kinds of food), products made of wheat flour, more animal proteins (especially meat and large, fresh fish), sweet products and lipids (Table 12.16). One of the Massa songs praises a warrior: 'His mouth only knows the Diamaor!' (one of the local brands of peanut oil). Products bought at the market are frequently mentioned and it is obvious nowadays that solving food insecurity cannot be undertaken totally through local self provisioning. In order to assert themselves, acquire prestige, reach a certain level of self appraisal and have a positive attitude in relation to their diet, traditional consumers need to be able to buy manufactured food products such as wheat flour, biscuits, salt, sugar, tea, sweets, alcohol, tinned tomato sauce, etc. To make food consumption a psychologically rewarding exercise, reaching a monetary income has become a must.

Abandoning the traditional style of life and its values has another consequence. To quote Trémolières & Baquet (1967) 'as the Golden Age

Table 12.16. *Frequency of consumption, preferences and dislikes among the Massa, Mussey and Fulani of three villages in Mayo Danaye (Cameroon)*

	Mussey (Gobo)	Traditional Massa (Bougoudoum)	Urbanised Massa (Yagoua)	Urbanised Fulani (Yagoua)
Frequently consumed	Early red sorghum Bulrush millet Late red sorghum Pulses Cucumbers Traditional vegetables	Early red sorghum Fish Okra Traditional drinks	Rice Early red sorghum White pricked sorghum Fish, pasta Doughnuts Traditional and imported drinks	Rice Late white sorghum Meat, fish and meat preserves Okra, doughnuts Carbonated drinks Syrups Coffee and tea
Preferences	Bulrush millet Rice, white sorghum White pricked sorghum Chicken, meat, fish Sesame, fatty foods Soft and alcoholic drinks	Early red sorghum Rice, meat Large fresh fish All alcoholic drinks	Early red sorghum Rice, meat Fish Soft drinks	Same as above plus fresh meat
Dislikes	Early red sorghum Bulrush millet Groundnuts Traditional dishes with an unpleasant smell Alcohol (traditional)	Bulrush millet Pricked sorghum Rice Early red sorghum Sorrel Wild leaves Catfish Wild cereals	Bulrush millet Pricked sorghum Early red sorghum Traditional drinks	Red sorghum Alcohol Pork
Prestige, wealth	White sorghum Rice, meat, fish Fatty and sweet foods Manufactured drinks and alcohol	Rice, pricked white sorghum, meat Fresh fish, fatty and sweet foods Manufactured drinks and alcohol	Rice, fish, pasta Meat, large fish European food Manufactured drinks and alcohol	Rice Meat

is considered to be in the future', traditional life appears backward and the bush a 'no man's land', dangerous and dirty. Seen through the eyes of outsiders, urbanites, and children who have been to school, resorting to bush and famine foods appears as behind the times and a sign of poverty, only fit for the youngsters when they watch the cattle grazing away from the village. Bush food, leaves, frogs, rodents and insects are scorned and labour in the bush is considered to be unsafe. In some cases, knowledge of the detoxification techniques of wild tubers has been forgotten, which resulted in mild accidents during the 1985 shortage when famine foods, although despised, played an important role in relieving hunger. This tendency will restrict the range of seasonal solutions available to the villagers.

Present evolution

Broader social environment

The general picture appears to be dark. We witness, as in other populations, the collapse of many parts of a system which based the solution to food problems on group solidarity and a certain discipline in the framework of a strong social structure, relying on traditional values and beliefs. The general conception of the world was based on a cyclic time and stressed homeostasis. One of the Mussey prayers is very explicit in this respect: 'Oh God! Let nothing happen to us!'

Nowadays we notice a tendency towards anomy, little social control is exerted, solutions are sought at the level of the nuclear family or even at the individual level. Money consciousness diminishes hospitality. Food has become a profane commodity which can be manipulated, or even stolen, with no special risk, but it does not yet rank very high among the many priorities to which money should be devoted. Today even the remotest communities are confronted with outside influences and the general flow of information. Traditional societies are perfectly conscious of the fact that their system of thinking and conception of life and the world have become largely obsolete. This does not bring them a feeling of security. What is happening to the Massa and the Mussey is a common process, they are simply moving into a market economy and being integrated into a larger community – their nation. They can no longer spend most of their life in a face-to-face group at local level. They are shifting from a warm kind of community to a more abstract type of social framework, moving from a *gemeinschaft* to a *gesellschaft* type of society, as Tönnies would have termed it (1944).

They have to deal with outsiders (considered in the old days as enemies or not human), who may represent abstract functions at the country level rather than being socially identifiable neighbouring groups. It is not possible for them to fall back on the traditional pattern of interaction, and precise rules of reciprocation may have to be invented before confidence is regained. The traditional social system as well as inter-individual ties are called upon to create an efficient network to face food and economic uncertainties. It is much broader than in the past, but they are doing their best to give it a personal, friendly touch and secure privileges similar to those to which they were accustomed at the village level.

Technical freedom

In the material field, the Massa and the Mussey are increasingly confronted with extension services and new techniques. In the area of agriculture, they are being trained to expand their cotton acreage, to manage pesticides and fertilizers, to use oxen to work in the fields. Very little is done to teach them how to increase their food crops, although this is recommended by the government. The collapse of the traditional taboo on the cultivation of *Sorghum durra* allows them to increase their crops and solve their seasonal food problem if land is available and if they have access to it. This is not exactly the case, owing to ethnic administration and political favouritism.

There is a tendency to focus development on cash crops, believing that access to cash will inevitably solve the food problem. As regards fishing, the constant use of modern nets owned by outsiders and the indiscriminate use of pesticides to kill fish in the river is steadily diminishing the resources. Cattle remain the main asset for success according to traditional channels. Their commercialization is steadily increasing, as well as the turnover of goats and sheep. This is a very dynamic field of action where resources are available to improve the seasonal food shortage and are more and more readily used. The emancipation of the younger generation and of women has positive consequences as they enjoy more freedom to prove their dynamism, they reach, on their own initiative, a better nutritional level through production and commercialization.

Spatial freedom

Although theft and even highway robbery are still practised, individuals are moving spatially more freely than in the past. There are no more threats from enemy tribes or muslim raiders. Land can be rented or even acquired outside the community of origin. Temporary migrations are

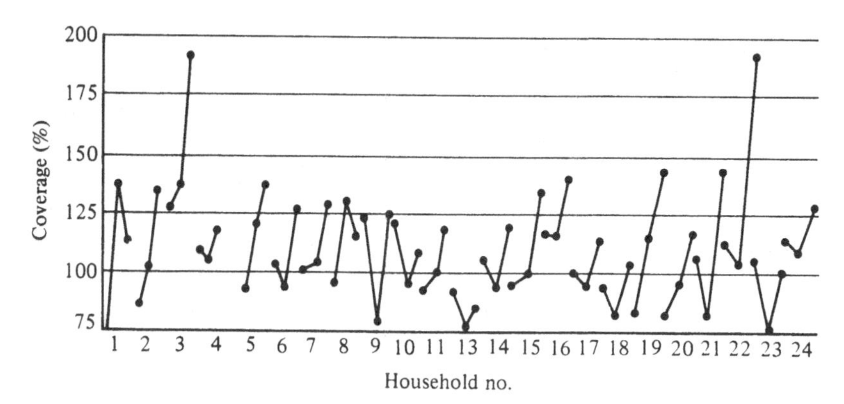

Fig. 12.3 Coverage of the energy needs of 24 households in Kogoyna during three periods in 1976: dry season (Feb–May), rainy season (Jun–Sep), crop season (Oct–Dec) (from Koppert, 1981).

more and more frequent. Permanent migrations can take place to a town or to other rural areas using the kinship and friendship network. In case of extreme demographic pressure or soil impoverishment, neighbouring groups, such as the mountain people of Mora or the Tupuri, have been resettled successfully through governmental action. The Massa and Mussey temporary migrations to Fulani land may also result in permanent settlement. In many cases, while keeping their traditional subsistence economy, the migrants settle to work as salaried agricultural labourers on Fulani or muslim estates.

Household and individual variability in the impact of the seasonal shortage

In the food consumption and anthropometric surveys we have carried out, we have recorded a wide difference between the various households at every period of the year. The households having consumed more during the rainy season than during the dry one are the most affluent in terms of food stores (Fig. 12.3). It is obvious from the literature that shortage periods provide a very good opportunity for the enrichment of the most clever or wealthy individuals (Watts, 1988). In Kogoyna, the 1985 difficult lean period enabled those who had the means to do so, to sell grain at the highest possible price and to obtain cattle, sheep and goats at the lowest possible one. As a matter of fact, one of those who exploited his fellow villagers during the shortage was later accused of sorcery. This illustrates one of the ways in which society is keeping a check on those who try to gain economic affluence too rapidly and not through the traditional channels.

As we have mentioned, traditionally wealth is achieved through the acquisition of cattle, sheep and goats to obtain women. Women in turn produce children; if they give birth to girls, these in turn will bring in more cattle; if they bear boys, cattle will have to be given away. As approximately the same proportion of boys and girls are born, the system is self-regulating. Today a number of individuals who have a permanent source of monetary income use it to accumulate wives and make the best use of the lean period to speculate on grain and cattle. As the traditional control on land by the Earth Priests and the elders of the community is diminishing, land can be hired or even sold to wealthy or politically powerful outsiders, who will pay labourers to work on it.

A rural bourgeoisie

We are witnessing the sprouting of a social-economic elite, usually related to the political power and having in many cases linked interest with the Fulani merchants they are so eager to imitate. The emergence of this rural 'bourgeoisie' should not appear as a new possibility to decrease the overall uncertainty of food supplies through generalized reciprocity, but rather as a means for individuals to enrich themselves at the expense of their neighbours without fear of magical sanctions or social retaliation. The ordinary villagers are witnessing the rise of these trend setters and are perfectly aware of the fact that the key to wealth and, among other things, food security is not in food production but in trading and, more precisely, in taking advantage of the seasonal shortage to speculate on grain and cattle prices. They have also come to understand that the best way of eliminating seasonal uncertainties is to obtain a regular salary, preferably in a non-manual job and, if possible, away from the rural area. There is a general tendency to break away from rural life in order to become a fully-fledged modern citizen according to the Fulani/muslim or negro-urban style of life. There are now opportunities to fulfill these dreams: education.

Education and the urban appeal

Massa and Mussey villagers are increasingly sending their children to school (although in 1985 this only concerned 10% of the male and 3% of the female population) and investing in them to carry out the best possible studies in order to draw heavily on them once they have a permanent job. In the village, receiving money subsidies from successful children is prestigious, and drawing on them during the lean periods an efficient formula. There is already a successful minority of educated

Massa and Mussey holding jobs locally and in various towns, including the capital. Visiting them provides villagers with adventure and is a good way of escaping food and health hardships in the village. The interaction between villagers and urbanites belonging to the same kinship, residential, village or simply ethnic group is likely to become one of the most adequate ways of restoring reciprocity links between the modern and traditional sections of a given population. Odeye (1985) showed the efficiency of the system in Senegal. It began in a rather negative way. As a Serer informant put it: 'We were fed up with being constantly begged at for food by our village relatives, but we could not let them die in the hunger period.' Among the Serer, going to town during the lean season and helping with domestic tasks was a common solution for young women. Nowadays, a two-way flow seems to be operating: villagers go to town to benefit from the economic security of the urbanites, but town dwellers send their children to the village for holidays and come themselves to obtain special foods which are no longer available in town, to relax from their urban stress and to receive traditional medical treatment. One may even foresee the rise of a renewed interest for the 'good old authentic life and food in the hospitable village atmosphere'.

The craving for education in order to raise their living standards has other positive consequences. The Massa and especially the Mussey are very open to new information and training. This can easily be used to reconcile, through educational programmes, nutritional needs and the traditional attitude, which is not very prone to favouring adequate diets for the vulnerable groups, and to establish a little more nutritional justice in the household.

Bridewealth: the blocking factor

However, the positive trends we have mentioned are hampered by a very negative feature which will, among other things, preclude the use of more money for food, especially among the Massa. They have not abandoned their bridewealth system, which monetization of the economy is making even more constraining but which remains the main path to social success and economic prosperity, provided one is lucky enough to give birth to more girls than boys. Obtaining a wife through a bridewealth of ten cows represents in cash about one thousand US dollars: approximately ten years' monetary income. The price is climbing as the members of the new local elite are ready to increase the rates in order to boost their prestige. Fear of remaining a socially immature bachelor does not predispose young men to use cattle, sheep and goats just for food. Providing all year round adequate food to all members of the household carries little weight compared with the desire to hoard cattle in order to become a prestigious polygynous family head.

Monetization and food anxiety

Although the evolution is slow, the Massa and the Mussey are heading towards a situation where seasonal food fluctuations might be reduced through the market economy and access to a sufficient income. This does not necessarily mean that they will enjoy a more rational diet, but that they may instead come to experience the same type of nutritional shortcomings as Western societies. Their tendency towards eating more abundantly, consuming more lipids, fat meats, sugar and alcoholic drinks as well as rejecting physical work give us strong clues in this respect. Reaching a steady income and constant over-eating may bring them the same kind of queries in relation to body image and health that we observe in modern industrialized society; this is already the case in urban Africa. They will exchange a seasonal food insecurity, responsible for mild temporary malnutrition and channelled through symbolic collective rituals, for a food cycle linked to the periodicity of their income, accompanied by a permanent uneasiness and social anxiety which they will have to negotiate at the individual level, sometimes through precisely those magical techniques of the traditional society they believe they are running away from. They are likely, however, to have lost sight of all the seasonal attitudes and celebrations which allowed them to attune their lives, even symbolically, to the course of the seasons.

Seasonal food uncertainty: possible positive aspects

The last aspect of this chapter is: has seasonal food uncertainty only negative consequences? The answer made by most of the specialists gathered in two recent symposia* appears to be affirmative. However, this issue needs careful examination and may not lead to such a negative conclusion if it is considered in terms of quality of the diet rather than of its calorie/energy value. The diet of the Massa, although satisfactory as a whole, is deficient in vitamin C during six months, especially during the dry season when only dried okra and amaranthus leaves are available (Table 12.2). It is precisely during the 'bad rainy season', when green leaves from the bush and cucumbers are available, that the diet reaches

* (i) 'Seasonal Dimension to Rural Poverty' organized by the Institute of Development at the University of Sussex, UK, 4–7 July 1978.
(ii) 'Coping with Uncertainty in Food Supply' organized by the Maison des Sciences de Homme and the Werner-Reimers Stiftung in Bad Homburg, Germany, 13–16 December 1982.

Table 12.17. *Food group consumption per season (g/d) in Kogoyna, Massa, 1976–80 (Garine & Koppert, 1988, p. 223)*

Group of foods		Annual average 1976	Dry season (Feb–May) 1976	Rainy season (Jun–Sep) 1976	Crop season (Sep–Dec) 1976	Crop season (Sep–Dec) 1980	Significance level of differences		
							dry vs rainy season	rainy vs crop season	dry vs crop season
1. Cereals	m	638	578	604	726	638	ns	***	***
	se	149	113	116	177	120			
2. Fresh fish	m	151	183	98	171	116	***	***	ns
	se	79	75	58	68	101			
3. Dried fish	m	11	11	3	17	6	**	***	*
	se	13	14	4	14	11			
4. Vegetables (fresh)	m	44	1	74	60	63	***	ns	***
	se	46	2	38	49	75			
5. Vegetables (dried)	m	8	16	1	7	7	***	**	***
	se	9	7	3	9	8			
6. Meat	m	11	18	10	7	22	ns	ns	ns
	se	26	39	14	22	36			
7. Milk	m	47	34	89	17	66	ns	*	ns
	se	12	99	160	76	145			
8. Salt (mineral & vegetable)	m	3	2	3	5	ns	ns	**	
	se	1	2	1	1	3			
9. Nuts and seeds	m	10	6	4	17	14	ns	*	ns
	se	23	12	8	35	36			
10. Roots and tubers	m	3	8	0	1	12	ns	ns	ns
	se	—	21	—	3	28			

m Mean.
se Standard error.
Significance level between means: ns, not significant; * $p<0.1$; ** $p<0.01$; *** $p<0.001$ (Student's t-test for paired observations).

Table 12.18. *Food consumed outside mealtimes, Massa (Kogoyna) 1976 (number of occasions)*

	Dry season (Feb–May)	Rainy season (Jun–Sep)	Crop season (Sep–Dec)	Total
Cereals	6	5	1	12
Nuts, pulses, berries	18	23	100	141
Tubers	13	7	22	42
Fruit	100	35	104	239
Cucumbers	0	33	7	40
Other vegetables	0	1	9	10
Sweet stalks	0	70	4	74
Green cereal cobs	0	60	0	60
Meat	1	4	4	9
Fish	6	7	1	14
Eggs	1	0	0	1
Sugar, confectionery	15	3	5	23
Sorghum alcohol	26	8	21	55
Sorghum beer	33	7	9	49
Fermented millet drinks	2	4	1	7
Kola nuts	26	8	19	53
Total	247	275	307	829

its best qualitative equilibrium (between June and October). Each season is strongly characterized. The Massa could be considered to be going on a 'fresh fish diet' in the dry season (January–May), a 'greens and vegetables' one in the rainy season (June–September), and a 'cereals, pulses and oil seeds diet' during the crop season (October–December) (Tables 12.17 and 12.18).

A group of ethnobotanists (Etkin & Ross, 1982) pointed out in northern Nigeria the adaptive function of the increase during the rainy season of plant consumption in non-grain dishes (because grain is lacking). Many of the plants used have medicinal properties and specifically combat the increase of gastro-intestinal disease (helminth infections and some forms of infectious enteritis).

All animals are submitted to seasonal variations in their diet, are adapted to their environment and reach an acceptable level of fitness. Why should it not be the case for Man, especially in traditional societies?

Whether we consider the !Kung Bushmen and the mongongo nut, the Eastern Tuaregs and dates or the Aka Pygmies and forest caterpillars, traditional societies demonstrate a tendency to gorge on food resources when they are available rather than to spread them out in order to reach a more permanently balanced diet (according to the nutritionist's views). Such behaviour has probably immediate as well as long-term biological consequences which are not often as carefully studied in Man's normal environment, his society, as they are in hospitals and clinical research. Besides approaching the influences of a seasonal factor in human nutrition and physiology through spontaneous selection of food at different periods of the year under laboratory conditions and referring it to vestigial physiology (Debry, Bleyer & Reinberg, 1975), it might be worth studying the short and long-term effects of seasonal changes in food consumption in a range of contemporary societies located in characteristic ecosystems.

The contemporary industrial urban society provides modern Man with a style of life in which he can abolish from his diet the time and distance factors. If he has the necessary economic means, he is free to fancy a very varied diet or to stick to a very monotonous one, satisfying his personal taste and his ego. The results he reaches biologically and psychologically are by and large not especially brilliant. The consequences of over-eating, malnutrition and lack of exercise in urban societies are well known; they demonstrate stress as well as well-being. It is possible that looking at seasonal food shortage in terms of an environmentally and culturally inbuilt cure or dieting session would provide a more constructive and refined approach towards understanding the profound effects of seasonal variations in human diet.

Acknowledgements

The studies on which this paper is based have been supported by the Centre National de la Recherche Scientifique, France. Since 1983 we have benefited from the support of the Institute of Human Sciences of the Ministry of Higher Education and Scientific Research of Cameroon, within the framework of the collaborative programme entitled 'Food Anthropology of Cameroonian Populations'. We acknowledge their help and wish to thank them.

References

Ancey, G. (1975). *Niveaux de Décision et Fonctions Objectifs en Milieu Rural Africain*. AMIRA (Groupe de Recherches pour l'Amélioration des Méthodes d'Investigation en Milieu Rural Africain) No. 3. INSEE (Institut National de la Statistique et des Etudes Economiques), Paris.

Annegers, J.A. (1973). Seasonal food shortage in West Africa. *Ecology of Food and Nutrition*, **12**, 251–8.

Bahuchet, S. (1985). *Les Pygmées Aka et la Forêt Centrafricaine*. SELAF (Société d'Etudes Linguistiques et Anthropologiques de France), Paris.

Bellisle, F. (1979). Human feeding behaviour. *Neuroscience and Behavioural Reviews*, **3**, 163–9.

Bennett, M.K. (1953). *The World's Food*. Harper & Brothers, New York.

Bois, C. du (1960). *The People of Alor – a Social Psychological Study of an East Indian Island*. Harvard University Press, Cambridge, MA.

Brookfield, H.S. (1972). Intensification and disintensification in Pacific aquaculture, a theoretical approach. *Pacific Viewpoint*, **13**, 30–48.

Brown, E.P. (1983). *Nourrir les Gens, Nourrir les Haines*, Société d'Ethnographie, Paris.

Brun, T.A., Ancey, G. & Bonny, S. (1979). Variations saisonnières de la dépense énérgétique des paysans de Haute Volta. In: *Environnement Africain* 14, 15, 16, Paris.

Cantrelle, P., Diagne, M., Raybaud, N. & Villod, M.-Th. (1967). Mortalité de l'enfance dans la région de Khombole-Thiénaba (Sénégal 1964–65). In: *Conditions de Vie de l'Enfant en Milieu Rural en Afrique*, ed. M.-J. Bonnal, P. Cantrelle, M. Diagne, I. Paul-Pont, N. Raybaud & M.-Th. Villod, pp. 134–9. Réunion et Conférences XIV, Centre International de l'Enfance, Paris.

Chambers, R. (1982). Health, agriculture and rural poverty: why season matters. *Journal of Developmental Studies*, **18**, 217–33.

Cook, S. (1973). Production, ecology and economic anthropology – notes towards an integrated frame of reference, *Social Science Information*, **12**, 25–52.

Currey, B. (1984). (ed.) *Famine as a Geographical Phenomenon*. Reidel, Boston, MA.

Debry, G., Bleyer, R. & Reinberg, A. (1975). Circadian, circannual and other rhythms in spontaneous nutrient and caloric intake of healthy four-year olds. *Diabète et Métabolisme, Paris*, 1975, **1**, 91–9.

Durkheim, E. (1893). *Division du travail social*. Sage, Paris.

Etkin, N.L. & Ross, P.J. (1982). Food as medecine and medecine as food – an adaptive framework for the interpretation of plant utilisation among the Hausa of Northern Nigeria. *Social Scientific Medicine*, **16**, 1559–73.

FAO (1958). *Food and Agricultural Development in Africa South of the Sahara: the state of Food and Agriculture 1958*. FAO, Rome.

FAO/OMS (1974). *Manuel sur les Besoins Nutritionnels de l'Homme*. FAO, Rome.

Fox, R.H. (1953). Energy Expenditure of Africans Engaged in Various Agricultural Activities. PhD Thesis, London University.

Friedman, J. (1979). Hegelian anthropology: between Rousseau and the world spirit. In: *Social and Ecological Systems*, ed. P. Burnham & R.F. Ellen, pp. 253–70. Academic Press, London.

Gardner, L.I. (1972). Deprivation and dwarfism. *Scientific American*, **227**, 76–82.

Garine, I. de (1960). *Le Prestige et les Vaches*. Communication au VI° Congrès des Sciences Anthropologiques et Ethnologiques, Paris 1960. In: Actes du Congrès II, 2° partie, Paris 1963–4, pp. 191–6.

Garine, I. de (1962). Usages alimentaires dans la région de Khombole (Sénégal). *Cahiers d'Etudes Africaines*, **10**, 218–65.

Garine, I. de (1964). *Les Massa du Cameroun – Vie Economique et Sociale*. Presses Universitaires de France, Paris.

Garine, I. de (1980). Approaches to the study of food and prestige in savanna tribes – Massa and Mussey of northern Cameroon and Chad. *Social Science Information*, **19**, 39–78.

Garine, I. de (1981). Greniers à mil dans l'arrondissement de Thiénaba, région de Thiès (Sénégal). In: *Les Techniques de Conservation des Grains à Long Terme*, ed. M. Gast & M. Sigaut, pp. 85–98. Centre National de la Recherche Scientifique, Paris.

Garine, I. de (1984). De la perception de la malnutrition dans les sociétés traditionnelles. *Social Science Information*, **23**, 415; 731–4. Sage, London.

Garine, I. de & Koppert, G. (1988). Coping with seasonal fluctuations in food supply among savanna populations: the Massa and Mussey of Chad and Cameroon. In: *Coping with Uncertainty in Food Supply*, I. de Garine & G.A. Harrison, pp. 210–60. Clarendon Press, Oxford.

Geertz, C. (1966). Religion as a cultural system. In: *Anthropological Approaches to the Study of Religion*, ed. M. Banton. Tavistock, London.

Grivetti, L.E. (1981). Geographical location, climate and weather and magic: aspects of agricultural success in the Eastern Kalahari, Botswana. *Social Science Information*, **20**, 509–36.

Guedeney, A. (1986). Les aspects psychosomatiques des malnutritions protéino–caloriques de la première enfance en milieu tropical, faits et hypothèses. *Psychiatrie de l'Enfant*, **XXIX**, 155–89.

Harrison, G.A. (1988). Seasonality and human population biology. In: *Coping with Uncertainty in Food Supply*, ed. I. de Garine & G.A. Harrision, pp. 26–31. Clarendon Press, Oxford.

Heath, D. (1987). A decade of development in the anthropological study of alcohol use: 1970–1980. In: *Constructive Drinking – Perspectives on Drink from Anthropology*, ed. M. Douglas, pp. 16–69. Cambridge University Press, Paris.

Heywood, P.F. & Nurse, G.T. (1980). *Regular Seasonal Hunger: a widespread phenomenon?* In X ICAES (International Congress of Anthropological and Ethnological Sciences) Series No. 2, series ed. L.P. Vidyarthi, ed. I.P. Singh & S.C. Tiwari, pp. 61–70. Concept Publishing, New Delhi.

Hunter, J.M. (1967). Seasonal hunger in a part of the West African savanna: a survey of body weights in Nangodi, north east Ghana. *Transactions of the Institute of British Geographers*, **41**, 167–85.

Hussain, M.A. (1985). Les variations saisonnières et la nutrition dans les pays en développement. *Alimentation et Nutrition*, **11**, 27–32.

Jelliffe, D.B. (1967). Parallel food classifications in developing and industrialising countries. *American Journal of Nutrition*, **2**, 273–81.

Kardiner, A. (1939). *The individual and his Society*. Columbia University Press, New York.

Koppert, G. (1981). *Kogoyna, étude alimentaire, anthropométrique et pathologique d'un village massa du nord Cameroun*. Département de Nutrition, Université des Science Agronomiques, Wageningen, Netherlands. Mimeo, 151 pp.

Koppert, G. (1988). Alimentation et culture chez les Tamang, les Ghalé et les Kami du Népal. Thèse de Doctorat de 3° cycle en Ecologie, Université de Droit, d'Economie et des Sciences d'Aix/Marseille, miméo 259 pp.

Kroeber, A.L. & Kluckhohn, C. (1952). Culture: a critical review of concepts and definitions. *Papers of the Peabody Museum of American Archaeology and Ethnology*, **47**, No. 1.

Laughlin, C.D. Jr. & Brady, I.A. (1978). Diaphasis and change in human populations. In: *Extinction and Survival in Human Populations*, ed. C.D. Laughlin Jr. & I.A. Brady, pp. 8–48. Columbia University Press, New York.

Lee, R.B. (1979). *The !Kung San*. Cambridge University Press, New York.

Linton, C.R. (1936). *The Study of Man*. Century, New York.

MacGregor, I.A., Rahman, A.K., Thompson, B. & Thomspon, A.M. (1968). The growth of young children in a Gambian village. *Transactions of the Royal Society of Tropical Medicine and Hygiene*, **62**, 303–14.

Mauss, M. & Beuchat, H. (1904). Essai sur les variations saisonnières des sociétés Eskimoes – etudes de morphologie sociale. *L'Année Sociologique*, T.IX, p. 39, réédité in Mauss, M. (1983). Sociologie et Anthropologie, 8th edition, pp. 389–477. Presses Universitaires de France, Paris.

Mauss, M. (1950). *Sociologie et Anthropologie*. Presses Universitaires de France, Paris.

Minnis, P.E. (1985). *Social Adaptation to Food Stress – a Prehistoric South-Western Example*. University of Chicago Press, Chicago.

Miracle, M.P. (1961). Seasonal hunger: a vague concept and an unexplored problem. *Bulletin de l'IFAN*, **XXIII**, Série B, 271–83.

Odeye, M. (1985). Les relations ville/campagne intra-familiales: le cas de Dakar. In: *Nourrir les Villes en Afrique Sub-Saharienne*, ed. N. Bricas, G. Courade, J. Coussey, P. Hugon & J. Muchnik, pp. 256–70. L'Harmattan, Paris.

Ogbu, J.U. (1973). Seasonal hunger in tropical Africa as a cultural phenomenon. *Africa*, **XLIII**, 317–32.

Pagezy, H. (1988). Contraintes nutritionnelles en milieu forestier équatorial liées à la saisonnalité et à la reproduction: réponses biologiques et stratégies de subsistance chez les Ba-Oto et les Ba-Twa du village de Nzalakenga (Lac Tumba), Zaire. Thèse pour le Doctorat d'Etat ès Sciences, Université d'Aix/Marseille III (miméo) 511 pp.

Platt, B.S. (1954). Food and its production. In: *The Development of Tropical and Sub-Tropical Countries, with special reference to Africa*, ed. A.L. Banks. Arnold, London.

Richards, A.J. & Widdowson, E.M. (1936). A dietary study in northeastern Rhodesia. *Africa*, **IX**, 166–96.

Sahlins, M. (1974). *Stone Age Economics*. Tavistock, London.

Schwartz, A. (1968). Calendrier traditionnel et conception du temps dans la société guéré. Cahiers de l'ORSTOM (Office de la Recherche Scientifique et Technique Outre-mer), *Série Sciences Humaines*, **V**, 53–64.

Shack, W.A. (1977). Anthropology and the diet of man. In: *Diet of Man: Needs and Wants*, ed. J. Yudkin, pp. 262–80. Applied Science Publishers, London.

Shostak, M. (1976). A !Kung Woman's memories of childhood. In: *Kalahari Hunter/Gatherers*, ed. R. Lee & I. de Vore, pp. 246–78. Harvard University Press, Cambridge, MA.

Silberbauer, G.B. (1981). *Hunter and Habitat in the Central Kalahari Desert*. Cambridge University Press, New York.

Spitz, P. (1982). *Models, Theories and Applications: Drought and Self-Provisioning*. Reprint from the regional workshop on Seasonal Variations in the Provisioning, Nutrition and Health of Rural Families, AMREF (American Medical and Research Foundation), Nairobi, Kenya, pp. 22–40.

Teokul, W., Payne, P. & Dugdale, A. (1986). Seasonal variations in nutritional status in rural areas of developing countries: a review of the literature. *Food and Nutrition Bulletin*, **8**, 7–10.

Tönnies, F. (1944). *Communauté et Société*. Presses Universitaires de France, Paris.
Torry, W.A. (1979). Anthropological studies in hazardous environments: past trends and new horizons. *Current Anthropology*, **20**, 517–41.
Trémolières, J. & Baquet, R. (1967). Le comportement alimentaire de l'Homme *Maroc Medical*, **508**, 761–2.
Turnbull, C.M. (1972). *The Mountain People*. Touchstone, New York.
Turton, D. (1977). In Torry, op. cit. *Response to drought: the Mursi of southwestern Ethiopia* in Human ecology in the tropics. Ed. J.P. Garlick and R.W.J. Keay, pp. 165–92. Symposia of the Society for Human Ecology 16.
Watts, M. (1988). Uncertainty and food security among Hausa peasants. In: *Coping with Uncertainty in Food Supply*, ed. I. de Garine & G.A. Harrison, pp. 260–90. Clarendon Press, Oxford.
WHO (1980). *Problems Related to Alcohol Consumption*, Technical Report Series No. 650. WHO Geneva.
WHO (1986). *Besoins énergétiques et besoins en protéines*, Série de Rapports techniques No. 724. OMS, Geneva.
Wiessner, P. (1981). Measuring the impact of social ties on nutritional status among the !Kung San. *Soc. Sci. Inf.*, **20**, 641–78.

13 *Food distribution, death and disease in South Asia*

BARBARA HARRISS

Reviews of evidence on the distribution of hunger within the household (Harriss, 1986) and of the literature on gender differences in mortality, morbidity and malnutrition in South Asia (Harriss, 1989) suggest a proposition which needs a critique. The proposition is exemplified in a generalisation of Amartya Sen's, in common with other feminists, that in South Asia 'the pattern of sex bias against women in the distribution of food has come through strikingly' (Sen, 1985). It has been concluded further from this that the sex bias may also be traced in disease and mortality and is the outcome of a particularly South Asian brand of patriarchy. Now if this is true, it has implications for policy in the area of targeted, 'reverse discrimination' by governments in order to bypass households practising such gender bias. Alternatively it is possible to bypass the inadequate practices inside households which are thought to lead to individual malnutrition for the female sex, particularly in certain age groups, by empowering women and by organising women to group together to demand their legal rights and/or more resources from the State. The generalisations made about sex bias are thus of considerable political importance.

I have been looking at sex bias in terms of three broad sorts of question. One is whether sex bias in mortality indicates sex bias in morbidity and sex bias in the allocation of food within the household. The second is whether sex bias in South Asia is uniformly distributed. The third is how sex bias varies economically and socially in a given region – how are the agrarian ecology and material conditions of life reflected in individual welfare? Here I shall summarise some evidence that has been considered at length elsewhere.

In response to the first question, whether sex bias in mortality translates itself directly into sex bias in morbidity and malnutrition, the short answer is no. In India it looks as though a sex bias in aggregate mortality has declined almost to the point of being non-existent. This is a very surprising fact for many people who have studied this subject. The life expectation at birth now is 52.5 for males and 52.1 for females and is on a path towards convergence (Dyson, 1987). Of course such a trend masks gender differences in mortality. There is still excess female mortality. This is maximised in absolute terms in the years one to four, and in terms relative to male mortality in the reproductive years 15 to 29. These imbalances are countered by excess male mortality in older age groups, which is increasing and at a progressively lower age throughout the subcontinent (Dyson, 1984). It is thought by some that the convergence of life-expectation is the result of excess male mortality rather than any improvement in female health or female nutritional status (Karkal, 1987). Quite what constitutes the driving force behind this convergence is still controversial, but I think it is clear that mortality decline brings with it changes in the causes of death. The disease environment in South Asia is also changing and may no longer be such as to translate discriminatory practice into gender differences in mortality rates. Mortality is less and less easily correlated with morbidity. On morbidity, the role of poor female health status in the reproduction of malnutrition is perfectly clear: malnourished women produce low birth weight babies who are particularly vulnerable to perinatal mortality (Karkal, 1987). On the other hand the evidence for morbidity which is unrelated to female reproduction is not so clear. Geraldine McNeill found no gender differences in the generalised incidence of disease in South India (McNeill, 1986) and Chen and his associates also reached the same conclusion in Bangladesh. There may, however, be gender differences in the duration and intensity of general morbidity (Chen, Huq & Huffman, 1981a). Yet the etiology of specific diseases provides counter intuitive examples. Nicholas Cohen (1987) considered eye diseases which may pauperise their victims and lead indirectly to their deaths. Here it may be the response to the disease rather than the incidence of disease which is gender specific, where there is a strong male bias in treatment (though, ironically, this may be iatrogenic). Alternatively the sexual division of labour (that is, the segregated lives and tasks of men and women, boys and girls) affects exposure, duration and intensity of eye disease. To date the evidence is insufficient for any general conclusion about gender differentials in mortality and gender differentials in morbidity. Data for the latter type of study are patchy, whereas data on mortality are derived from censuses and are much more thorough. There has been a recent rash of village studies in North India which show unbiased child feeding practices coexisting with excess female child mortality (Visaria, 1987; Warrier, 1987). Explanations hypothesised are of two sorts: first that the excess female child mortality may be the result of male biased health expenditure or gender differentials in aspects of expenditure other than feeding (see de Garine, Chapter 12 of this volume).

Alternatively Monica Das Gupta has suggested that excess female child mortality may be very highly specific. She posits a continuum between generalised neglect and individualised murder and suggests that, in the villages she is studying in the Khanna region of Punjab, the fact that the second or third daughter has a 70% greater probability of dying before five years than the first daughter is a very specific form of demographic manipulation. Excess female child mortality can then coexist with no generalised neglect of the female sex at all (Das Gupta, 1987). So for the question of the closeness of the relationship between mortality, morbidity and malnutrition, the data do not allow us yet to make very strong inferences about positive associations. Further, as mortality declines and the standard of living increases, mortality, morbidity and malnutrition will become increasingly less good indicators of sexual discrimination. For evidence of discrimination, one will have to look (as in western countries) at education, earning capacity, control over domestic budgets and household decision, including fertility decisions etc.

The second question is whether South Asia has uniform patterns of age and gender bias. By way of background, the average sex difference in mortality is four per thousand, aggregate mortality is declining at the rate of one per thousand per three years. Regional variations in mortality are much more pronounced and interesting. Female child mortality rates vary between Kerala in south (the low end) and Uttar Pradesh in the north (the high end) by 58 per thousand. It has been argued that it is regional variations rather than gender variations that ought to be the more important political issue (Caldwell & Caldwell, 1987).

There is a constellation of demographic variables which are thought to vary from the north west of South Asia to the south east. The sex ratio varies from masculine in the north west to feminine in the south east. Fertility varies from high in the north west to low in the south east. Aggregate excess female mortality, excess female mortality in reproductive years, and excess child mortality in relation to other age groups again all vary from high in the north west to low in the south east, as does the probability of widowhood. Dyson (1987) has concluded from this clustering of demographic variables that sex variations in mortality are the key to other dimensions of demographic variation which are possibly more important. On the other hand little research concentrates on the problem of explaining the regional geography of child deaths (see Fig. 13.1).

The sample registration surveys of the early 1980s, spliced in with the 1981 census, show that excess female mortality is still very high – it is maximised in Punjab, Haryana, Rajasthan and Bihar. In other words there is a belt of high female child mortality across the north of India and now it is only in these regions where life expectation at the age of five

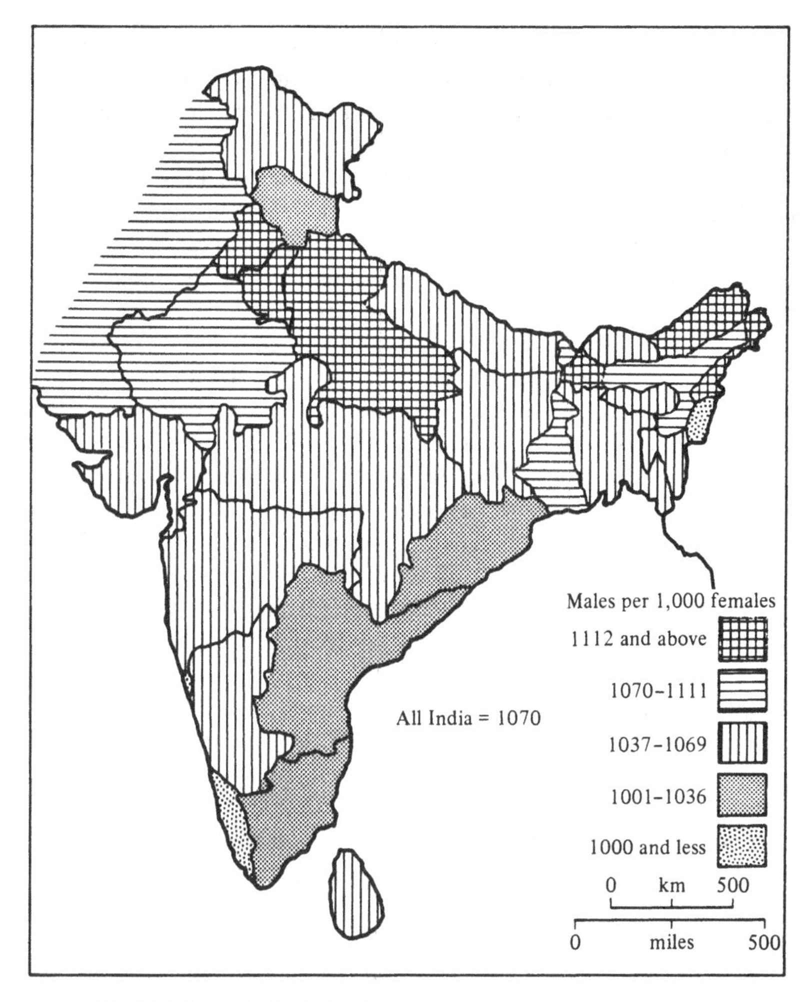

Fig. 13.1 Sex ratios in Indian states, 1981. (Source: Padmanabha (1983), Table 7.)

favours men; elsewhere it favours women. There is no excess female child mortality in Tamil Nadu and Kerala in the south, nor is there any in the hill states of the north east and in Assam. It is very low in Karnataka as well as in Andhra Pradesh, so rather than showing a gradient from north west to south east it is equally valid to conceptualise it as a belt of female disadvantage across the north, surrounded by increasingly less female disadvantage in all directions. What happens on the other side of the Himalayas in Tibet and in Soviet Central Asia is not known. Thus controversy surrounds explanations of mortality in the north west part of the Asian subcontinent. If the north west is part of a continuous trend, then its demography can be explained in the same way as other parts of the subcontinent. If Punjab and Haryana and eastern Uttar Pradesh are anomalous, they require a separate and unique analytical treatment. They do seem to me to be anomalous because they have low overall mortality yet high excess female child mortality, high life expectation yet an excess male mortality which begins very late compared with the rest of the subcontinent. Excess male mortality starts in this region at above the age of 40, whereas everywhere else it occurs in the late 20s and 30s.

How do nutritional status data fit in with this geography? The answer is: not very well. Anthropometric status is a combined and proximate outcome of the interaction between food, disease and health care, and it cannot be regarded as the outcome of any one of them without information on the other two. As with data on morbidity, our map of anthropometric status has to be constructed from spot data. This data is not supportive of a direct causal connection between relatively low female anthropometric scores and either excess female child mortality or excess female aggregate mortality. Interesting anthropometric results in Bangladesh and Uttar Pradesh show both discrimination against girls and lack of it (compare Chen *et al.* (1981b) with Abdullah (1983) on Bangladesh for contrasting examples). Anthropometric studies in South India (Pereira, Sundaraj & Begum, 1979; Pushpamma, Geervani & Usha Rani, 1982; McNeill, 1986) on children, adolescents and adults do not evince gender bias. Yet it is quite possible that these average results mask countervailing trends according to class position, or income and poverty. Certainly these issues need a great deal more research.

With regard to energy intakes, my reading of the debate about adaptation would lead me to be very cautious in interpreting relatively low individual nutrient shares as causing physiologically costly malnutrition, if these data are unsupported by clinical or sociological evidence. Fig. 13.2 is derived from 24 studies in the subcontinent, all surveys of individual nutrient intakes within households. Unfortunately, these 24 studies are not standardised in terms of methods of observation, period of observation, sample size and sample selection, or in terms of the classification of the age groups of people inside the household and the number of nutrients which have been analysed, so one is forced to make some heroic assumptions about their consistency. I created a sharing index on the individual observations of each household (Harriss, 1986). Each observation was indexed in relation to the male household head and that index itself was related to the Indian Council of Medical Research nutrient recommendations transformed in like fashion into an index. Thus actual sharing practices are compared against ideal shares in

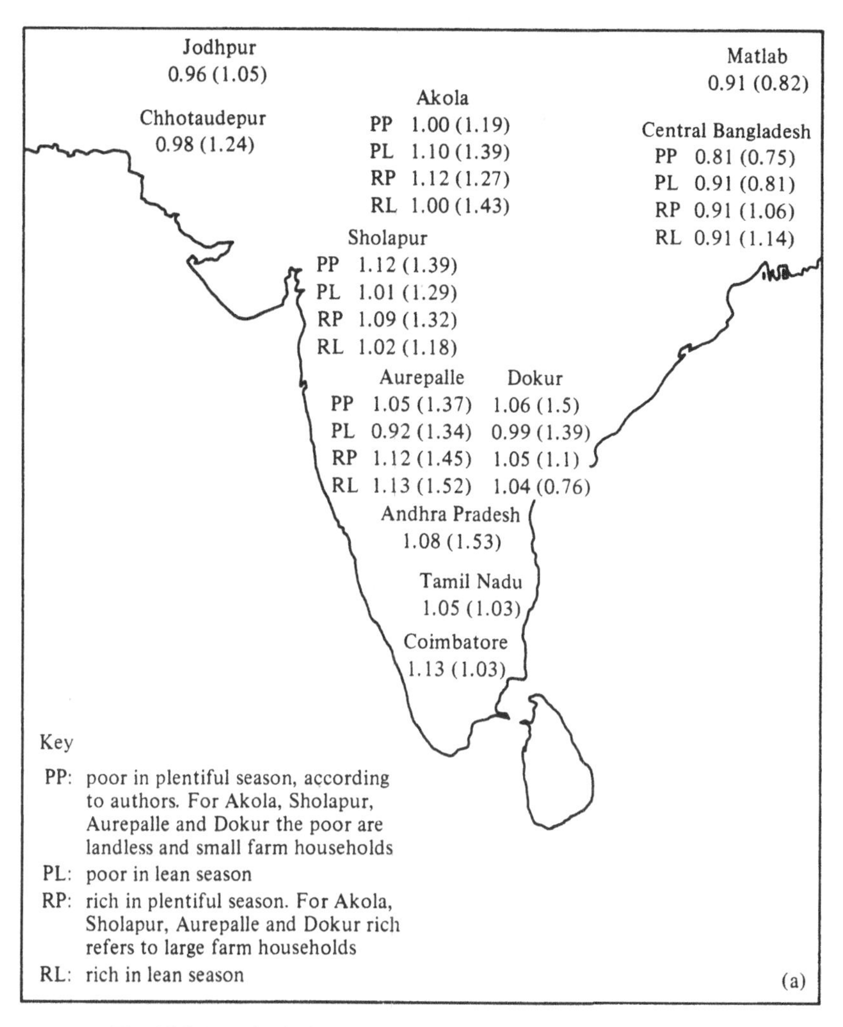

Fig. 13.2 Actual calorie allocation indices as proportions of the ICMR norm for (a) adult women and (bracketed) adolescent girls (aged 13–18), (b) the elderly and (c) children under five. Sources: central Bangladesh, Abdullah (1983); Matlab, Chen *et al.* (1981b); Jodhpur, Sharma (1983); Chhotandepur, Gopaldas *et al.* (1983); Akola, Sholapur, Aurepalle and Dokur, Ryan *et al.* (1984); Andhra Pradesh, Pushpamma *et al.* (1982); Coimbatore, Sadasivan, Kasthuri & Subramaniam (1980); Tamil Nadu, Cantor & Assocs. (1973).

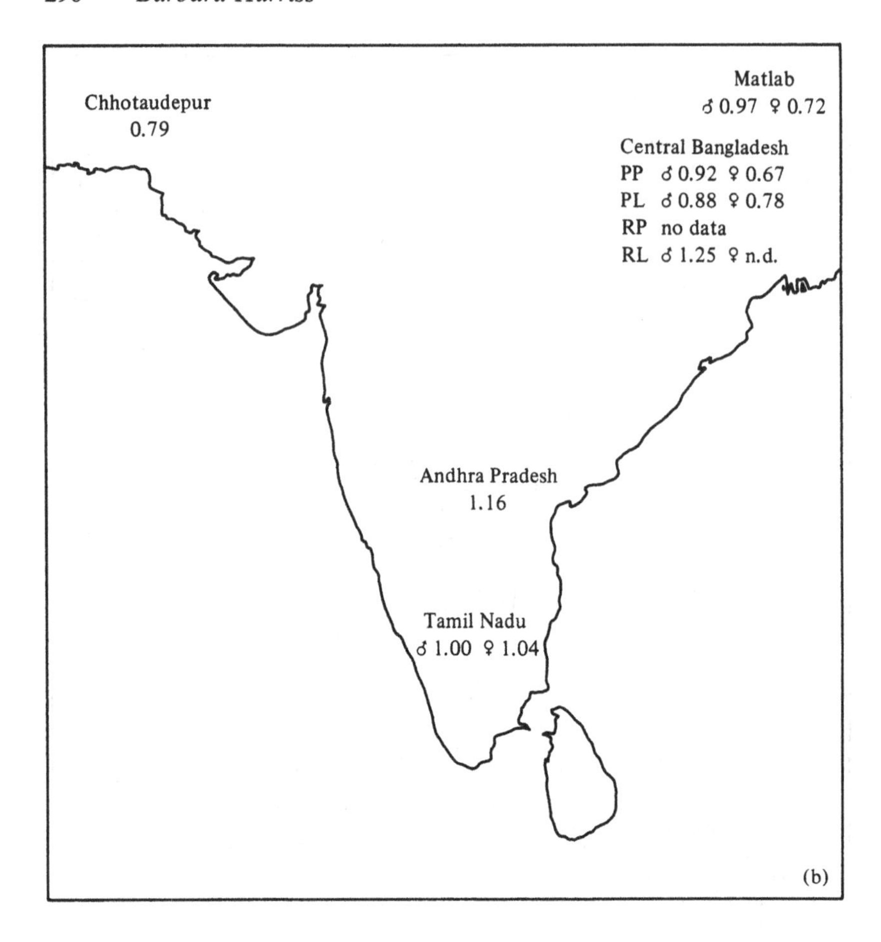
Chhotaudepur
0.79
Matlab
♂ 0.97 ♀ 0.72
Central Bangladesh
PP ♂ 0.92 ♀ 0.67
PL ♂ 0.88 ♀ 0.78
RP no data
RL ♂ 1.25 ♀ n.d.
Andhra Pradesh
1.16
Tamil Nadu
♂ 1.00 ♀ 1.04
(b)

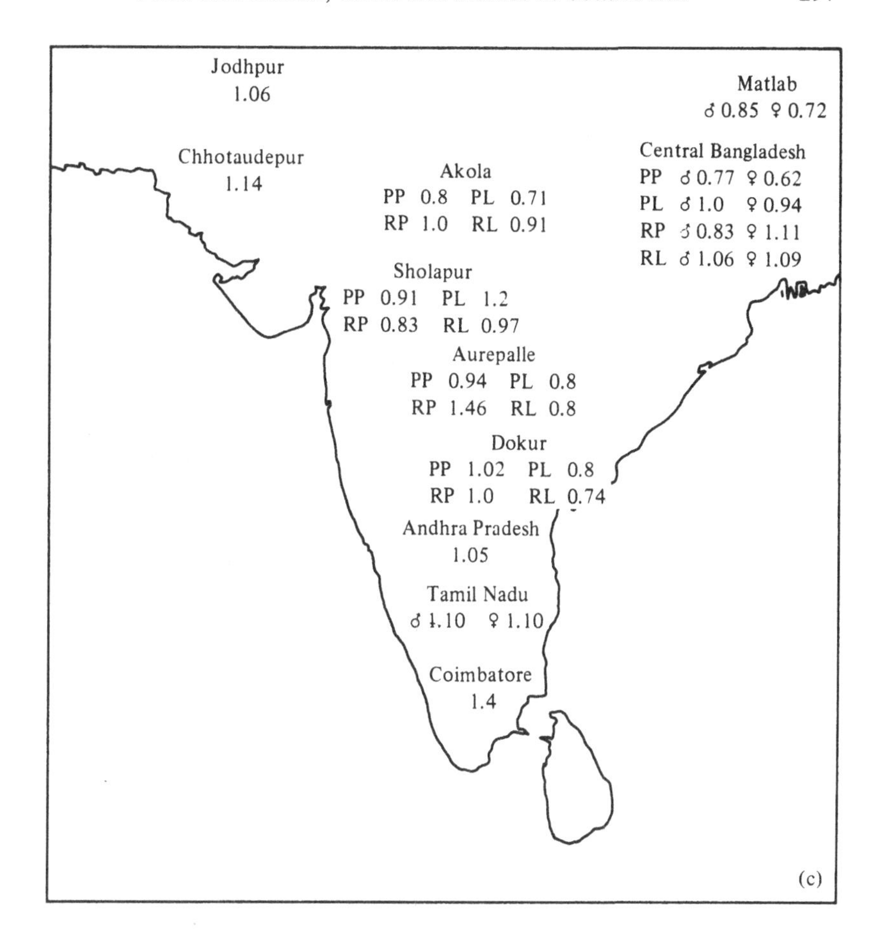
Jodhpur
1.06
Matlab
♂ 0.85 ♀ 0.72
Chhotaudepur
1.14
Akola
PP 0.8 PL 0.71
RP 1.0 RL 0.91
Central Bangladesh
PP ♂ 0.77 ♀ 0.62
PL ♂ 1.0 ♀ 0.94
RP ♂ 0.83 ♀ 1.11
RL ♂ 1.06 ♀ 1.09
Sholapur
PP 0.91 PL 1.2
RP 0.83 RL 0.97
Aurepalle
PP 0.94 PL 0.8
RP 1.46 RL 0.8
Dokur
PP 1.02 PL 0.8
RP 1.0 RL 0.74
Andhra Pradesh
1.05
Tamil Nadu
♂ 1.10 ♀ 1.10
Coimbatore
1.4
(c)

the absence of scarcity (another unstandardised aspect of the data). Some surveys are of poor people in time of scarcity, and some of unpoor people and not in times of scarcity. Anyway, for what it is worth, these 24 studies show that gender differences in nutrient distribution are not as dramatic as was supposed. The absolute calorie intake of the male household head explains about two thirds of the absolute intakes of women and children where we have enough data to run regressions. Thus about a third of the absolute intakes of women and children must be explained either by error or by intra-household factors or by other factors such as activity. Discrimination in energy and protein allocation still appear to be greater in the north and east than in the south and centre. In the north they appear to be least fair for the very young of both sexes and very elderly widows and possibly for women with special needs, as in pregnancy and lactation, but we do not have the activity or behavioural data to investigate this rigorously. In the centre and south the absolute aggregate energy intakes are about 400 calories lower and nobody knows yet whether this is a measurement error or whether these populations are subsisting on considerably less energy than the sub-continental average. Certainly the shares are relatively low in comparison to ICMR recommendations for preschool children of both sexes and possibly for adult men themselves. It may be concluded that the crude pattern of sex bias is not reproduced by detailed data. Nutrition intake by itself does not seem to be a good indicator of female disadvantage. It is certainly not a good predictor of excess female mortality. So my conclusion to the second question is that sex bias, whether measured by mortality, morbidity or nutrition in its various forms, does not have a straightforward geography in South Asia. Yet rather straightforward explanations have been put forward for the regional geography of misallocations and individual illfare. South Asia is a battleground between those exploring material explanations for discrimination and those exploring cultural explanations.

Material explanations for gender bias turn round the proposition that this discrimination is an efficient survival strategy, reflecting and perpetuating gender differences in the value of men and women. Male wages virtually invariably exceed female wages, irrespective of skills and productivity (Binswanger & Rosensweig, 1984). While the demand and supply of male and female labour is found to be the determinant of this difference in wages, it is obvious that it is the sexual division of labour (the allocation of tasks between men and women in South Asia), which conditions the demand for female labour. The gender division of labour is reasonably well fixed for a given region, but it varies between regions. Transplanting rice, for example, is a male task in the north east, but a female task elsewhere. Evidence has been quoted already in this symposium and I would add to that the research of Deolalikar (1984) which showed that both male wages and productivity vary with anthropometric status whereas female wages do not seem to vary in the same way. Men are more valuable than women in South Asia not only because of wage differentials, but also because of the old age insurance they bring to their parents (Cain *et al.*, 1979; Cain, 1981). The male can bring future insurance to parents along with current security, both by his waged labour, by the dowry which is commonly settled on the groom and the groom's kin at the time of marriage, and by remittances consequent to male migration. Women do not supply security to their natal kin in the same way by their cash flows.

One other material aspect of gender bias is that males prove to be more costly to rear than females, so that the loss of investment is greater for sons than for daughters. By contrast for girls in South Asia, not only are their relative wages for paid work lower, but so too are contributions that they make to their household through post-harvest processing, food preparation, child care etc. because they cease after marriage. Women typically marry and migrate and have no further economic value to their parents. Moreover the accumulation of dowry very often involves labour on the part of the males of the household, or the contracting by them of debt. In such ways, mortality differentials have been related to the phenomenon of the economic undervaluation of women. In turn this is thought to depend on two institutional factors: one is the labour demand of the agrarian system and the associated levels of female participation in agriculture, in other words the commercialisation of female labour power. The other is the control of property manifested in the gender distribution of inheritance rights and in the exchange value of women at marriage (expressed in dowry), which reflects the commercialisation of their bodies.

Testing these explanations for South Asia as a whole has been one of the many research projects of Pranab Bardhan at Berkeley. Bardhan (1987) tried to show that the ratio of male to female earnings is negatively related to indices of son preference and to the ratio between female and male morality. State-level data generate significant, but not very high, correlation coefficients for these associations. Bardhan tried to investigate the relationship between female mortality and property rights quite ingeniously, by estimating the proportion of households that are assetless and arguing that in an assetless household there is no adverse effect from patrilineal inheritance systems because there is nothing to inherit. Correlating assetlessness with female child mortality and with the index of son preference, again generates negative correlations. Some comments are in order: there is actually no statistical or empirical relationship between the variables chosen for the commercialisation of women's labour power and the variable chosen to proxy for the commercialisation of female bodies. About 70% of the co-variance between these variables is left unexplained. Then the restriction of this explanatory effort to wage work ignores the unwaged work of women, that is work on domestic agricultural production, in post-harvest processing (which is also a productive activity), and in all the aspects of household reproduction (child care, food preparation, finding water etc.). Such work is unvalued and it is impossible to assign market prices to this kind of work. Then, interestingly, the gendering of economic participation may relate to norm rather than to actuality. Female participation may be low, not because it is low at all, but because ideas about the gendering of tasks have it that all agricultural tasks are male irrespective of the sex of those active in the labour process. Then there are regional anomalies in labour demand, as is evident from Fig. 13.1. The wet rice agriculture of Bengal should have high female participation because it is a highly labour intensive system of production. Yet in West Bengal there is low female participation and in Bangladesh female participation is virtually non-existent. Also the dry wheat agriculture of Rajasthan should have low female labour participation, whereas in fact the reverse is the case.

It is argued that culture is omitted in economic explanations, but it can be counter-argued that the material explanation is in fact a cultural one. Female economic status is conditioned by the demand for female labour. What conditions the demand for female labour? First the agrarian ecology and the labour demands of crop combinations. Second the gender division of waged tasks has no biological inevitability about it at all. It is a profoundly cultural artefact and it varies regionally. In the same way, systems of property ownership are cultural in origin. To say as Bardhan (1987) has done that matter and culture interact in a mutually reinforcing way is another point altogether, but one with which it is hard to find fault.

The cultural explanation for gender differentials in demographic variables, in morbidity and in nutrition recognises that in the north of South Asia there tends to be village and kin-group exogamy, that is daughters leave their village on marriage and no longer have very much contact with their parents. This practice has two implications: one is that the young bride is very isolated and may have low social status in her new family and may be maltreated and discriminated against, in a way which brooks no reprisals. The inter-generational reproduction of such behaviour may translate itself into gender bias in the treatment of children. Further the loss to her natal locality of the outgoing bride means that she is a net drain on household resources. Also in North India women are excluded from property ownership. The south is purported to differ in all these respects, resulting in fewer gender differences in illfare, in morbidity, in anthropometric status, in food allocation and mortality (see Dyson & Moore, 1983). Anthropologists such as Karve (1953) have also identified major regional differences between north and south in language, caste, and practices of seclusion, which can also affect status. Such cultural dichotomisation may also be criticised. Existing ethnographic evidence shows that the north–south cultural regionalisation is exaggerated. Demographic anthropologists such as the Caldwells (1987), who have looked at the culture of illness and nutrition, have concluded that there is less rigour in the social relations affecting power and freedom for women in the south and in the far north east than there is in the north west; but to say that there are two sharp demographic divisions related to Aryan and Dravidian culture is quite misguided. Then the cultural regionalisation proposed by Dyson & Moore (1983), among others, is unhistorical and ignores very rapid cultural change, exemplified by the colonisation of the practice of dowry towards the south of the subcontinent and downwards through society. The basis of the cultural regionalisation can be disputed not only in terms of the clarity of the north–south contrast but also in terms of its boundary.

Alternative regionalisations might distinguish Kerala, the southern tip of Tamil Nadu and Sri Lanka on the grounds of geographical isolation and of the relative power of women and the virtual non-existence of sex bias in medical and demographic variables. The Caldwells have set up an intriguing hypothesis which is where I shall leave this aspect of the explanation. The hypothesis is that the regionalisation of gender has less to do with Aryan or Dravidian culture, wheat ecology, or rice ecology than it has to do with the difference between the heartland of a peasant civilisation and its periphery. According to the Caldwells (1987), old settled peasant societies with closed land frontiers have a 'definite internal logic' based on command, and based on segregation and hierarchy, and that this is associated with high mortality (even in boys) and is unrelated to material economic levels. By contrast the periphery of peasant society is not only less rigid, but also slowly adopting the dominant model as remnants of matrilineality and polyandry are submerged. This hypothesis now needs testing in South Asia and in other regions where there is a long-settled peasant heartland and a peripheral fringe. The conclusion is that neither 'material' nor 'cultural' explanations are very satisfactory. The ethnographical evidence does not support either incontrovertibly. Simple explanations are being sought for phenomena which we have seen are actually very complex.

Recent village research by anthropologists and demographers in south Asia has been concerned with explanations for oddness, results which are not predicted from these macro-regionalisations. The third question concerning the role of socio-economic status in gender bias has been examined in the northern region, where excess female mortality and gender bias in morbidity and nutrition is most strongly pronounced. Within the north west of India there is quite contradictory evidence as to whether girls in pauperised households are more at risk than girls from the propertied classes and there are two contradictory arguments about class position which have been advanced for this evidence. One is that the relative economic valuation of women and girls is highest, and the patrilineal control over property is lowest, among the assetless so that less gender bias would be expected. This has been recorded among poor Muslims (Jeffery, Jeffery & Lyon, 1987) and poor tribal families (Warrier, 1987). The opposite argument is that it is among the poorest that the opportunity cost of health care, in terms of income foregone by illness and by searching for health care and then the direct costs of treatment are greatest, and that a given level of discrimination may be more fatal (Das Gupta, 1987; Wadley & Derr, 1987). So with respect to our third question, there is evidence both ways.

There is also contradictory evidence for the cultural variable that has excited those attempting to explain gender bias in mortality – that is female autonomy. Greater female autonomy may increase women's control over household resources, but it almost certainly also increases female work burdens. The evidence in relation to female autonomy shows that the several factors which affect it do not act without contradiction and that a given factor does not have an unambiguous outcome. Education for instance may reduce the infant mortality rate, but it does not appear to reduce sex bias in infant mortality. Seclusion may have absolutely no effect on whether or not a woman works. The nuclear family which might be expected to increase female autonomy may also increase the power of husbands over resources and decisions. Clearly, trying to seek an explanation for the geography and sociology of sex bias, either in an economic variable like wage work participation or a cultural variable like autonomy, is not proving very easy.

Another point made by these village studies is that it looks as though the operation of patriarchy within the household means that some girls are more at risk than others (which qualifies the value of sibling controls in research on morbidity and malnutrition). Monica Das Gupta (1987) found a 70% greater risk of death among the second daughters than among the first. Other work has shown that some households, irrespective of caste, class or income, appear to put little girls more at risk than do

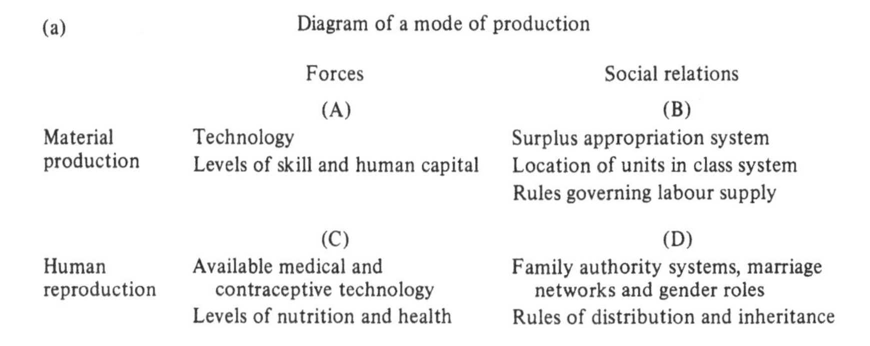

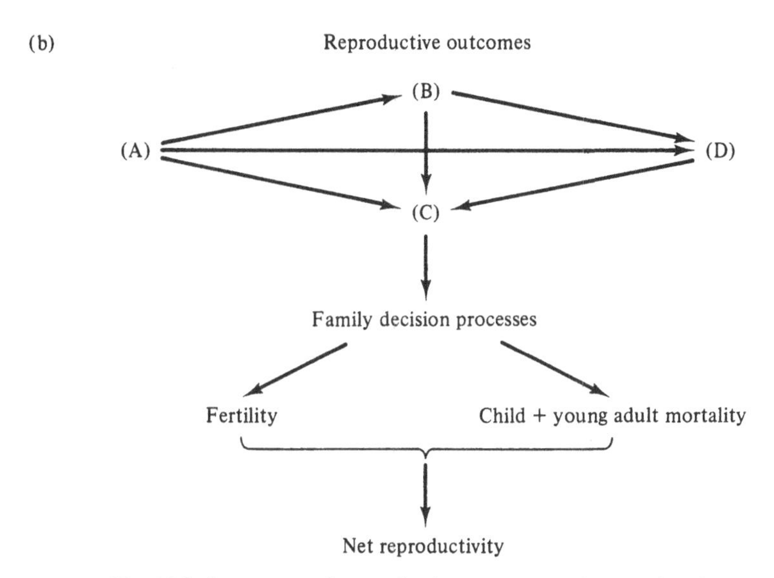

Fig. 13.3 Structure and reproductive outcomes of a mode of production. (Source: Clark, 1987, Fig. 1).

others. Jeffery *et al.* in Uttar Pradesh (1987) found the probability of female deaths maximised, irrespective of caste or class, in households where there were no living sisters and where one daughter had died already. There are certain killing kinds of households.

To conclude, Alice Clark has recently argued that excess female child mortality and gender differences in morbidity and in the allocation of food are best conceptualised as part of reproductive strategy practised by social units which are larger than that of a household. In other words, reproductive strategy may be specific to social classes (Clark, 1987). Discriminatory acts which lead to disease or to malnutrition may be affected by factors other than patriarchy (which as we have seen is the main cultural variable used to explain sex bias) (Fig. 13.3). Social class relations, gender relations and relations between the generations within the household may be much more important than the kind of nutrition–infection relation first modelled by Chen and his colleagues (1981b) as being the crucial determinants of mortality and malnutrition. The proximate determinants beloved of nutritionists and clinicians may be being mediated by material relations of production such as technology and technical skills and by material relations of reproduction, that is medical and contraceptive technology and nutrition. The age and the gender distribution of mortality, morbidity and malnutrition is then to be explained by the interaction between class position and the cultural institutions of patriarchy in South Asia: the division of labour, the way production is distributed, inheritance practice, marriage networks, and reproductive technology, including access to health care. This would seem a fruitful conceptual framework for the future.

References

Abdullah, M. (1983). Dimensions of intra household food and nutrient allocations: a study of a Bangladesh village. PhD Thesis, Faculty of Medicine, London University.

Bardhan, P. (1987). *On the Economic Geography of Sex Disparity in Child Survival In India: A Note.* Paper to American Social Science Research Council Workshop on Excess Female Mortality and Health Care in South Asia, Dhaka, Bangladesh.

Binswanger, H.P. & Rosenweig, M. (ed.) (1984). *Contractual Arrangements, Employment and Wages in Rural Labour Markets in Asia.* Yale University Press, New Haven.

Cain, M.T., Syed, S.K. & Nahar, S. (1979). Class, patriarchy and women's work in Bangladesh. *Population and Development Review*, **5**, 3, 405–38.

Cain, M.T. (1981). Risk and ignorance: perspectives on fertility and agrarian change in India and Bangladesh. *Population and Development Review*, **7**, 3, 435–74.

Caldwell, P. & Caldwell, J. (1987). *Where there is a Narrower Gap between Female and Male Situations: Lessons from South India and Sri Lanka.* Paper to American Social Science Research Council Workshop on Excess Female Mortality and Health Care in South Asia, Dhaka, Bangladesh.

Cantor, S.M. & Associates (1973). *Tamil Nadu Nutrition Study* (6 vols.). Haverford, P.A.

Chen, L.C., Huq, E. & Huffman, S.L. (1981a). A prospective study of the risk of diarrheal diseases according to the nutritional status of children. *American Journal of Epidemiology*, **114**, 284–00.

Chen, L.C., Huq, E. & D'Souza, S. (1981b). Sex bias in the family allocation of food and health care in rural Bangladesh. *Population and Development Review*, **7**, 1, 55–70.

Clark, A. (1987). Social demography of excess female mortality in India: new directions. *Economic and Political Weekly*, **27**, 7, WS12–WS21.

Cohen, N. (1987). *Sex Differences in Blindness and Mortality in the Indian Subcontinent: Some Paradoxes Explored.* Paper to American Social Science Research Council Workshop on Excess Female Mortality and Health Care in South Asia, Dhaka, Bangladesh.

Das Gupta, M. (1987). Selective discrimination against female children in rural Punjab, India. *Population and Development Review*, **13**.

Deolalikar, A.B. (1984). *Are there Pecuniary Returns to Health in Agricultural Work?* Progress Report no. 66, Economics Program, International Crops Research Institute for the Semi Arid Tropics, Hyderabad.

Dyson, T. (1984). Excess male mortality in India. *Economic and Political Weekly*, **19**, 422–6.

Dyson, T. (1987). *Excess Female Mortality in India: Uncertain Evidence on a Narrowing Differential.* Paper to American Social Science Research Council Workshop on Excess Female Mortality and Health Care in South Asia, Dhaka, Bangladesh.

Dyson, T. & Moore, M.P. (1983). On kinship structure, female autonomy and demographic behaviour in India. *Population and Development Review*, **9**.

Gopaldas, T., Saxena, K. & Gupta, A. (1983). Intrafamilial distribution of nutrients in a deep forest dwelling tribe of Gujirat, India. *Ecology of Food and Nutrition*, **13**, 69–73.

Harriss, B. (1986). *The Intrafamily Distribution of Hunger in South Asia.* Paper for the Conference on Poverty, Hunger and Economics, World Institute of Development Economics Research, Helsinki.

Harriss, B. (1989). Excess female mortality and health care in South Asia Journal of Social Studies, **44**, 1–123.

Jeffery, P., Jeffery, R. & Lyon, A. (1987). *Domestic Politics and Sex Differences in Mortality: a View from Rural Bijnor District U.P.* Paper to American Social Science Research Council Workshop on Excess Female Mortality and Health Care in South Asia, Dhaka, Bangladesh.

Karkal, M. (1987). *Differentials in Mortality by Sex.* Paper to American Social Science Research Council Workshop on Excess Female Mortality and Health Care in South Asia, Dhaka, Bangladesh.

Karve, I. (1953). *Kinship Organisation in India.* Deccan College, Poona.

McNeill, G. (1986). *Energy Nutrition in Adults in Rural South India.* Report to Ford Foundation, London School of Hygiene and Tropical Medicine, London.

Padmanabha, P. (1983). *Census of India Series 1*, Paper 1 of 1981 Provisional Population Totals, UN Demographic Yearbook.

Pereira, S.M., Sundaraj, R. & Begum, A. (1979). Physical growth and neuro integrative performance of survivors of protein energy malnutrition. *British Journal of Nutrition*, **42**, 165–71.

Pushpamma, P., Geervani, P. & Usha Rani, U. (1982). Food intake and nutritional adequacy of the rural population of Andhra Pradesh, India. *Human Nutrition: Applied Nutrition*, **36A**, 293–301.

Ryan, J.G., Bidinger, P.D., Prahlad Rao, N. & Pushpamma, P. (1984). *The Determinants of Individual Diets and Nutritional Status in Six Villages of Southern India.* Research Bulletin no. 7, International Crops Research Institute for Semi Arid Tropics, Hyderabad.

Sadasivan, S., Kasthuri, R. & Subramaniam, S. (1980). Nutritional survey in a village of Tamil Nadu. *Indian Journal of Nutrition and Dietetics*, **17**, 245–50.

Sen, A.K. (1985). *Women, Technology and Sexual Divisions*. UN, Geneva (UNCTAD TT/79).
Sharma, S. (1983). Food distribution pattern in drought affected farm families of Rajasthan. MSc Thesis, University of Rajasthan, Udaipur.
Visaria, L. (1987). *Sex Differentials in Nutritional Status in a Rural Area of Gujarat State, India*. Paper to American Social Science Research Council Workshop on Excess Female Mortality and Health Care in South Asia, Dhaka, Bangladesh.
Wadley, S. & Derr, B. (1987). *Child Survival and Economic Status in a North Indian Village*. Paper to American Social Science Research Council Workshop on Excess Female Mortality and Health Care in South Asia, Dhaka, Bangladesh.
Warrier, S. (1987). *Daughter Disfavour, Women, Work and Autonomy in Rural West Bengal*. Paper to American Social Science Research Council Workshop on Excess Female Mortality and Health Care in South Asia, Dhaka, Bangladesh.

14 *The cultural context of diet, disease and the body*

RICHARD BURGHART

It is a commonplace of human nutrition that people are selective in their use of the environment for dietary purposes. Nutritionists have evaluated the energy and nutritional values of numerous substances that can be assimilated in the human body, but the cross-cultural research of anthropologists shows, if nothing else, that man's dietary habits are considerably more narrow than the potential indicated by nutritionists.

Both nutritionists and anthropologists share this observation, yet they interpret it differently. Nutritionists take the human body as a given entity and focus on culturally specific diets as a matter of taste. Admittedly there are minor racial differences in the human body, plus life span and gender differences, but the human body is taken to be a more or less standard item of equipment as indicated in the universal applicability of the so-called *Recommended Dietary Allowance* that continues to inform so much research on nutritional deficiency. Assuming that the political economy does not constrain the availability of local foods, the human body and its basic needs remain universal; it is dietary preference that varies.

The image of the individual who, on the basis of certain rational calculations, affective predispositions and biological tolerances, selects a diet has some plausibility, as does the view that cultural factors influence those calculations and predispositions. Yet from an ethnographic point of view this selectiveness is not necessarily a fact in itself, for many people do not consider their essential diet to be a matter of 'selection'. Moreover, the model of personal preference may work in the West with its class differences in eating habits and where 'taste' itself becomes a means of discrimination between the cultured and the vulgar. This is especially the case in the intra-class dynamics of bourgeois society in which the bourgeois distinguishes himself from the *parvenu* only in his exercise of superior taste. Yet this model breaks down in other societies where taste lacks such social power of discrimination and where the basic diet of the rich and the poor, even if they could 'select' it, would not differ significantly in its choice. Moreover, there are sound ethnographic reasons for querying the universality of this concept of the person as the seat of human subjectivity. From a pious Hindu point of view, for example, the image of the individual giving free rein to his appetites – choosing his diet according to taste – seems a blueprint for a licentious society, not a free one.

My aim, in this paper, is to discuss dietary differences in the light of varying conceptions of the human body, and to explore the relation between diet and disease as mediated by such conceptions. Ethnographers have recorded considerable diversity in people's bodily images and practices, as well as their diet. It only stands to reason that there is some relationship between the two, because different concepts of the body entail different definitions of what it takes to sustain the body, and hence different ideas of what constitutes food. My attempt to relate dietary practices to the body has two parts. The first entails a consideration of food as a weighted rather than distributive category. If edible things become food in their capacity to nourish the body, then one would expect the category of food not to comprise a class of items equal in value; rather there are ontologies at work that weight some foods as being 'real' food. The second entails a consideration of the identification of the body with individual selfhood. Such an identification has common sense validity yet it restricts one's understanding of other people's experience of a 'social body', relegating the concept to the realm of the metaphorical and overlooking the significance of the social experience for a person's well-being. For both these parts of the essay to succeed as ethnography, we shall have to disprivilege expert biological conceptions of the body. Yet having qualified the category of food and relativized the body, some sense can be made of the distribution and meaning of disease in different societies. It is to this theme that I shall turn in the conclusion of this paper.

'Real' food, nourishment and the human body

Ethnographers have noted that not everything in the class of edible things called 'food' is of equal value. Some foods are thought to be more 'food-like' than others. Conklin (1957, p. 29) recorded in the Phillipines that among the Hanunoo only cooked food is considered 'real food'. Not too far away in Fiji Pollock (1985) observed that 'real food' (*kakana dia*) is starchy (e.g. taro, breadfruit and yams). This category contrasts with the relishes and side dishes (*i coi*) that accompany the starchy staple. Whereas questions of taste might govern the selection of side dishes, the consumption of the staple is not a matter of taste (p. 202):

> Even though *kakana dina* is only one kind of edible, it is food in the sense that it is the only edible which must be consumed daily in order to

> bring a sense of well-being, emotional security and a sense of being full or repleteness. Without a starchy component, a Fijian can say in his view he has not eaten.

Croll (1983) recorded the distinction between *fan* and *cai* in China, the former denoting cereals and the latter vegetables and meat. Of the two, cereals are fundamental and indispensable. 'Without *fan*, it is said, an individual would not be full, while the absence of *cai* merely makes the meal less tasteful' (p. 20). Similarly, among southern Bantu peoples cattle and wild game are real food, the vegetables, porridge and millet beer – that make a more significant contribution to nutritional health – are not.

These few examples illustrate that in geographically dispersed societies some foods are said to be more 'real' than others. These foods would also seem to be less subject to intra-cultural variability in taste. The cross-cultural data are insufficient to demonstrate why this should be so, but I would suggest 'real' foods are thought to constitute the body and hence only they can nourish and sustain it. In order to advance this argument, I shall resort to the ethnographic material on Hindu peoples, both in south Asia and overseas, for which there is sufficient literature to argue the case.

In south Asia two related ideas seem to be at work: first, the human body assimilates the qualities inherent in food; and second, the body is a system which refines food into energy. These two ideas, stemming from different dietary purposes, do not contradict each other. For example, a lactating mother might be encouraged to drink lots of milk to produce milk. On the other hand, a person might be encouraged to drink milk because this substance is dense in those things which, upon assimilation, give energy to one's mind and body. Although food should be pleasing to the taste, one's diet should not be subservient to taste. Rather food is thought to influence the body and mind; and in this measure becomes subject to moral constraint. To return to the first sense of the body, certain food is appropriate for a certain type of person – that is, the person who can assimilate its qualities without injury because the food is morally appropriate for the eater. The second sense of food enables the eater to refine himself by opting for a diet most amenable to refinement – that is, foods that can increase one's energy, are not injurious and that enhance the mind's capacity to control the body.

Both these ideas support the Hindu view that 'you are what you eat'; that is to say, the body is a 'food-body'. For many Hindus, especially the higher castes who speak on behalf of civilized values, the human body is constituted of grain (*anna*). The term is a general one that includes wheat, millet and barley, but the primary reference is to rice. Rice is valued as the source of life, and the body that it nourishes is referred to auspiciously as that which is 'composed of grain' (*annamaya*). Conversely the expression 'to exhaust one's rice' is synonymous with death (Inden & Nicholas, 1977; pp. 5–6; Cantlie, 1984). In Assam the term is used exclusively with reference to boiled rice (*bhat*) which is regarded not only as staple food, but as the only food (Cantlie, 1984, p. 194):

> Anna is our life (*pran*). Life is called 'life depending on anna' (*anna pran jiva*). As we sometimes say anna is god (Brahma), so we think anna is our life. Boiled rice is our life.

Similarly Greenough (1983, p. 36) recorded that Bengalis consider boiled rice to be the principal source of well-being and bodily substance. Boiled rice should be consumed at least once per day. Most other foods are classified as snacks, no matter how filling or copious they may be. In sum, rice is 'food'; and, hence, it is fundamental to any culturally relevant definition of basic need in south Asia.

Not all castes, however, are known as rice-eaters. Certain untouchable castes, such as Halkhor and Musahar, may eat rice, but they eat other things as well that constitute their being. The Musahar, a Maithil caste of earth diggers, are reputed to eat the rats and field mice (*musa*) from which their name is derived. As a caste, they are known for their brute strength, slowness in thought and ritual impurity. They are what they eat and therefore are said to have certain abilities lacking in the higher castes: e.g. to dig ditches all day without feeling any pain or to digest unleavened millet bread, something the more sensitive upper castes claim they are unable to do themselves. This is not to say that the Musahar would not also like to sit down to a meal of rice, if they could afford it. It does mean, though, that caste differences are naturalized by dietary means. Boiled rice is not only 'real' food, it is also a food of bodily refinement and civilized values. The higher castes who espouse Hindu values observe that the high and low castes are what they are by virtue of what they eat. Conversely, by virtue of being what they are, low castes can eat the crude and disgusting things they do.

To these observations one must add 'water' as the other dietary essential. Tea, coffee, milk, soft drinks, fruit sorbet and alcoholic beverages may be drunk with snacks, but water is taken with one's meal. Water is not drunk in the European manner, in sips throughout the meal. Rather after the entire meal is eaten one drinks down a glassful or two of water. Water is thought necessary for digestion to take place. Additionally, it is thought to animate the body, giving suppleness to movement and freshness to the complexion. As Maithil people say, 'where there is water, there is life'. It is like the sap in a tree, for which dryness is akin to ageing and eventual death. In a similar vein rickets in children is referred to as *sukhaniya*, literally the 'drying out (illness)'; and women at menopause sometimes express their morbid feelings by remarking that they are 'dried out'.

The relation between 'real' food and the body can be adduced further in the weekly and occasional fasts undertaken by pious Hindu women for the welfare of their family. Despite the 24 hour period of abstention, women frequently nibble on fruit and snacks. Such edibles are not, however, 'food' and hence their ritual vow remains unbroken. Similarly, a fast of Hindu ascetics, undertaken for a proverbial 12 years, entails limiting one's diet to milk and fruit (*phalahar*). Since the body is constituted of rice the fast entails a starvation of sorts. Hindu ascetics, who renounce transient existence, claim to be dead to this world. One of the deathly images they cultivate is that of a 'dried out piece of wood'. The image goes back to the notion 'where there is water, there is life'. Whereas in European languages one might refer to a lifeless, standing tree as being 'dead', in Indo-Aryan languages one would say that such a tree was 'dried (out)'. An alive, but dried out body is a living death of sorts. For the Hindu ascetic – unlike the menopausal women – such death betokens transcendence.

So far my discussion implies that the body simply takes on the qualities of food and water. The body, however, also refines these qualities in a system of transformation. Here the human body is understood as a system composed of a seven stage transformation which is likened to an alchemical process, in which heat transmutes food into increasingly refined substances, called *dhatu*, until energy, the '*elixir vitae*' is produced. The process begins with the ingestion of food. Rice and water collect in the stomach where they are heated by digestive fire and transformed into style. The style is transported to the liver where it is heated in the second stage and transmuted into blood. The digestive process continues with blood being transmuted by heat into flesh, flesh into fat, fat into bone, bone into marrow and marrow into either semen or menstrual blood depending upon the sex of the person. As in any alchemical action dross is produced at each stage, starting with urine and faeces in the transformation of food into style; phlegm (*kapha*) in the transformation of style into blood; bile (*pita*) in the formation of flesh; finger and toenails in the formation of bone, and so on (Kutumbiah, 1962, pp. 37–44). To complete the alchemical analogy, semen and uterine blood are thought of as an *elixir vitae*, giving physical vitality, mental energy and strength of character. Longevity, and by implication health, entail on the one hand the refinement of food into this *elixir vitae* and on the other hand, the elimination or counterbalancing of dross within the body. The integration of the body as a digestive system turns the body into a circulatory system of sorts in which semen and uterine blood provide the energy to digest food during the first stage of the cycle. Without such energy coming back into the system, a vicious cycle builds up in which less and less energy is available for the digestion of food such that even less energy is produced.

The centrality of the digestive system in the Hindu concept of the body underscores the importance of dietary considerations in monitoring one's health. One cannot highlight the importance of digestion with reference to morbidity figures, for they are based on the bio-medical understanding of the body. Instead I shall offer several illustrations of people's morbid feelings, for such feelings give insight into Hindu perceptions of physical vulnerability. In south Asia males of weak disposition are preoccupied by the loss of semen in intercourse as well as the condition, known in Indian English as 'nightfall', for which they seek medical treatment (Carstairs, 1961). Less well documented are women's complaints of weakness. The weakness is possibly caused, or at least exacerbated, by overwork and anaemia, but is attributed to the loss of blood during menstruation. One of the classic male explanations for the 'fact' that there are no great female ascetics is that women cannot accumulate energy as men do by virtue of sexual abstinence. Rather they lose their sanguine source of energy every month. Alternatively, as a symptom of physico-spiritual greatness, any woman who is recognized by her fellow nuns as an accomplished ascetic is said to have ceased menstruating upon receiving initiation. The importance of ridding the body of impurities also explains the response of Maithil people during my research in Nepal on the relation between the bacteriological quality of water and the incidence of diarrhaeal diseases. Upon my arrival in Nepal local people informed me that I had come to the wrong place; their problem was constipation which they attributed to the presence of iron in the water. They were more preoccupied by the retention of stools than by any looseness in their bowels. When suffering from diarrhaea the dross is at least being evacuated from the body, albeit at a more rapid rate than desirable.

Having argued that the consumption of 'real' food and water are not matters of taste but of need, it is in the side dishes that likes or dislikes of food items are expressed. I shall not enter here into a detailed discussion of the aesthetics of a formal Hindu meal in upper India, except to note that there are six tastes – sweet, sour, salty, pungent, bitter and astringent – and that each taste should complement the other in a harmonious whole of counter-balanced dishes. More important for the present argument is that aesthetic preoccupations mirror those of health. Each taste varies in the degree to which it is cooling or heating, light or heavy, dry or wet, in its effects upon the body. Moreover, each taste has its own influence upon the digestive process (detailed in Kutumbiah, 1962). Sweetness increases the *dhatu* (blood, flesh, fat, marrow, semen, etc.); acidity initiates digestion and develops the body; salinity facilitates digestion and removes the dross of wind (*bat*) and phlegm (*kapha*); pungency provokes the digestive fire; bitterness sharpens the appetite and assists the digestion of as yet undigested food; and astringency restores harmony.

The aim of the eater is that each dish should counter-balance the other thereby maintaining the bodily equilibrium. If, however, the body is in some state of disequilibrium, a particular dish is given on its own to restore the balance. Boiled rice is cool and wet and thereforc is best eaten at midday in winter, but not in the evening at which time the cool, damp air exacerbates the effect of the rice. Oranges, lemons and bananas are cooling and should similarly be taken in moderation during the cold winter months, and avoided altogether by persons suffering from a cold. Papayas and mangos are heating and should be eaten in moderation during the summer months. Women are often said to be fond of sour foods, such as mango pickle, but men are advised to take such foods in moderation since acidity thins semen and debilitates the organism. Heating foods, such as garlic, onions, chillies and meat, warm the blood and agitate the mind. Such foods may be appropriate for the 'brash' Rajput, eager to test his strength in the battle field, but not for the ascetic in search of the mental stability that comes from stillness. However, whether vegetarian or non-vegetarian food has been consumed, it is usual to round off the meal after several heating dishes with something cooling, such as sweetened milk or curds. Unless, of course one suffers from constipation, in which case one might let the heating foods take their effect. In pregnancy women avoid heating foods in fear that the warmth will abort the foetus, but postpartum they eat heating foods, prepared with pepper, chillies and ginger, for the warmth loosens the stale, dangerous blood of birth and causes it to drain from the body. In short, the connection between food and the body provides a rationale whereby people work out their diet; think about it in relation to their well-being, solicit advice about it and test that advice in practice. The range of interpretation is diverse, varying from context to context. All that is constant is the idea of equilibrium that integrates the particular interpretation.

Since these dietary practices are contextual in application, there is no end to their description. Their general implications, however, can be briefly summarized. First, food and water constitute one's being. Dietary practices habituate the eater to his social and natural environment.

Having become so naturalized, sudden deprivation – even of a bad habit, such as cigarette smoking – may cause shock and be more harmful than the perpetuation of the habit. Too strong a cure is worse than the disease. Second, the Hindu universe is highly relative to time, place and person; and illness manifests itself in disequilibriums. Diet-related diseases have no ontological existence outside the body, as they do in European folk ideas of illness. Third, it would be difficult to institute the code of Recommended Dietary Allowances in Hindu culinary terms, for the human body does not work in a standardized manner. Some foods are tolerated by high castes, others by low. Some by men, others by women. What is an appropriate diet in one place may be inappropriate in another. In western India, vegetarian Gujarati mothers cook vegetarian food for their infants. In London, by contrast, vegetarian mothers counteract the cold northern climate by serving processed meat baby foods to their children. No dietary recommendation is absolute or context free. Fourth, by virtue of one's dietary regimen one may come to improve the mind and body. Food is profoundly cultural and, like health, becomes part of a civilizing process. From this perspective the Western fad of associating healthy living with natural foods appears curious to some Hindus and erroneous to others. One wants to eat unadulterated foods, but not ones that are natural. 'Food' is cooked food. 'Natural' food is eaten only by tribal people and animals; it is associated with savagery, not health.

In closing these remarks, it is important to stress that the concept of the body does not cause people to look upon certain edibles as food. Rather the two form part of a native rationale, which explains dietary practices as purposive action. Such a rationale does not explain all dietary behaviour. For example, food taboos, construed as non-actions, may have no rationale; native explanation may not pertain to why someone observes the taboo, but what would happen if they did not observe the taboo. Moreover, the rationale that relates 'real' food to the concept of the body is supported by productive relations and cultural values. 'Real' food often privileges men who are identified with its cultivation or collection. In south Asia it also privileges the civilization of lowland people with irrigation systems at the cost of rustic highlanders and of 'uncultured' tribals who survive by hunting and gathering. Furthermore, 'real' food may derive an ontology from metaphysical notions and divine origins. For example, in Hindu south Asia paddy fields are identified with one's ancestors. Therefore, to eat the rice of one's fields creates a link with one's forebears to whom one owes one's bodily existence. Poor is the man who must nourish himself with anonymous rice purchased in the bazaar. Blessed is the man whose granary overflows with paddy. In the same way that a paddy-filled granary is a symptom of wealth, so a plump body is a symptom of health. Both health and wealth are signs of a household blessed by Lakshmi, the Goddess of Fortune.

Multiple embodiments of the self

Throughout the first section we have suspended expert belief in order to take on faith local understandings of the relation between food and the body. Nonetheless we have grounded our faith on certain pre-suppositions that have led us to take for granted that the body is quite substantial and that it is also the seat of selfhood. These pre-suppositions lead us in English society to treat as metaphors other usages of the term body, as in public bodies, the body politic, or a business corporation. In other societies the family, lineage, caste or some other social group may also see itself as a social body, in which case one is likely to treat their linguistic reference as metaphorical. But is the reference necessarily a metaphor for the people concerned? To take an example, the 'heat of passion' is a metaphor in English, but a literal statement in Indo-Aryan languages as evidenced in *ayurveda* where the heat of sexual love is one of the sixteen types of fever. Presumably the heat of passion is no more metaphorical than that of typhus.

One way of addressing this problem is to begin with the symbolism of food. I do not query the view that food can acquire symbolic significance, but one can never be certain from the ethnographic literature whether the symbolism is native in origin or an artefact of the ethnographer's disbelief. Typical of this quandry is the work of Laderman (1981) in her understanding of the Malaysian category of poison (*bisa*) as it affects Malay food avoidances. Some poisonous foods are 'really' poisonous because there is empirical evidence that such is the case. Other foods are not empirically poisonous, yet are believed to be so. To spare the Malays the slur of irrationality, Laderman concludes that these empirically non-poisonous foods must be symbolically poisonous.

Yet from Laderman's material it would seem that the Malays do not discriminate in their behaviour between the empirical (read 'real') and the symbolic (read 'imaginary'); and her use of the term symbolic emerges not out of an empirical study of the Malay theory of signs, but out of the ethnographer's field encounter with the natives. An example might clarify the matter. Several years ago politicians and journalists spoke of the British economy as being 'ailing', 'sick', 'weak', etc. The Japanese were reluctant to invest in Britain because labour disputes were the 'English disease'. Inflation was a 'symptom' of wage claims that exceeded productivity gains. The balance of payments deficit was also a 'symptom' of a 'weak' economy. Yet the government of the day did not call in the medical experts to cure the economy; instead they called in economists. Clearly the government, in its behaviour, was treating its medical diagnosis of the economy as a metaphor. Yet the Malay do not treat their category of poisonous foods in a metaphorical way; instead they act in the same way to empirically 'real' poisons as they do to the so-called symbolic ones. In other words, their symbolism – if one will – is also real.

The forced nature of the distinction between literal and metaphorical foods and poisons stands to reason when one considers the ethnographic material on Hindu theories of illness. Although Hindus distinguish between physical, mental and moral domains, their theory of contagion brings these three domains together as mutually related influences. To take an example, a lactating mother transmits both her positive and negative qualities to her infant through the medium of her milk (see Reissland & Burghart, 1988). Hence if a mother becomes ill, the symptoms of her illness become manifest in her child. What is of interest in the transmission of these qualities is that the physical, mental and moral domains are not different types of experience, but different aspects of the same experience. If a woman – because of her easy virtue – should contract venereal disease and then wet-nurse a child, she will transmit through her milk not only her venereal disease, but also her loose moral character. To return to the symbolic poisons of Malaysia, if symbolic reality is something mental and empirical reality something physical, then it could be that the Malay theory of contagion does not require different sorts of practical behaviour for each of the two domains to influence the other.

The Hindu theory of contagion leads us, however, to a further point. If society is a moral system, then the mutual influence of moral, mental and physical domains implies that moral relationships themselves may be physical in their effects. This, in turn, opens the possibility that morally constituted social groups may be seen as social bodies in which moral obligations and shared experiences are seen to have a physical basis of existence. Given the inherent implausibility of such a belief in England, it may be helpful to argue the case on native grounds, without recourse to exotic peoples. In England, after the birth of a child, it is customary for the relatives of the parents to visit the maternity hospital for their first look at the newest member of the family. The entire family gathers round the cot, expresses their delight in the offspring and promptly begins to dissect the body. The child is found to have its mother's eyes, its father's hair, grandmother's mouth, etc. The entire family – as a collectivity – comes to see itself embodied in that one infant. In those early days the infant is not yet an individual; rather it becomes one in the course of many years, integrating diverse family influences in such a way as to construct some measure of personal autonomy.

If the family comes to see itself in one body, why cannot the family itself be a body? There are in English society certain bodily practices that indicate the potential for just such a view. The repulsion that one feels for having someone else's saliva in one's food or hair in one's soup hardly exists for members of one's own family. Furthermore sexual otherness can only be encountered outside the family. The terms of the incest taboo imply that the family is one large self constituted of several individuals. In short, the structure of our physical revulsion and our social taboos indicate the potential for a family in England to see itself as an embodied self. The point is, of course, we do not. The values of English society together with the legal philosophy of civil society militate against such a reality. The social body is not real in England; and we find it difficult to imagine it otherwise.

In other societies, however, the self is variously embodied and individual identity is achieved in sorting out these various embodiments. To continue with the theme of birth, in Hindu south Asia the newly born is an extension of a seven generation deep patrilineal descent group (and in some regions one's maternal ancestors five generations deep), known literally as the 'body' (*sapinda*). The baby's first nourishment – honey-sweetened water to wish it a sweet life – comes from the women of this group (father's sister, father's mother, or father's elder brother's wife), not from the mother. Until the mother's milk comes on the second or third day, the infant's patrilateral female relatives continue to feed it, traditionally with goat's milk. The rationale for this is as follows. Out of all domesticated mammals, only the goat is said to be omnivorous. By virtue of having eaten and refined all manner of foods, goats produce a milk that cures all manner of illnesses (unlike the milk of man, cows, buffalo, etc. which by virtue of dietary selectiveness is partial in its healing potential). As a member of the *sapinda* group, the infant's blood is identical to that of his forebears; hence the neonate may come down with hereditary diseases transmitted by blood from parent to child at the time of conception. The virtue of goat's milk is that it destroys inherited taints thereby giving the child an autonomous existence, separate from the medico-moral history of its patrilineage. In sum, for Hindus the self is variously embodied and physical vulnerability concerns as much membership in a social body as it does the individual body.

Regardless of how real one credits the social body, its existence has implications for food, both as a medium of contamination and an item of distribution. With regard to the former implication it has already been mentioned that the patrilineage, seven generations deep, constitutes a social body. All members of this body are said to share the identical particles; it is for this reason that the death of one member of the group pollutes all the other co-members. Furthermore marriage must take place outside this group and sexual relations within the group are classified as incest. Notions of group identity and vulnerability come out most clearly in feasting. In cooking rice, the cook adds water to the raw rice in the pot, brings the water to the boil and throws away the excess water after the rice has been cooked. The transformation of the rice from the raw to cooked state is likened to an act of pre-digestion, for the rice is softened such that it can be assimilated in the human body. The metaphor – if it is one – is reversed in the act of digestion. One consumes the rice and several glassfuls of water and then the rice is cooked, that is digested, in the stomach heated by the digestive fire (see Cantlie, 1984, pp. 194–202). Because the cooking of rice is an act of pre-digestion, the persons with whom one cooks and shares food transfer their moral qualities, through the medium of food, to the members of the social body. One may tolerate pre-digested food from persons with superior moral qualities, but not persons with inferior qualities. There is considerable anthropological literature on the caste implications of commensality. Pious members of high castes do not accept boiled rice from lower castes, for they remain vulnerable to the inferior physical and moral qualities of the lower castes. In a similar vein celibate ascetics avoid corruption by sexual desire by not accepting boiled rice from married couples; and itinerant ascetics avoid corruption by material attachments, by not accepting boiled rice from sedentary, monastery-dwelling ascetics. In sum, pre-digested food is liable to contamination and the preparation and serving of cooked food provides a powerful means of discriminating between social groups.

With regard to food as an item of distribution, a second sense of social body structures the Hindu family hierarchically. Here the bodily image depicts the relation between the head of household and those members of the family whose labour and loyalty he commands. The head of household is the agent, likened to the mind; the spouse, unmarried daughters, sons and their spouses and children are the instruments of the agent's will – likened to the limbs of his body. There is no equality in the Hindu family; rather there is seniority within generations and superiority between generations. The head of household commands and coordinates the labour of the other family members. He provides for their welfare and upbringing and expects, in return, their unquestioned loyalty, respect and service. Agency is expressed by manual passivity and stillness; the head of household speaks minimally to those whom he commands. His wife may also command the labour of her daughters and daughters-in-law. She may do physical work inside her courtyard, but she would be ridiculed by neighbours if they saw her fetching water from the neighbourhood well or collecting brush for kindling a fire. They would want to know why her daughter or daughter-in-law was not carrying out such manual work.

The hierarchical organization of the family places a premium on obedience and service; yet this does not necessarily mean that obedience is the norm, only that it is seen to be the norm. Those who are commanded have minds of their own, which they express often in irony – making it difficult for their superiors to distinguish between flattery and sarcasm. One cannot disagree with one's superior's command, but one can say yes and then procrastinate. A woman would be disrespectful to disagree with her husband, but she can dissemble feelings of weakness to delay an unwanted command until the matter is forgotten. With regard to meals, food is prepared by the women and then served first to the men in order of precedence. The women eat what remains after the men have eaten; yet there is ample opportunity to snack on the food in the kitchen before it reaches the superior's plate.

The notion of the family as a social body emerges clearly in accounts of food distribution in times of famine. In European society there is the notion that every member of the family is an individual, motivated by a sense of self-preservation. Women and children, however, remain vulnerable in their individuality in that their physical strength or immaturity disadvantages them in self-preservation. Hence in situations of distress women and children receive special entitlement. In Hindu society, however, the family is a social body whose essence is identified with the agent who is head of household. The instruments – the women and children – are dependent upon and derivative of the agent. In situations of food scarcity women and children lose their entitlement before the agent. The head of household's behaviour may look like self-preservation, but it is the family, not the individual, that must be preserved; and in that preservation he is the essential person (see Greenough (1983) on the Bengal famine of 1944–5). Furthermore if the family splits up and the women and children fend for themselves in cities or at workcamps, they lose their entitlement to be reconstituted within the family after the calamity (Greenough, 1983). By virtue of no longer being commanded by the head of household, a woman's virtue cannot be guaranteed and hence her suitability for preserving her husband's patriline is cast into doubt. Instead the head of household remarries. One of the main social problems after the Bengal famine was the continuing provision of shelter and food for married women who had been rejected by their husbands.

Diet, disease and the body

The Hindu concept of the body is fundamental to understanding both the assimilation and refinement of food as well as the importance of food as a medium of contagion and an item of distribution. It goes without saying that the body is also central to understanding illness and vulnerability. Furthermore Hindu dietary practices are invested with a rationale that is seen to be health promoting, even though the nutritional and medical evidence may be somewhat at variance with local views. Unfortunately there is insufficient inter-disciplinary research to explore this variance in any detail, yet there are three areas of research where nutritional and anthropological pre-occupations can be brought together, if not in the field, at least in the library.

The first pre-occupation is the cultural refinement of tribal or low caste people. The dietary implications of cultural refinement have already been described in the first section above. One opts for a meal that does not agitate the mind nor injure the body. This entails consuming rice as the staple, cutting out alcoholic beverages and abstaining from most, if not all, varieties of flesh, fowl, fish and eggs. Since 'food' is cooked, there is less emphasis on raw vegetables, leafy greens and fruits (see Kharve, 1976a, b for details). Gopaldas, Gupta & Saxena (1983a, b) reported on the mixed nutritional implications of dietary refinement. They investigated the nutritional status of a tribal people in Gujarat, some of whom had improved their social status and refined their diet and others who had not. Differences in nutritional status between the two groups emerged at weaning. Tribal toddlers were better off than their 'civilized' cousins in all respects except for protein and calcium. In the pre-school group (4–6 years) the 'civilized' group were better off nutritionally for protein, calcium, thiamin and riboflavin; the tribal group – due to their consumption of fresh fruit and vegetables – were better off for iron, ascorbic acid, retinol, and niacin. With regard to nutritional disorders the tribal people suffered from essential fatty acids and vitamin B complex deficiency; the 'civilized' group from protein – energy malnutrition, anaemia, vitamin A and ascorbic acid deficiencies. In interpreting these results, one might also bear in mind that a social group in the process of cultural refinement is often one whose economic situation has improved. Hence one refrains from eating 'jungle' food which demean one's status (losing out on iron, vitamin A and ascorbic acid) and yet one can afford a diet richer in milk products, grain and tolerable meats (mutton and chicken rather than cow, pork, squirrel or monkey). It could be that tribal people – could they afford it – might also purchase such refined foods without abandoning their fresh fruit and vegetables. On the other hand, if they could afford it, they might also want to refine themselves culturally in which case fresh fruit and vegetables 'from the jungle' might be abandoned.

A second area of mutual concern by nutritionists and anthropologists is the special diet of women during pregnancy and puerperium (Eichinger Ferro-Luzzi, 1973a, b, 1974; Homans, 1983; Reissland & Burghart, 1988). There are few dietary recommendations in pregnancy. Pregnant women, however, fear miscarriage and complications in delivery; and hence they may avoid heating foods (said to cause miscarriage) as well as food, or amounts of food, that unnecessarily fatten the foetus (and thereby make it difficult to squeeze through the birth passage). One might add that pregnancy cravings are recognized, but they are said to be those of the infant in the womb, not the mother; and the foetus may be treated with some indulgence in its dietary requests. Most domestic dietary lore concerns the postpartum period, which is managed almost entirely by foods and tonics that serve to expel the blood of birth, heal the birth passage, produce milk, restore strength, etc. Because the qualities of the mother pass through the medium of her milk to her child, particular care ought to be given to the maternal diet postpartum. Studies by nutritionists indicate that home remedies and tonics may be beneficial (Mital & Gopaldas, 1985), but many women suffer from nutritional deficiencies. Poverty may be an important cause of malnutrition, but even among the better off the nutrient intake is substandard, except for such nutrients as calcium and riboflavin that come from a milk and cereal diet. Many fresh fruits and vegetables are avoided in pregnancy or during lactation, as are meat and eggs. Anaemia is common (Vijayalakshmi, Jacob & Devadas, 1975; Devadas, Vijayalakshmi & Chandy, 1980; Gupta & Sharma, 1980; Luwang & Gupta, 1980; Rao, 1985, 1986a, b).

A third area of mutual concern lies in the epidemiological research carried out on Hindus overseas in which some attempt has been made to relate morbidity and mortality to diet. South Asian immigrants to Britain are found to exceed expected mortality with regard to infective, parasitic disease (especially tuberculosis), endocrine diseases (diabetes), circulatory diseases (ischaemic heart disease) and digestive diseases (cirrhosis). Mortality because of malignant neoplasms is less than expected (Balarajan *et al.*, 1984; Marmot, Adelstein & Bulusu, 1984). One is tempted to interpret these figures in the light of the 'healthy' Hindu diet, based around cereal staples, milk and ghee and vegetable oils used to fry the side dishes. The beneficial aspects of a rice or wheat staple and a relatively low prevalence of smoking helps to explain the less than expected deaths from malignant neoplasms. Setting aside the cases of tuberculosis, many of which stem from infected cow's or buffalo's milk consumed in South Asia, the other causes of death are of interest in the investigation of the cultural context of diet and disease.

Reports of rickets among Asian youth in Glasgow, and later of other cities in Britain, attracted considerable medical research (Goel *et al.*, 1976). An early hypothesis was that phytic acid, found in chapathi flour, impaired the absorption of dietary calcium. This led to the proposal that south Asians be put on enriched leavened bread (Ford *et al.*, 1972). The hypothesis, however, was not widely accepted; and comparative research in the United Kingdom and northern India indicated that the limited exposure to sunlight was the probable reason for the higher prevalence of rickets, holding income and dietary practices constant, among Asians in Britain than among the people of north India (Hodgkin *et al.*, 1973). Researchers in Glasgow noted that the recommendation to switch from ghee to clarified, fortified margarine and to avoid chapathis was untenable; far better to enrich the chapathi flour with vitamin D (Pietrek, 1976). This policy was thought to have unwanted or unnecessary consequences for families in which there is no deficiency, especially in view of the fact that there are important regional differences in diet: e.g. Gujaratis have a vitamin D deficiency, but the Goanese do not (Hunt *et al.*, 1976). It was later discovered that vitamin D deficiency is not just a child problem but afflicts adults as well, and especially women who are more likely to remain at home such that their skin synthesizes insufficient vitamin D (Shaunak *et al.*, 1985). By this time the rickets campaign had become a political issue in the Indian community. While there was widespread concern for the health of the community, many Indians were disturbed to find themselves problematized. Hindu dietary practices are inextricably linked with cultural values and personal refinement – in a word, with self-esteem. Hence the public attack on the traditional diet was treated by Hindus as not being solely a medical matter. Rather they felt that they were being stigmatized by racist public health workers. Much nutritional research and local policy was subsequently caught up in, and eventually stymied by, this political imbroglio.

The other two diet-related diseases on which some research has been carried out are diabetes and ischaemic heart disease. The prevalence, adjusted for age, of diabetes in Britain is 3.8 times greater for Asians than Europeans; and in the 40–64 age group five times greater (Mathger & Keen, 1985). Ethnographically little is known about this disease except that it is stigmatic and is locally diagnosed. In rural north India it is called the 'sugar disease' (*ciniya bimari*) and is diagnosed by observing whether ants gather at the place where one has urinated. As for ischaemic heart disease, British citizens and Gujaratis have among the highest mortality rates in the world; higher still are the rates for Gujaratis in Britain. Medical opinion is divided on the relation between diet and ischaemic heart disease. Despite the fact that plumpness is a sign of health for Gujaratis and that vegetable oils and ghee are thought to be nourishing foods, proof of the dietary factor is difficult to obtain; and family history, environmental conditions and other factors all contribute to coronary heart disease. Moreover, not everyone is convinced of the primary importance of diet. Sociologists, as well as some Asian opinion (Coronary Prevention Group for the Confederation of Indian Organisations, 1986), argue that an important factor in explaining Asian heart disease in Britain is the stressful social environment in which many coloured immigrants live. Finding fault with the Asian diet is said to 'blame the victim' of racism. Again expert opinion and official policy have been caught up in political recriminations.

This brief summary of research on diet-related illnesses among three transitional social groups indicates where the health-promoting rationale of the south Asian diet is at variance with the nutritional and medical evidence. Health educators may be tempted to think that by means of education Hindus could become better informed of the relation between food and health such that they might preserve their health by changing their diet. Here there is the idea that items of food, previously selected according to taste, might now be selected according to their value in promoting health. Yet the ethnographic material, inadequate as it is in some cases, indicates how unsuccessful such a policy might be. The nutritional knowledge of the health educator does not replace the ignorance of local people. Rather Hindus themselves already have their own knowledge of the relation between diet and disease, which is embedded in their concept of the body – both individual and social bodies. Their dietary rationale makes sense because its logic and pre-suppositions are reinforced in other aspects of their personal experience. Moreover, diet and digestion are so central to the cultural, social and physical experience of Hindus that any attack on the Hindu diet might, as the British example shows, be interpreted as an attack on Hindu civilization itself. At the very least the ethnographic material shows that diet-related diseases and disorders are a medical problem only for the medical profession; for Hindus there may not be a problem. Or if there is a medical problem then it cannot be readily divorced from the social experience of well-being.

References

Balarajan, R., Bulusu, L., Adelstein, A.M. & Shukla, V. (1984). Patterns of mortality among migrants to England and Wales from the Indian subcontinent. *British Medical Journal*, **289**, 1185–7.

Cantlie, A. (1984). *The Assamese: Religion, Caste and Sect in an Indian Village*. Curzon, London.

Carstairs, G.M. (1961). *The Twice-Born: a Study of a Community of High Caste Hindus*. Indiana University Press, Bloomington.

Conklin, H. (1957). *Hanunoo Agriculture*. FAO Forestry Development Paper, 12. FAO, Rome.

Coronary Prevention Group for the Confederation of Indian Organisations (1986). *Coronary Heart Disease and Asians in Britain*. CPGCIO, London.

Croll, E. (1983). *The Family Rice Bowl: Food and the Domestic Economy in China*. Zed Press, London.

Devadas, R.P., Vijayalakshmi, P. & Chandy, A. (1980). Nutritional status of expectant mothers and their offspring. *Indian Journal of Nutrition and Dietetics*, **17**, 275–80.

Eichinger Ferro-Luzzi, G. (1973a). Food avoidances at puberty and menstruation in Tamilnad. *Ecology of Food and Nutrition*, **2**, 165–72.

Eichinger Ferro-Luzzi, G. (1973b). Food avoidances of pregnant women in Tamilnad. *Ecology of Food and Nutrition*, **2**, 259–66.

Eichinger Ferro-Luzzi, G. (1974). Food avoidances during puerperium and lactation in Tamilnad. *Ecology of Food and Nutrition*, **3**, 7–15.

Ford, J.A., Colhoun, E.M., McIntosh, W.B. & Dunningan, M.G. (1972). Biochemical response of late rickets and osteomalacia to a chupatty-free diet. *British Medical Journal*, **3**, 444–7.

Goel, K.M., Logan, R.W., Arneil, G.C., Sweet, E.M., Warren, J.M. & Shanks, R.A. (1976). Florid and subclinical rickets among immigrant children in Glasgow. *Lancet*, i, 1141–4.

Gopaldas, T., Gupta, A. & Saxena, K. (1983a). The impact of Sanskritization in a forest-dwelling tribe of Gujarat, India: ecology, food consumption patterns, nutrient intake, anthropometric, clinical and hematological status. *Ecology of Food and Nutrition*, **12**, 217–27.

Gopaldas, T., Gupta, A. & Saxena, K. (1983b). The phenomenon of Sanskritization in a forest-dwelling tribe of Gujarat, India: nutrient intake and practices in the special groups. *Ecology of Food and Nutrition*, **13**, 1–8.

Greenough, P. (1983). *Prosperity and Misery in Modern Bengal: the Famine of 1943–1944*. Oxford University Press, New York.

Gupta, R. & Sharma, I. (1980). An overview of the dietary consumption pattern of pregnant and lactating mothers of Haryana region, Hissar. *Indian Journal of Nutrition and Dietetics*, **17**, 13–19.

Hodgkin, P., Hine, P.M., Kay, G.H. & Lumb, G.A. (1973). Vitamin D deficiency in Asians and at home in Britain. *Lancet*, ii, 167–72.

Homans, H. (1983). A question of balance: Asian and British women's perceptions of food during pregnancy. In: *The Sociology of Food and Eating*, ed. A. Murcott, pp. 73–83. Gower, Aldershot, Hants.

Hunt, S.P., O'Riordan, J.L.H., Windo, J. & Truswell, A.S. (1976). Vitamin D status in different subgroups of British Asians. *British Medical Journal*, **2**, 1351–4.

Inden & Nicholas (1977). *Kinship in Bengali Culture.* University Press, Chicago.
Kharve, R.S. (1976a). *The Hindu Hearth and Home.* Carolina Academic Press, Durham, NC.
Kharve, R.S. (1976b). *Culture and Reality: Essays on the Hindu System of Managing Foods.* Indian Institute of Advanced Study, Simla.
Kutumbiah, P. (1962). *Ancient Indian Medicine.* Orient Longmans, Madras.
Laderman, C. (1981). Symbolic and empirical reality: a new approach to the analysis of food avoidances. *American Ethnologist*, **8**, 468–93.
Luwang, N.C. & Gupta, V.M. (1980). Anaemia in pregnancy in a rural community – influence of dietary intake in the multifactorial aetiology. *Indian Journal of Nutrition and Dietetics*, **17**, 414–17.
Marmot, M.G., Adelstein, A.M. & Bulusu, L. (1984). *Immigrant Mortality in England and Wales 1970–1978.* Office of Population and Census Statistics Studies of Medical and Population Subjects, 47. HMSO, London.
Mathger, H.M. & Keen, H. (1985). The Southall diabetes survey: prevalence of known diabetes in Asians and Europeans. *British Medical Journal*, **291**, 1081–4.
Mital, N. & Gopaldas, T. (1985). Habit survey of a culturally acceptable mother food in Gujarat, India. *Ecology of Food and Nutrition*, **16**, 243–52.
Pietrek, J. (1976). Prevention of vitamin D deficiency in Asians. *Lancet*, i, 1145–8.
Pollock, N. (1985). The concept of food in a Pacific society: a Fijian example. *Ecology of Food and Nutrition*, **17**, 195–203.
Rao, M. (1985). Food beliefs of rural women during the reproductive years in Dharwad, India. *Ecology of Food and Nutrition*, **16**, 93–103.
Rao, M. (1986a). Diet and nutritional status of pregnant women in rural Dharwad. *Ecology of Food and Nutrition*, **18**, 125–33.
Rao, M. (1986b). Knowledge, attitude and practice regarding nutrition among pregnant women in rural Dharwad, Karnataka, India. *Ecology of Food and Nutrition*, **18**, 197–208.
Reissland, N. & Burghart, R. (1986). The quality of a mother's milk and the health of her child: beliefs and practices of the women of Mithila. *Social Science and Medicine*, **27**, 461–9.
Shaunak, S., Colston, K., Ang, L., Patel, S.P. & Maxwell, J.D. (1985). Vitamin D deficiency in adult British Hindu Asians: a family disorder. *British Medical Journal*, **291**, 1166–8.
Vijayalakshmi, R., Jacob, M. & Devadas, R.P. (1975). Relationship between diet during pregnancy and nutritional status of the new-born. *Indian Journal of Nutrition and Dietetics*, **12**, 233–42.

Index

Page numbers in italics refer to figures and tables.